Modern Techniques in Metrology

MODERN TECHNIQUES IN METROLOGY

Edited by
Paul L Hewitt

World Scientific

Published by

World Scientific Publishing Co Pte Ltd.
P O Box 128
Farrer Road
Singapore 9128

ISBN 9971-966-45-X
9971-966-46-8 pbk

Printed in Singapore by Richard-Clay (S.E. Asia) Pte. Ltd.

Preface

"Modern Techniques in Metrology for Developing Countries" is essentially the Technical Proceedings of the Regional Workshop on Metrology for Developing Countries, which was held at the CSIRO National Measurement Laboratory (headquarters of the Division of Applied Physics) from 30 August to 10 September 1982. The Workshop was organized by the Regional Coordinator, Mr T. Jones, and financial support was given by the Australian Development Assistance Bureau (ADAB), the Commonwealth Fund for Technical Cooperation and the Commonwealth Secretariat through the Commonwealth Science Council (CSC), the United Nations Education Scientific and Cultural Organization (UNESCO) and the Australian Department of Science and Technology for the Associatiion of Science Cooperation in Asia (ASCA). In all, over 60 participants from 16 member countries of the Asia/Pacific Metrology Programme, international agencies and national standards laboratories attended.

The objectives of the Workshop were to examine the essential elements in the development of a national measurement system, and the role of regional cooperation in this regard. The Workshop also provided intensive training in four measurement fields: physical metrology, engineering metrology, ther mometry and electrical metrology.

Paul L Hewitt
Editor

CSIRO Division of Applied Physics
National Measurement Laboratory
Sydney Australia 2070

REGIONAL WORKSHOP ON METROLOGY FOR DEVELOPING COUNTRIES, SYDNEY, SEPTEMBER 1982

Squatting (Left to Right): J D Thane, J C Macfarlane, D R R Gowdie, M Beavis

Front Row: D Fawzy Othman, Md. T Sarkar, Tum Tham Yee, J C Park, Herudi Kartowisastro, Tay-Sum Kit Mun, M N G A Khan, C Kurz, G Ram, T V P Upadhyaya, K C Joshi, R B Dacuyan, P N Puntambekar, G M Putera, H L R W Madanayake, Sun Jhitkrikroun, Teo Nam Kuan, C C Franchimon, W R Blevin

Back Row: M H Molla, B S Harrap, R B Frenkel, P J Campion, U S Kuruppu, E S Juan, J Tapping, R B Hurst, Siri Nandhasri, T J Quinn, B Tyler, H M P Stock, B D Inglis, Pi Jiajing, Yu Datong, A G Driver, N Bignell, K Chandra, J J Connolly, D B Prowse, R E Bentley, N B Tabuyaqona, T P Jones, J B Patterson, M J Puttock, P J Sim

CONTENTS

Modern Techniques in Metrology

Australia's Measurement System

W.R. Blevin
CSIRO Division of Applied Physics, Sydney, Australia 2070

Abstract

In Australia, as in other countries, government has recognized a responsibility to establish a structured measurement system. The objective is to provide for uniform and reliable measurements thoughout the nation, as far as possible in conformity with measurements made overseas. Although there are specific legislative requirements concerning units and standards for the measurement of physical quantities, and particularly concerning measurements in trade and commerce, the Australian measurement system is loosely structured and dependent on goodwill and cooperation between the participating organizations. Since Australia has a federal system of government, the administration of the measurement system is shared by the Commonwealth, States and Territories.

1. Introduction

Measurements play a vital role in the great majority of human activities, in trade and commerce, manufacturing, environmental protection, health care, military technology, and so on. It has been estimated that in the United States, for example, about 6% of the gross national product is spent on measurements.

The great majority of measurements are made by end-users - in factories, in the market-place, on construction sites, in testing laboratories, etc. These measurements would be of much lower value, however, if they were not supported by a national and indeed international infrastructure that ensures their compatibility with measurements made elsewhere. Recognition of the economic significance of this point has led nations to establish national measurement systems and to develop mechanisms for promoting international uniformity in measurements.

In Australia, as in most other countries, the national measurement system is not the result of one, all-embracing, planning process. Instead it has evolved over many years as a result of changing social, economic and technical factors. It is not necessarily the most appropriate system for Australia today, and is certainly not being put forward as an ideal model for other countries. Nevertheless, delegates to this Asia/Pacific metrology conference in Australia might find some value and interest in looking at the local system in a little detail.

2. Background Information on Australia

Although Australia is a country of large area, 7 682 300 km^2, it was very sparsely populated prior to the beginning of European settlement in 1788. The aboriginal people, estimated to have numbered about 300 000, were a nomadic, non-agricultural race and apparently had no system of weights and measures. During the period 1788 to 1900 Britain established six colonies in Australia, and in due course granted each of them responsible government for their internal affairs.

The Commonwealth of Australia became an independent nation in 1900, and at present comprises six States (the former colonies) and two Territories (the Australian Capital Territory and the Northern Territory), with a total population of about 15 000 000. It has a federal form of government, the powers and functions of government being shared by the Commonwealth (or national) parliament and by State parliaments.

3. Weights and Measures Legislation

The governmental responsibilities for Australia's measurement system are shared by the Commonwealth, States and Territories.

The Commonwealth Constitution makes one specific reference to weights and measures. Its section 51(xv) states simply that the Commonwealth Parliament shall have power to make laws with respect to weights and measures. The first Act to be passed under this power was the Commonwealth Weights and Measures (National Standards) Act of 1948, which because of certain weaknesses was subsequently repealed and replaced by the current Act of 1960. This Act created a new statutory authority, the National Standards Commission (NSC), whose responsibilities include advising the national government with respect to weights and measures generally. It also placed obligations for maintaining the national physical standards of measurement on the Commonwealth Scientific and Industrial Research Organization (CSIRO), which is a statutory authority that carries out the major part of Australia's government-sponsored scientific research.

Each State has its own Weights and Measures Act, going back for many years in some cases. These Acts are primarily concerned with measurements in trade and commerce and therefore deal chiefly with measurements of length, mass and volumetric capacity. The sale of electricity and gas are dealt with in other State Acts, and an exhaustive survey of all State legislation would reveal references to the measurement of a very wide range of physical quantities.

4. Legal Units of Measurement

The objects of the Commonwealth Act are to provide for the establishment and use throughout Australia of uniform units and standards of measurement for physical quantities. One of the responsibilities of NSC is to advise the government which physical quantities should have prescribed units and what those units should be. At present the Regulations under the Act prescribe units for 30 different physical quantities (length, mass, density, time interval, force, electric current, temperature, luminance, radioactivity, etc.), and they are currently being revised to include certain acoustical and magnetic quantities as well.

When units are prescribed for a physical quantity, those units become the sole legal units for that quantity. For example, the only prescribed units for the measurement of force are the newton and combinations of the newton with certain SI prefixes. This does not mean that it is necessarily an offence to use some other unit of force such as 'kilogram weight', but it does mean that a clause in a contract that used that unit would be declared void if tested in court.

Australia has recently undergone a metric conversion program with considerable success, but it will require further time for the last vestiges of the former British Imperial System of Units to be completely phased out. Meanwhile, for some physical quantities such as mass, both SI and Imperial units are prescribed as legal units. In many applications, however, food-packaging being a good example, other legislation requires that only the metric legal units be used.

5. Physical Standards of Measurement

As already stated, the Commonwealth Weights and Measures (National Standards) Act obliges CSIRO to maintain Australia's primary standards of measurement for all physical quantities for which legal units are prescribed. CSIRO is given the option, however, of either maintaining all the standards in its own laboratories or appointing agents to maintain some of them. In practice it maintains the great majority of the primary standards itself, in its Division of Applied Physics. It has appointed the Australian Atomic Energy Commission near Sydney and the Australian Radiation Laboratory in Melbourne to serve as agents, however, for the standards relating to ionizing radiations.

The CSIRO Division of Applied Physics has its headquarters and most of its facilities at the National Measurement Laboratory in Lindfield, a suburb of Sydney. Small branch laboratories, principally concerned with tests and calibrations, are located in Adelaide and Melbourne. Although the maintenance of standards and the provision of appropriate calibration services are central functions of the Division, there is also a considerable amount of research undertaken in other areas of applied physics.

The present Division of Applied Physics had its origins in the establishment in 1938 of the National Standards Laboratory as it was then known. It is thus a considerably younger institution than most of the major standards laboratories of the northern hemisphere. Initially most of its 'primary' standards were indeed secondary standards that had been calibrated overseas. By now, however, the Division has established most of its standards *ab initio*, though there are a few exceptions to this. As well as meeting its mandatory obligation to maintain standards for all physical quantities having prescribed units, the Division also maintains standards and testing facilities for several other quantities, examples being hardness, refractive index, and humidity.

Calibration services are provided to enable the Australian community to relate its measurements to the national standards. CSIRO does not unquestioningly accept items for calibration simply because the service is sought. Broadly speaking, its policy is to undertake the minimum number of tests and measurements that is consistent with providing the basis for an effective calibration hierarchy throughout the community, with appropriate traceability to the national standards.

Much attention is paid to ensuring that Australia's standards are as consistent as possible with those of other nations. To this end Australia has been a signatory to the Metric Convention since 1947 and works closely with the International Committee of Weights and Measures and its specialist advisory committees. The increasing regional collaboration through the Asia/Pacific Metrology Programme is also highly valued.

6. Measurements in Trade and Commerce

The provision of a framework to ensure that measurements made in trade and commerce are reliable is a joint responsibility of the Commonwealth and the States.

Before a new model of measuring instrument can be legally used in Australia for trade purposes its pattern must be examined and approved. Since 1966 this has been the responsibility of NSC, which maintains in the Sydney suburb of North Ryde a pattern-approval laboratory specifically for this purpose. During the examination of a pattern particular attention is paid to reliability and accuracy of the instrument, and also to the susceptibility of the instrument to fraud. The approval criteria are based largely on recommendations by the International Organization of Legal Metrology, but these are sometimes modified to take specific account of Australian requirements. NSC represents Australia on this International Organization and contributes to its work.

The States and Territories employ teams of inspectors to regulate measurements in the market-place. They ensure that only pattern-approved instruments are used, and periodically carry out tests to verify that the instruments remain accurate within prescribed limits. Their responsibility is not restricted to retail selling direct to the public, but extends to measuring devices used internally within certain industries, such as oil-flow provers, weighbridges, milk tankers, and continuous coal loaders.

Inevitably different requirements arise amongst the various States, and industry sometimes finds that an instrument or practice acceptable within one State may not be acceptable in another. One of the roles of NSC is to minimize this type of occurrence by promoting the adoption of uniform legislation in the States and Territories.

7. Accreditation of Laboratories

For over three decades Australia has had in operation a national system for the assessment and recognition of testing, measurement and calibration laboratories. The system is supervised by the National Association of Testing Authorities (NATA). It covers all technical disciplines and is open both to privately and publicly owned laboratories. At 30 June 1982 there were 1287 registered laboratories, distributed over nine different fields of testing - acoustic and vibration measurements, biological, chemical, electrical, heat and temperature measurement, mechanical, metrology, non-destructive testing, optics and photometry.

NATA is incorporated as a company and derives its income chiefly from a Commonwealth government grant and the annual subscriptions of its member laboratories. It employs only a small staff and relies on a large body of honorary specialists to evaluate laboratories. Recently it has begun to reinforce its well-established assessment techniques with a comprehensive program of proficiency tests.

A significant number of the registered laboratories offer calibration services and thus provide an effective extension service to the CSIRO Division of Applied Physics. They thus form part of the hierarchy giving traceability to the national standards of measurement. Also, by fulfilling many of the everyday calibration requirements of the community, they allow CSIRO to concentrate on calibrations at the first level.

Over the past decade many other countries have followed NATA's lead and established a national system of laboratory accreditation. This has been followed by the holding of International Laboratory Accreditation Conferences at appropriate intervals. NATA is working with these Conferences towards internationally accepted criteria for laboratory approvals, and has already signed, with its New Zealand counterpart, an agreement that provides for mutual acceptance of Australian and New Zealand accreditations.

8. Defence Measurements

One of the basic requirements in the quality control of military hardware is a strong measurement capability, both in the armed services themselves and also in the relevant areas of manufacturing industry. This had been recognized in Australian defence circles before 1938 when CSIRO first entered the standards field. From 1923 until 1978 the defence authorities operated their own first-level standards laboratory at the Melbourne suburb of Maribyrnong, in the establishment now known as the Materials Research Laboratories. In 1978, however, it was decided to terminate this standards function at Maribyrnong since it largely duplicated work being performed in CSIRO. It was formally agreed between the Department of Defence and CSIRO that the CSIRO Division of Applied Physics would henceforth be responsible for first-level calibration requirements in the defence area. Within the defence ambit, however, there continue to be a wide range of other laboratories providing calibration and measurement services at various levels in the hierarchy.

9. Conclusion

The above discussion has shown that the principal responsibilities within Australia's measurement system are apportioned between several organizations, as follows.

CSIRO	- national physical standards of measurement; first-level calibration service; international collaboration under the Metric Convention
NATA	- accreditation of testing, measurement and calibration laboratories; collaboration with International Laboratory Accreditation Conferences

NSC	- legal metrology; collaboration with International Organization for Legal Metrology; pattern approval of trade instruments
States and Territories: Weights & Measures Authorities	- regulation of trade measurements; verification of trade instruments.

Success of the measurement system as a whole depends on a strong commitment, and willingness to cooperate, by all the participating organizations. None of them has over-riding authority over the whole system. In the past the system has worked tolerably well. In recent years, however, there have been increasing difficulties in some areas, caused both by the harsher economic climate and by decisions of some of the organizations concerned to give a lower priority to measurement matters.

The BIPM and the Asia/Pacific Metrology Programme

T.J. Quinn
BIPM, Pavillon de Breteuil, Sèvres, France

Abstract

A brief outline is given of the history of the Treaty of the Metre and of the foundation and present-day activities of the BIPM. This is followed by a description of the organizational structure of the International Committee of Weights and Measures and its eight Consultative Committees. It is shown how this provides the framework for international agreements on matters related to the definition and practical realization of units and within which international comparisons of units can be arranged and carried out. Finally, the growing links between the various regional metrology organizations and the BIPM are discussed and prospects for the future suggested.

1. The Treaty of the Metre - 1875

Starting about the middle of the nineteenth century, the rapid growth of world trade, combined with the increasing complexity of technology, led to a general acknowledgement of the need for some sort of international agreement on weights and measures and units of measurement. Both in Great Britain and in continental European countries, moves were made to try and establish such uniformity of measurement. The British Association for the Advancement of Science took early initiatives in the field of electrical measurement, while the International Geodesic Association at its second General Conference held in Berlin in 1867 made definite proposals aimed at unifying length measurement throughout the whole of Europe. One of the proposals was for the creation of a European Bureau of Weights and Measures.

The metric system based upon the metre as the unit of length equal to one ten millionth of the quadrant of the earth, and the kilogram as the unit of mass equal to the mass of a cubic decimetre of water had by then become firmly established in many European countries. By the time the need for a universal system of weights and measures had become pressing, it was clear that the metric system was the only serious candidate. Already at the Great Universal Exhibitions, held in London in 1851 and 1862 and in Paris in 1855 and 1867 various resolutions had been put forward urging some sort of international cooperation on weights and measures. Finally in 1869, following the proposals of the International Geodesic Association, supported by the Academies of Science of St Petersburg and Paris and the French Bureau des Longitudes, the French Government proposed the creation of a Commission charged with preparing the ground for making the metric system international. Twenty-four countries accepted the original invitation to send delegates to Paris but eventually, after an interruption caused by the Franco-Prussian war of 1870, a total of thirty countries participated in the discussions. The result was the Treaty of the Metre (Convention du Mètre) signed by 17 nations in 1875 (and now adhered to by 45 nations) under which members undertook to establish in Paris, at their common expense, a Bureau International des Poids et Mesures (BIPM) where would be deposited new international standards of the metre and the kilogram.

Among the original tasks of the Bureau was to be the comparison of national prototype standards of length and mass with the international prototypes. The Treaty of the Metre, as well as founding the BIPM, set up an organizational structure within which international agreements on units could be formalized and the administrative and financial affairs of the BIPM be maintained. The Comité International de Poids et Mesures (CIPM) was established, made up of 14 scientists (increased to 18 in 1921) each from a different member nation, to administer the affairs of the BIPM and to make Recommendations to periodic Conférences Générales des Poids et Mesures (CGPM), to which all member nations are entitled to send delegates. The CIPM is now advised by eight Consultative Committees of experts. Under the Treaty of the Metre it is the CGPM alone that has authority to adopt proposals concerned with new units or re-definitions of old units. The CGPM now meets once every four years in Paris, the most recent having been the 16th CGPM in 1979.

2. Present role and activities of the BIPM

The activities of the BIPM have evolved considerably since its foundation in 1875. In addition to the original tasks related to the custody of the International Prototypes of the Metre and Kilogram (which included thermometry), the maintainance of electrical units was added in 1921, and photometric units in 1939. In 1960 a major extension to our activities was the establishment of a section concerned with units and standards of ionizing radiations. Finally, since 1973 the BIPM has participated in the realization of International Atomic Time in collaboration with the Bureau International de l'Heure at the Paris Observatory.

The major advances in science that have taken place since the early days of the BIPM have resulted in far-reaching changes in the way that most units and standards of physical measurement are defined and realized. No longer, for example, is the metre defined in terms of the distance between two lines engraved on a particular bar of Pt-Ir. Instead it is defined as a given number of the wavelengths of a certain sort of light emitted by a krypton atom. Even this is likely to change soon and the metre will probably be defined in terms of the distance travelled by light in a very small fraction of a second, the fraction being given by (1/c) where c is the speed of light. This latest change results from the invention of the laser and all of what is now possible in terms of direct frequency measurement of such optical sources. As with all changes in definitions, it is designed to enable yet higher accuracies to be obtained, but in this case it will also lead to a considerable simplification and ease of realization of the unit at the more modest levels of accuracy such as 1 part in 10^8. But this is not the place to enter into a detailed discussion on the proposed new definition of the metre. It is mentioned as an example of the way modern science is reflected in the way we define our units of measurements. Another example, and a particularly appropriate one to mention here, is the invention in this laboratory of the calculable capacitor which has allowed the ohm to be realized directly in terms of the metre and the permittivity of free space (which includes the speed of light) to an accuracy approaching a part in 10^8. Also in the electrical field is the application of the Josephson effect to the maintenance of the unit of voltage in terms of fundamental constants. Atomic constants are also employed in the definition of the unit of time, the second. We now use a time scale, TAI, which is based upon the properties of the caesium atom rather than one based upon the movement of the earth about the sun, although for obvious practical reasons the atomic time scale is not allowed to depart too much from the rising and setting of the sun!

The only base unit of the International System of Units (SI) which has remained unchanged is that of mass, the kilogram which remains defined as the mass of the International Prototype Kilogram kept at the BIPM. The difficulty in trying to put the unit of mass on an atomic basis is simply that the mass of an atom is too small - by some twenty orders of magnitude - for practical use. Such a definition would require a very large, specified, number of atoms. The practical problem of counting such a large number or making a known volume to contain a calculable number of atoms, does not seem solvable to the accuracy required - at least 1 part in 10^8. There seems no prospect, at present, of being able to measure the volume of an object to better than a few parts in 10^7. There may, of course, be other ways around the problem. For example, since we can measure 'g' to a few parts in 10^9, if we could measure a force to the same accuracy we could deduce a mass. It is not inconceivable that an electrical method of developing a calculable force could be devised which might have the required accuracy. For the time being, however, the kilogram remains the last of the base units defined in terms of a particular material object. Thus, through periodic calibrations of national prototype kilograms, mass measurement throughout the world is traceable to BIPM.

The role of BIPM in fields other than mass is different but nonetheless essential. The fact that a unit is defined in terms of an atomic transition or fundamental constant might be imagined to remove all doubt from its practical realization. This is far from being the case. In any practical realization of a unit there are a multitude of small sources of error that may combine to lead to a value very far from what is expected. The only way that such errors may be discovered is by comparison with realizations of the same unit carried out elsewhere, preferably by a different method. The very fact that the new atomic-based definitions of units allow a higher accuracy to be achieved leads inevitably to the requirements for accuracy on the part of the user to be increased. The same advances in science and technology that lead to units of length being defined in terms of laser wavelengths or frequencies, or units of voltage in terms of h/e, lead just as quickly (and sometimes even more quickly) to advanced-technology industries demanding increased accuracy in calibrations.

The role of the BIPM in these other fields is thus to provide an international, independent means by which national realizations of units may be compared and hence verified. This takes place in two ways; directly, by means of comparisons carried out in the laboratories of BIPM and indirectly, through comparisons organized within the framework of the eight Consultative Committees of the CIPM.

Taking first the direct verification of units by comparison at BIPM, leaving aside the field of mass; we find that, more and more, BIPM is acting as the central laboratory for the comparison of laser frequency and wavelength standards. Lasers are brought to BIPM and compared, usually by beat-frequency techniques, with BIPM lasers. The results of all of these comparisons are published and taken together provide a comprehensive network showing the differences between the various types of lasers. In recent discussions on the measurements of the velocity of light, which preceded the proposed new definition of the metre based upon a fixed, conventional value of c, differences in laser frequencies uncovered during BIPM comparisons allowed corrections to be made that led to an improvement in overall agreement. In the field of electrical units, our Josephson voltage equipment allows us to monitor the drift of standard Weston cells and enables comparisons to be carried out which have an overall uncertainty of about 20 nV. Unfortunately the behaviour of most commercial standard cells during transport is poor and an uncertainty of about 0.5 μV (0.2 μV under special circumstances) is about the best that can be achieved in international comparisons of voltage standards in this way. The major task outstanding is to develop a means of transporting voltage standards that can take advantage of the very high accuracy now possible using Josephson standards.

In addition to the complex and time-consuming comparison of lasers and similar activities, we also carry out calibration work for those national laboratories that do not have the resources to work at the highest level in all fields. For example, in addition to standard-cell enclosures, we calibrate electrical-resistance standards, platinum-resistance thermometers, thermocouples, radioactive sources and so on. The amount of such calibration work that we are able to do is very limited because of the very small number of staff. Nevertheless we recognize that it is one of the essential tasks of the BIPM.

Coming now to the indirect ways in which BIPM works. At the beginning the CIPM by itself dealt with all matters related to units, both technical and administrative. This soon ceased to be possible as the complexity of metrology increased, and a number of Consultative Committees were set up. These now number eight and are made up of experts drawn for the most part from the national standards laboratories. These Consultative Committees meet at BIPM when there is some need in the field such as a new definition or more usually, new methods that must be checked by international comparisons at the highest level of accuracy or questions related to the realization of the unit. In practice, some of the Consultative Committees meet at two-yearly intervals, others of three or even four-yearly intervals. Because of the small size of BIPM (a total staff of 56) intercomparisons of derived units must usually be carried out at the national laboratories rather than at BIPM. Such intercomparisons are organized and overseen by the appropriate Consultative Committee or one of its Working Groups. Also in certain cases it is necessary for comparison of some base units to be made in this way. All of this is part of the tasks of the Consultative Committees. It is clear that one of the essential tasks of the BIPM is to maintain the scientific expertise and capability required to carry out such intercomparisons in the principal areas related to the base units. The scientific staff of the BIPM should be able to participate in discussions at the Consultative Committees on a level of equality with their counterparts from the national laboratories.

The areas of work looked after by each Consultative Committee is clearly shown by its title in the following list:

Consultative Committee for:
- Mass and Related Quantities (CCM)
- The Definition of the Metre (CCDM)
- The Definition of the Second (CCDS)
- Electricity (CCE)
- Thermometry (CCT)
- Photometry and Radiometry (CCPR)
- Standards of Ionizing Radiations (CCEMRI)
- Units (CCU)

3. The BIPM and the Asia/Pacific Metrology Programme

Having outlined the way in which the BIPM operates I now come to more specific questions of participation in BIPM activities of Asia/Pacific countries through the Asia/Pacific Metrology Programme. First of all at the level of the CIPM; among the 18 members, each of a different nationality, there are at present Dr Blevin from Australia, Prof. Wang Daheng from the People's Republic of China, Dr Verma from India and Dr Kawata from Japan.

Between General Conferences to which all 45 member nations send delegates, the CIPM has the responsibility of overseeing the activities of BIPM and the Consultative Committees. The existence of regional metrology organizations makes this task easier for the CIPM. As at present constituted the CIPM members originate from a wide enough range of countries to ensure that at least one member, and usually more, is aware of the needs and problems proper to any regional metrology group. In this way the regional metrology organizations provide a strong support for BIPM while at the same time allowing a rapid and efficient exchange of information both to and from BIPM and CIPM*. Not only is this a convenient way of

---

* I would like to draw your attention to the ways in which we publish the results of our activities. The Comptes Rendus of General Conferences and Procès-Verbaux of the meetings of the CIPM are published by BIPM. We also publish Reports of Consultative Committee Meetings and a wide variety of BIPM Reports on the experimental work carried out in the laboratories of BIPM. The international journal of scientific metrology 'Metrologia' is now edited from BIPM and carries an annual account of the activities of the CIPM and Consultative Committees under the title 'News from the BIPM'. Copies of all of these publications and reprints of News from the BIPM are available free of charge on application to BIPM.

---

representing the interests of all countries, but it is probably one of the best. Direct representation by all member nations would lead to a Committee which would be far too cumbersome to be effective and it would be necessary to have a smaller group in which technical decisions could be taken. Such a transformation would not be a move in the direction of greater efficiency.

The participation in Consultative Committees of experts from one or more laboratories of nations included in a regional metrology organization is also available. It allows other members of such regional organizations to be linked with and have direct access to the results of the international comparisons carried out under the auspices of BIPM. Finally, I should perhaps comment on the meaning that we now attach to the words 'Weights and Measures' in the context of BIPM and the other organs of the Treaty of the Metre. The activities now carried out in the laboratories of BIPM and which are the concern of Consultative Committees are probably better described by the term 'scientific metrology'. The journal Metrologia is the journal for scientific metrology. In the field of industrial metrology BIPM and the Consultative Committees do not have special competence; it is not our role to advise or participate in this area. The modern meaning of the term 'weights and measures' is one related to commerce and above all to trading standards. There are, of course, very direct legal implications in the term 'trading standards' and in most countries the weights and measures services are regulated by law. The international aspects of such legal metrology is the concern of the International Organization of Legal Metrology (OIML), established in 1955. Thus we find that the role of BIPM and the other organs of the Treaty of the Metre is one strictly limited to the field of scientific metrology. This includes the definition and realization of the base units and some of the derived units at the highest levels of accuracy of the International System of Units (SI), (those calling for some special scientific cooperation) the development of the SI itself, and of course the practical work, which I have already mentioned, related to the intercomparison of standards and calibration. Thus, although the activities of BIPM may not be felt directly by some of the members of regional metrology organizations, the indirect results certainly are, and the activities of the regional metrology organizations provide a strong support to BIPM and CIPM.

Laboratory Accreditation - Its Role in National and International Trade

J.A. Gilmour
National Association of Testing Authorities, Australia, Chatswood 2067

Abstract

National Association of Testing Authorities, Australia (NATA) is an established element in the Australian National Measurement System combining with the Division of Applied Physics, the National Standards Commission and the Standards Association of Australia. NATA operates a comprehensive national laboratory accreditation system which permits integration of laboratories of demonstrated competence into a coherent system irrespective of ownership or geographic location. Many other countries are developing similar systems. Australia's international trade commitments require that its technical conditions of trade in harmony with those required by GATT, OECD and other international treaties and the existence of a laboratory accreditation system will assist in this regard. This paper reviews the place of NATA in the Australian context and describes some of these recent international developments which very much broaden its activities.

1. Introduction

The International Organization for Standardization (ISO) defines laboratory accreditation as being: a formal recognition that a testing laboratory is competent to carry out specific tests or specific types of tests.

Ever since the industrial revolution there have been forms of laboratory approval or accreditation systems. Such schemes have, however, always been directed to special areas of technology or industry or to serve the special purposes of particular users. For instance, defence services throughout the world have some form of a laboratory approval system, motor vehicle manufacturers have elaborate systems of approved suppliers, and there are many other examples.

Australia established its national laboratory accreditation system, National Association of Testing Authorities (NATA) in 1947 and NATA's unique contribution to the field of laboratory accreditation has been to bring a multitude of systems together into one comprehensive scheme for the recognition of testing laboratories to meet the needs of all users of laboratories - government, industry, commerce and trade and which covers all areas of science and technology.

Its range of activities extends from tests on abrasives and aircraft fuels to testing of zinc rich primers and, alphabetically, everthing in between.

Other countries are now following Australia's example in developing comprehensive national accreditation systems.

In most countries measurement systems have developed from weights and measures considerations and very few attempts have been made to integrate the various elements into a coherent whole. Newer nations, such as Australia, are fortunate in that they have been able to bring some order into what is, in many older societies, a very complicated or even chaotic situation.

In these newer countries two broad approaches are evident - centralised and decentralised system. Australia has opted for a system which coordinates decentralised activities while countries such as Israel, South Africa and I believe Singapore, have adopted a very centralised approach which involves maximum utilisation of a few "official" laboratories.

This paper concentrates on the decentralised approach which relies on the recognition of competence of a wide range of laboratories throughout the country in which direct government ownership or influence is minimal.

2. Australia's National Testing and Measurement System

Australia's testing and measurement system has two main streams with traceability to the international system of units and quantities; namely legal metrology (weights and measures), and other testing and measurement. The various components are illustrated in abreviated forms in Figs. 1 and 2.

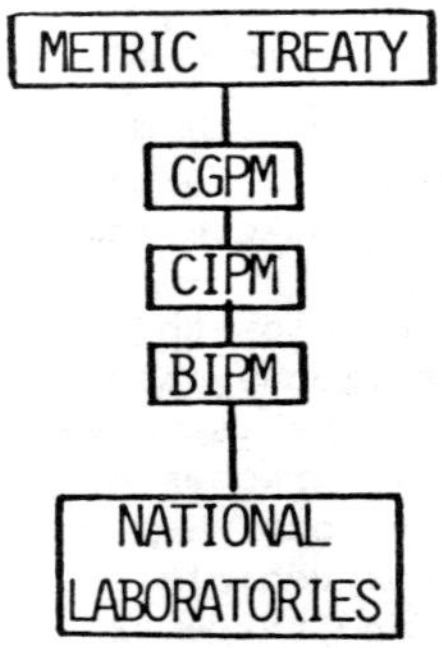

Fig. 1. International standards of measurement

National physical standards or national standards of measurement are fundamental to any national metrology system. They provide the apex of the national calibration system and the vital link to international traceability. For countries to participate effectively in international trade and technology transfer, accurate calibration of, and precise agreement between, national standards is of the utmost importance.

Australian national standards are defined in terms of SI units and are traceable to international standards of measurement maintained at the International Bureau of Weights and Measures (BIPM) or to standards maintained by other laboratories in conjunction with BIPM.

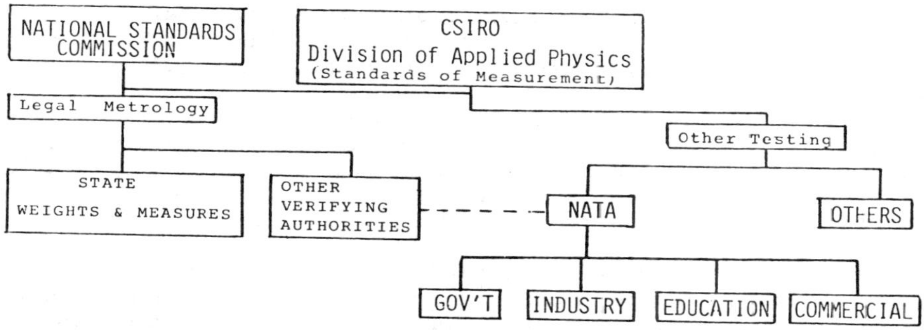

Fig. 2. The Australian System

Within Australia, primary standards of measurement are maintained by CSIRO Division of Applied Physics, formerly known as the National Measurement Laboratory. The Australian system is relatively simple and coherent and its unique feature is the integration and recognition of laboratories from all sectors of the community through the process of laboratory accredition by NATA. It is this element which is being studied by many other countries.

3. Development of National Laboratory Accreditation Systems

(a) Australia

The operation of NATA was the subject of a paper by Mr H.F. Monaghan at the Wellington conference in 1979 and therfore I will mention NATA only briefly.

The idea of establishment of a national laboratory approval system for Australia was first proposed in the early 1920s, at the time of formulation of the nation's first Science Policy. The original concept envisaged a network of government-operated laboratories which would provide for any necessary official testing. For various reasons, no action was taken until the early 1940s, when, as an emergency measure, the Approved War-Time Test-House Scheme was introduced. This scheme permitted routine testing of war materials in laboratories other than the official Defence establishments and other government laboratories which had become severely overloaded as war-time production expanded. This scheme was in every sense a laboratory accreditation system.

After the war, the Munitions Supply Department acknowledged that the Approved War-Time Test-House Scheme had enabled it to exercise adequate control of the quality of materials brought in or manufactured by its own plants. The success of the scheme demonstrated the possibilities of a broadly based laboratory accreditation system incorporating industrial and commercial laboratories as well as government establishments.

So, in 1947, in Australia, an organisation called National Association of Testing Authorities (NATA), embarked upon an experiment in laboratory accreditation which had a number of novel features.

Firstly, this Australian system was designed to provide a national testing service which would serve the needs of government, industry, and commerce, and would have no scientific, technological, or industrial boundaries. If the community had a need for a certain type of laboratory service, then the system would provide the necessary accreditation.

Secondly, there would be no geographic limitations within Australia. The system had to operate uniformly throughout a country approximately the size of the United States and comprising six independent states and two territories.

Thirdly, the system was to be based on a unique method of assessment of the competence of laboratories by a comprehensive examination of each laboratory's operations by independent experts. The assessment was to take the form of a true peer review with emphasis on evaluation of management.

For the next twenty years the rest of the world looked on with interest but considerable scepticism. NATA was regarded as an interesting curiosity, practical only in a homogeneous country like Australia.

By the mid 1960s however, the concept was no longer novel and it began to appear to Australia's overseas friends and trading partners as more and more sensible and very practical.

(b) New Zealand

In terms of development of secondary industry, New Zealand has tended to lag behind Australia by about 15 to 20 years. It wasn't until the early 1960s, therefore, that New Zealand seriously began to broaden its industrial activity to complement its efficient and profitable agricultural industries.

Government scientific and technological advisers realised that in any modern industrial environment, quality control by manufacturers must be an integral part of all production units and that the market place was showing more and more interest in the quality of goods. These same advisers were quick to perceive that NATA had had a significant impact on the development of measurement science and quality-control practice in Australia.

By the early 1970s there was widespread agreement that a NATA-style organisation would be valuable to New Zealand.

In 1972, New Zealand established its Testing Laboratory Registration Council (TELARC) which is almost a replica of NATA - technically exactly equivalent but with some administrative differences.

New Zealand does have a national standards laboratory known as the Physics and Engineering Laboratory of the Department of Scientific and Industrial Research. The formation of TELARC meant that New Zealand had a very similar national measurement system to Australia - at least in concept. Over the past 10 years the growth of TELARC has actually increased this similarity.

(c) Denmark

Quite independently, and with little contact with NATA, Denmark established a national accreditation system in 1973 with the institution of the Danish National Testing Board (STP) which again is almost a pure replica of the Australasian systems - remarkably close to the New Zealand pattern in terms of administration.

(d) United States

The United States is one of the world's most complex societies and it is not surprising to find that there are approximately 100 identified laboratory accreditation systems. There may be many more. This situation has led to some laboratories being subject to many different approval systems and needing to demonstrate compliance with the different requirements of each one. Obviously an undesirable state of affairs.

In the late 1960s the U.S. Department of Commerce sought ways to improve this situation and in 1976 the National Voluntary Laboratory Accreditation Program (NVLAP) was launched. This program is now operated within the National Bureau of Standards (NBS).

It is now fully operational although not yet fully comprehensive in the same way as the older systems, but it is being extended each year.

(e) Other systems

Table 1 lists existing national accreditation systems currently in operation, but the U.K. situation is worthy of note as it reflects the general pattern being developed in Europe.

In 1966 the U.K. established an organisation called the British Calibration Service (BCS) which took ideas from NATA and applied them to calibration only while retaining the criteria and technical requirements of NATA. It has been quite successful.

Table 1. List of National Accreditation System

System	Date Formed	Country
National Association of Testing Authorities (NATA)	1947	Australia
British Calibration Service (BCS)	1966	U.K.
Testing Laboratory Registration Council (TELARC)	1972	New Zealand
Danish National Testing Board (STP)	1973	Denmark
National Voluntary Laboratory Accreditation Program (NVLAP)	1976	U.S.A.
National Network for Testing (RNE)	1979	France
The National Testing Laboratories Accreditation System (SINALP)	1980	Mexico
National Testing Laboratory Accreditation Scheme (NATLAS)	1980	U.K.
Accreditation Program of the Standards Council of Canada	1981	Canada

In 1980 the National Testing Laboratory Accreditation Scheme (NATLAS) was launched to complement the work of BCS by accrediting laboratories engaged in general testing and measurement.

Both BCS and NATLAS are operated as independent sections within the National Physical Laboratory.

This separation of calibration from other testing and measurement functions seems to be the preferred route for Europe in general with at least France, West Germany and the Netherlands following it.

More recently, Canada through its Standards Council and the French through its Ministry of Industry have developed frameworks for the accreditation of testing laboratories on a national scale. Such developments bring a degree of rationalisation into individual national measurement systems and there is a high degree of awareness of Australia's initiative in this area.

In the very near future there will be at least another dozen acronyms added to the standards language, all relating to laboratory accreditation.

4. Why

It's a remarkable fact that between 1947 and 1970 national laboratory accreditation was regarded as a curiosity and yet from the early 1970s onwards, there has been an almost spontaneous development of national systems in countries varying in size and industrial development from the U.S. to Papua New Guinea and it is therefore an obvious question to ask why this would have happened.

Communities, and organisations within communities, operate laboratories for many reasons and which include:

(i) Governments

enforcement of regulations
forensic work
technical investigations of national
weights and measures

(ii) Purchasing agencies

to demonstrate compliance with specification

(iii) Producers and manufacturers

quality control
product development
failure investigation
data for advertising
certification of compliance with customer specification

(iv) Health Services

public health
medical diagnoses

(v) Consumers organisations

(vi) Product certification bodies

(vii) General contracting parties

These are usually internal, national needs and there appears to be a growing awareness of the cost of testing and the economies that can be achieved by minimising testing by ensuring that laboratories of demonstrated competence do the testing. Some form of laboratory accreditation is therefore seen as desirable.

Internationally, new factors and pressures have arisen.

Trade in many commodities is becoming more technical with test documentation a more and more important element in that trade. Even traditional products such as wool are now sold almost entirely on the basis of objective testing. With sophisticated technical products such as chemicals, this trend is even more pronounced.

Traditional trade arrangements and agreements are changing and Governments are taking a more active interest in the negotiations that take place in the commercial world. The Multilateral Trade Negotiations have resulted in the GATT Agreements which will, over the next few years, have a very pronounced affect on conditions of trade, even in countries that are not signatories to the Agreements. The Agreement that affects laboratory accreditation specifically is the GATT Standards Code and this is referred to again later in this paper.

Not only are Governments showing increasing interest in normal commercial relationships, they are also producing more and more regulations which include technical requirements to control imports and exports. The GATT codes are designed to reduce the possibility that such technical requirements will become trade barriers.

There is also an increasing flow of test data of all kinds around the world such as results of testing of pharmaceuticals for toxicity and food additives for safety. The Organization for Economic Cooperation and Development (OECD) work on chemicals is one of particular significance to the chemical industry and over the next few years new chemicals will be subject to much more rigorous screening than previously before they will be available in any market place. Accreditation of testing laboratories is to be a element in this control.

The past 20 years have seen a dramatic increase in product liability litigation in many countries and although presently Australia and other South Pacific countries are largely isolated from this activity it should be considered in a paper such as this since exporters are subject to legal action in other countries.

The United States has the dubious distinction of leading the world in the numbers and dollar volume of suits filed against manufacturers and distributors for damages to persons and property allegedly caused by defective products. The 50 states have each enacted legislation which increasingly places more and more of the responsibility on manufacturers and distributors for compensating injured parties, regardless of who may have been at fault in the use or misuse of the products.

Instead of traditional legal doctrines of negligence, with offsetting provisions of comparative negligence (i.e. differing degrees of negligence contributing to an injury caused by plaintiffs and by defendants), a newer doctrine has come into being which imposes what is called "strictly liability" on the parties responsible for placing allegedly defective or injury-causing products in commerce. The social theory behind this newer doctrine is that whenever a product causes an injury to someone there should be a legal remedy available to compensate the injured party for his damages, and the party that should provide such compensation should be the one which can best absorb the cost of the damages or possibly pass it on to others by increasing the cost of its products. Generally speaking, that party is considered to be a product's manufacturer or distributor. This strict liability doctrine has also taken root in Europe, and conceivably may spread to other nationsl the world over. It should be kept in mind, moreover, that the doctrine is not limited to U.S. manufacturers. Exporters whose products reach the U.S. will find themselves subject to the same product liability laws as are U.S. nationals.

In some jurisdictions "strict liability" can be distinguished from the harsher legal penalty which may be termed "absolute liability". In many strict liability jurisdictions there is frequently a requirement imposed upon plaintiffs that a product has to be proven defective and unreasonably dangerous when used as anticipated. If the seller or other defendant can prove that the product was not defective - and also was not unreasonably dangerous, the defence may prevail.

A point to keep in mind is that in litigation where the decision on liability may depend on whether a product is "defective and unreasonably dangerous", competent technical proof will be an essential factor. If tests are made by a laboratory which has been duly accredited as competent to make those tests, by an accrediting organization whose own status is beyond any question of credibility regarding the competency of its own procedures, the courts may have little difficulty in accepting the test data and other findings by such a laboratory.

Unfortunately this is supposition at this stage since the U.S. accreditation system is in its very early days but the above reflects some very respectable legal opinion.

Another international movement which is being promoted strongly in Europe is that of "product certification". There are various forms of such systems but the most common one is described as "third party certification" and the Standards Association of Australia Certification Mark Scheme and certification systems operated by other standards bodies are examples of such systems.

The subject is worthy of a separate paper but for the purposes of this presentation only a few brief comments will be made.

Product certification involves assessment of an organisation's total quality control and assurance procedures including laboratory facilities. Laboratory accreditation should therefore be an integral part of all certification systems. Laboratory accreditation does however have other roles outside product certification and there is more to quality control than simply laboratory testing. The two concepts overlap in that area.

The complexities of such systems do mean that they restrict themselves to limited ranges of products and these have been generally related to products involving safety. The IECQ system for certification of electronic components is an example of an international system of great interest to our part of the world.

Product certification, despite these limitations, is a growing feature of national and international trade and must be recognised as such.

All these international factors just mentioned have resulted, at long last, in the beginnings of some action in various forums such as GATT, ISO, IEC, OECD and the International Laboratory Accreditation Conference (ILAC).

5. ILAC

It is a fact that over the last 5 years there has been much international interest in the whole concept and philosphy of laboratory accreditation.

In 1977 the first international conference on the subject was held in Copenhagen and that conference, now known as the ILAC Conference, was attended by 17 national delegations plus three international organisations, namely, ISO, RILEM and UN/ECE.

The 1978 conference was held in Washington and was attended by 24 national delegations with 9 international organisations participating. The new international bodies included IEC, EEC, GATT and OECD.

Sydney hosted the 1979 conference which was attended by 21 delegations and 7 international organisations. The slight reduction of numbers was due, not to a drop in interest, but to the rather long distances that had to be travelled.

Subsequent conferences have been held in Paris (1980), Mexico City (1981) and Tokyo will host ILAC 82. To date 42 countries, including some from Eastern Europe, and 12 international organisations have at various times taken part. From the Asia/Pacific area, Australia, Hong Kong, India, Indonesia, Japan, Korea, New Zealand, Papua New Guinea, Peoples Republic of China, Taiwan and Thailand, have all participated in ILAC conferences.

The conference has as its objectives to:

exchange information,
advance the state of the art, and
promote international recognition of national accreditation systems to facilitate acceptance of test data.

The major achievements of the conference so far have been

(a) identification of legal obstacles to recognition of test results produced in foreign countries and other legal impediments to trade based on technical matters.

(b) production of a directory of laboratory accreditation systems that exist around the world (not just national systems)

(c) identification, development and codification of criteria used by different laboratory accreditation systems which, hopefully, will lead to the harmonisation of those criteria with the long-term objective of bringing about recognition of laboratories from country to country.

(d) a bibliography of the literature on the subject.

Further work is underway on:

calibration systems,
laboratory quality systems, and
multilateral agreements between accreditation systems

From Australia's point of view, NATA has achieved international recognition and acknowledgement from countries ranging from the major trading blocs, Japan, the United States and the EEC to countries such as Jamaica, India and Papua New Guinea. All countries have come to ILAC with international trade in their mind, but have gone away with an appreciation of the usefulness of such systems for a nation's internal purposes.

6. GATT Code

The GATT Code has already been mentioned and the subject is worthy of further comment.

The General Agreement on Tariffs and Trade is the product of international discussions at government level. The discussions began soon after World War II and cover trade and conditions of trade in great detail. Only Western countries have chosen to participate in the discussions and agreements so far.

From the point of view of laboratory accreditation and standards the GATT Standards Code is of most significance.

The purpose of this Code of Conduct is to prevent technical matters being used as trade barriers. Specifically, standards, testing and certification should not be used in this way. In the past Europe and the U.S.A. have used technical barriers to great effect, as has Australia also.

Specifically the Standards Code, Article 5.1, calls upon the Parties to "ensure that, in cases where a positive assurance is required that products conform with technical regulations or standards, central governments apply the following provisions to products originating in the territories of other Parties:

> "5.1.1. Imported products shall be accepted for testing under conditions no less favourable than those accorded to like domestic or imported products in a comparable situation: (et seq.)
>
> "5.2. However, in order to facilitate the determination of conformity with technical regulations and standards where such positive assurance is required, Parties shall ensure, wherever possible, that their central government bodies:
>
> > accept test results, certificates or marks of conformity issued by relevant bodies in the territories of other Parties; or reply upon self-certification by producers in the territories of other Parties;
>
> even when the test methods differ from their own, provided they are satisfied that the methods employed in the territory of the exporting Party provide a sufficient means of determining conformity with the relevant regulations or standards."

The Code is therefore an international treaty aimed at giving recognition in country of import to tests performed in country of origin.

This provision is admirable in its intent. If effective, it could eliminate a major technical barrier to international trade, and at the same time reduce the cost of such trade by eliminating the routine duplication of tests of products by exporters and by importers. What is needed is a mechanism to give a high degree of confidence to both parties and some form of international laboratory accreditation is seen as essential.

The EEC, Japan and the U.S.A. have signed the Code, as has Korea, New Zealand, Pakistan, The Phillipines and Singapore. Australia has not yet signed but will do so in the near future.

It is clear that test certificates, in one form or another, are therefore going to play an increasingly important part in international trade in the future.

What is evolving is the concept of mutual recognition of national accreditation systems and several such agreements have already been completed.

7. International Agreements

To date, progress on international arrangements has been slow. This is inevitable and perhaps even desirable. There will not be one single agreement or treaty but rather a matrix of bilateral agreements which may in the long term end up as a single agreement but that is many years away. The following agreements illustrate this trend:

for some time NATA and TELARC have unilaterally accepted certificates from BCS accredited laboratories;

in 1981 NATA and TELARC concluded a mutual recognition agreement;

in 1982 NATA signed a Memorandum of Understanding with the National Standards Council of Papua New Guinea to operate a laboratory accreditation system on their behalf;

before the end of 1982 NATA will complete a Memorandum of Understanding with the U.S. Bureau of Standards for mutual recognition between NATA and NVLAP under very similar terms to the TELARC agreement;

earlier this year the British Calibration Service signed an agreement with the German calibration service and the other calibration systems of Europe may soon follow;

TELARC is currently negotiating an agreement with NATLAS in the U.K.;

NATA is currently involved in a consultancy to the Hong Kong Government which may result in some form of recognition agreement.

The pattern is clear therefore. International acceptance of test documents will be enhanced by mutual recognition of laboratory accreditation systems and this activity is being actively pursued in various countries.

8. Conclusion

Australia's development of an integrated national measurement system using voluntary accreditation of laboratories in all sections of the community has been used as a model by many other countries in establishing their own systems. National standards of measurement laboratories have had formal and informal recognition agreements for many years and in more recent times recognition of second and third echelon laboratories is being developed through international activity in laboratory accreditation.

NATA has always been essentially an organisation serving Australia's internal needs for credibility of test data but its role is now expanding to meet external needs as well.

These international activities will lead to some changes in NATA's current operations but these will only improve the already high status of the organisation, at least internationally.

It seems clear that international trade will be subjected to increasing amounts of government influence and regulation particularly where health, safety and the environment are possibly involved.

Within Australia there will be more and more pressure on raw material suppliers to test and to certify products as part of a growing awareness of total quality assurance in the manufacturing industry and manufacturers will become increasingly demanding of their suppliers.

There will also be more specification of goods and services and certification required by more users of technical products.

References

Inglis, B.D. Metrology and Quality Assurance in Australia and other papers presented at Asia/Pacific Metrology Programme, Kuala Lumper April 1982.

Forman, H.I. The Impact on International Trade of: The GATT Standards Code and International Laboratory Accreditation, Quality Assessment and Product Certification and Domestic Product Liability Laws.

Paper presented at 50th Jubilee Conference of Standards Association of NZ, Wellington 1982.

Gilmour, J.A. Laboratory Accreditation - Its Concept and Developments. Paper presented in Mexico 1981.

ILAC Task Force C. Report on Needs, Objectives, Effects and Consequences of Laboratory Accreditation. Presented at ILAC 1979.

Work of the Mass-and-Related-Quantities Group

D.B. Prowse
CSIRO Division of Applied Physics, Sydney, Australia 2070

Abstract

This paper gives a summary of the activities of the Mass and Related Quantities Group. It consists of a brief outline of the six areas in which the group is working - mass, volume, density, pressure, force and gas flow.

1. Introduction

This paper is a brief overview of the activities and capabilities of the Mass Group. There are six areas in which the group is working, these being mass, volume, density, pressure, force and gas flow. Some aspects of the work relating to mass, density and pressure are described in detail in related papers. This paper paints with a broad brush the work of the group, with a little more emphasis on those areas not described in detail in other papers.

Because this is an overview, instruments, techniques and facilities only are described. Their use in calibration is not given. The measurement capabilities of the Group are illustrated by descriptions of the range over which measurements can be made and the uncertainties associated with them. Besides particular references, general references are given for the topic where these are available.

2. Mass

The Group maintains the Australian standard of mass, which is Copy No. 44 of the platinium iridium prototype at the BIPM at Sèvres in France. All other masses in Australia ultimately derive their value from copy no. 44. Prowse (1982b) describes this process in more detail.

Six stainless-steel 1 kg masses have been especially set aside for comparison with no. 44. These then form the basis of the calibration of other masses of both higher and lower denominations. These 6 masses are compared with no. 44 on the 1 kg standardising balance made by Stanton to an NPL design. This balance has a resolution and repetition of about 5 μg and a period of swing of 2 minutes. It is a tedious balance to use and so is used only for weighing 1 kg masses when the highest accuracy is required.

All other masses are weighed on a series of 2-pan balances which have an uncertainty of about 1 part in 10^7 - for balances up to 10 kg capacity. These balances have capacities of 10 kg, 1 kg, 200 g, 20 g and 2 g. Recently a Sartorius ultramicro balance of capacity 3 g and readability 0.1 μg has been obtained.

For weighings above 10 kg we have 30 kg, 100 kg and 1000 kg balances. The latter balance works on the substitution principle and has flexure pivots in place of knife edges. Each of these balances has an accuracy of better than 1 part in 10^6.

Although all reference masses are stainless steel, most of the larger working standards (above 1 kg) are of aluminium bronze.

Because all masses are immersed in a fluid (air) it is necessary to know both the density of the air and the density of the masses for accurate work. Stainless steel mass standards have a density of about 7800 kg/m^3, while brass, which was once used for masses, has a density of 8400 kg/m^3. It has become generally agreed throughout the world that most masses in general use have an assumed density of 8000 kg/m^3. Thus in use masses are assumed to have this density and to have been weighed in air of density 1.2 kg/m^3. Under normal laboratory conditions the errors introduced by making these assumptions is less than 3 parts in 10^7.

For accuracies greater than this it is necessary to work on what is termed the 'true mass basis'. In this case the mass is calculated taking into account the actual densities of the objects being compared, and the measured density of the air.

When comparing Pt-Ir kilograms with stainless-steel masses it is a knowledge of the density of the air that determines the ultimate accuracy. Most standardising balances used for these weighings have a precision of 5 μg or better. To obtain actual mass values to this accuracy it is necessary to know the air density to 5 parts in 10^5. The normal method of determining the air density is to measure the pressure, temperature, relative humidity and CO_2 content (or assume the composition). Jones (1978) states that with state-of-the-art measurements the air density can be determined by this means to 2 parts in 10^4.

Recently the BIPM produced the best formula for calculating the air density (Giacomo 1982). This formula should eventually replace the many variations used around the world.

In our Group a different approach has been made to the air-density problem. The aim is to make direct and continuous measurements of the air density. This is done by measuring the buoyancy difference between two objects of equal mass and surface area but differing in volume by about 1000 cm^3. The objects are a glass sphere (which can be evacuated) and a glass ring. These are weighed on a two-arm balance which has been modified to give an electrical readout. This readout is fed into a computer to program the calculations and give a continuous readout of the air density. This work is described further by Prowse (1982a).

3. Volume

In general all measurements of volume are made by weighing the amount of liquid (usually water) either contained or delivered. Most volumes can be determined with an uncertainty of about 1 part in 10^4 or 10^5.

The calibration of volumetric glassware is a well-established technique and can in many cases be done by the user provided a balance of the appropriate capacity is available. This Division will no longer calibrate such apparatus on a routine basis; however, the facilities are still available for special calibrations. A set of tables giving the corrections due to temperature for vessels of different materials has been prepared, (Humphries 1979). For further information on volumetric calibration see NPL (1955).

Because we have some large-capacity balances (up to 1000 kg) we are able to provide a calibration facility for large-capacity volumetric measures and mechanical displacement provers. The calibration of these is of great importance to the oil industry. We do not calibrate flow-metering devices as these can be measured in terms of the volumetric measures or provers by commercial firms.

4. Density

The method of establishing density standards has not changed since Archimedes had his fateful bath - the methods of measurement have merely become more refined.

In this Division the density standard is a sphere of approximately 75.8 mm diameter made of ultra-low-expansion (ULE) glass. This sphere has been polished to be as nearly spherical as possible and a large number of diameters have been measured by means of a special interferometer. From this it is possible to determine the volume with an uncertainty of 0.4 parts in 10^6. Using this sphere it is possible to determine the density of liquids, and solids by transfer via a liquid.

Over the past few years the main research in density has been in determining the density and thermal dilation of air-free water over the range 4 to 40°C. This has been done both by measurements with the ULE sphere and with a ULE dilatometer which contained the water (Bell 1982).

Measurements of the effect of dissolved air on the density of water have been made with a magnetic float. The position of the float is detected optically and the current required to hold the float in position is a measure of the density (Bignell 1982a). Measurements over the range 4 to 20°C indicate that the density decreases with dissolved air, the magnitude being about 2.5 parts in 10^6 at 20°C. This indicates that care is required in the use of water as a density standard when uncertainties of less than 5 parts in 10^6 are required.

Since the definitive work by Cook (1961) the density of mercury has been considered to be known to 1 part in 10^6. However some recent evidence has indicated that different samples of mercury may differ by more than 3 or 4 parts in 10^6. Cook found that samples of pure mercury should not differ from his value by more than 1 part in 10^6. Work has commenced in the laboratory to compare the density of different samples of mercury to accuracies approaching a few parts in 10^7. This will be done by weighing a polished cylinder of tungsten carbide in different samples.

Most of the routine work of the Group consists of the calibration of hydrometers. This is referred to further in the paper by Patterson (1982).

5. Pressure

Pressure standards are maintained over more than 11 orders of magnitude from 1 mPa to 300 MPa. The pressure range can be divided into 3 sub ranges

(a) Vacuum - 1 mPa to 1 Pa
(b) Atmospheric - 1 Pa to 100 kPa
(c) High pressure - above 100 kPa

The divisions are somewhat arbitrary and are made merely for convenience. The boundaries between the ranges are a grey area and change according to a number of factors.

Because the range covered is so large, no single instrument can hope to cover it and often the sub ranges are divided further.

(a) Vacuum

Work is in progress to establish vacuum standards in the range 10^{-1} to 10^{-4} Pa. At present the system consists of a stainless-steel vacuum chamber which can be baked to 300°C. The chamber is pumped by a turbomolecular pump through a 100 mm bakeable valve. Ion gauges calibrated at NPL will be used as standards in the initial stages of the project. The method requires the connection of the standard gauge and the test gauge to the chamber, then evacuation to a factor of 100 below the lowest calibration point whilst baking the whole system. The valve to the pump is closed and the readings on the gauges are compared at various pressures (obtained by leaking gas into the chamber).

Ultimately it is intended to use this chamber as the final stage of a static expansion system. By this means it will be possible to calibrate the ion-gauge standards without recourse to other laboratories. The static-expansion method of calibrating vacuum gauges works by taking a small measured volume of gas at a pressure sufficiently high that it can be accurately measured, and then allowing this to expand into a larger known volume. From the known geometry the pressure in the final stage can be calculated. Generally 3 or more expansion stages are used. For further information on this method see Berman (1979).

It is quite difficult to provide adequate calibration from 1 Pa to 100 Pa. It is an important range because it is covered by a number of sensitive electronic transducers which require calibration. There are also a number of pressure transducers that measure differential pressure in this range. These can be calibrated by precision mercury or water manometers, but that is a different problem and will not be discussed here.

Calibrations in the 1 Pa range are done with a McLeod gauge, which operates in an inverse fashion to the expansion method described above. Gas from a known volume is compressed into a small capilliary tube using mercury and the length of the gas column is a measure of the pressure in the original volume. (Leck 1964). A simple method for calibrations in this range is a modification of the expansion method described by Prowse and Caw (1980). This uses atmospheric pressure for the reference and because the pressures measured are relatively high outgassing is not a problem.

(b) Atmospheric

Pressures from 1 Pa to 100 kPa can be measured with manometers or barometers. In particular the interferometric manometer (Harrison et al 1976) has a precision of 0.1 Pa and can be used over this range. This and similar instruments are described by Bignell (1982b).

(c) High Pressure

In this range the pressure standards are generally free-piston pressure gauges, using either oil or gas as the medium. These are described in detail by Bignell (1982b). In general the accuracy obtained with these instruments is about 1 in 10^4.

6. Force

The Group maintains the force standards for Australia, which extend from zero to 10 MN. There are three instruments to provide these standards.

(a) Forces up to 5.6 kN

This consists of a small dead-load machine, made by Morehouse, in which the loads are applied manually by the operation of levers. Loads from 50 N to 5.6 kN may be applied in increments of 10 N in either tension or compression. The weights in the machine are known to within 1 part in 10^5 of their nominal value.

(b) Forces in the Range 5.0 kN to 550 kN

These are obtained by a dead-load machine built by Avery. There are 4 stacks of weights which are manipulated hydraulically under computer control. The machine as constructed has the value of the weights in pounds; these are listed in the table below. It should be noted that the minimum force that can be applied is limited by the weight of the loading frame - 1100 lbf (4.9 kN). The next force is 2000 lbf (8.9 kN) because the 900 lbf (4.0 kN) pan is the next load applied. After that forces may be incremented in units of 200 lbf (0.9 kN).

It is intended to convert the machine to newtons but unfortunately the pound values do not lend themselves to easy conversion to powers of ten. The proposed values are shown in the table below. The conversion will be achieved by 'hanging' weights of about 21 kg on each of the 2000 lb (907 kg) weights and proportional amounts on the other weights.

To overcome the gap between 1100 lbf (4.9 kN) and 2000 lbf (8.9 kN) it is proposed to add a fifth stack of weights. This will also enable any force in increments of 200 N between 5.0 kN and 550 kN to be achieved.

Weight Stack	Number of weights in each stack	Total force in each Weight stack lbf	Value of each weight after conversion to metric kN
1100 lbf loading frame	1		5.0
4000 lbf pan	1		18.0
2000 lbf pan	1		9.0
900 lbf pan	1		4.0
4000 lbf stack	14 (28x2000 lbf)	5100 + 56000	18.0
2000 lbf stack	14	3100 + 28000	9.0
1000 lbf stack	9	2000 + 9000	4.5
200 lbf stack	10	2000 + 2000	1.0

The Avery machine operates in a manner which may be considered to be the reverse of the usual operation of such machines. Normally to increase the load extra weights are 'hung' onto the crosshead which transmits the load to the instrument being tested. The Avery machine has a loading frame which is rigidly attached to the crosshead. In the rest position all the weights sit on the scalepans and to obtain zero load the weights are lifted off by the hydraulic system. This method gives great flexibility enabling any number of weights from any stack to be applied at any instant without having to go to zero load. Also any swinging or vibration in the hanging stack is transferred to the hydraulic support system, not the instrument under test.

All the weights and components contributing to the force have been weighed on the 1000 kg balance described in section 2 and have been adjusted to be equal to their nominal value to within 1 part in 10^5. This eliminates the need to apply corrections.

The machine is controlled by a HP 9835 computer which receives its input signals from potentiometers operated by the raising and the lowering of the stacks. Considerable effort has been expended in the programming to protect the instrument and the machine from failures, power blackouts or operator mistakes. This occupies most of the 120 k of computer memory available. At the conclusion of the calibration the report on the instrument is printed by the computer.

(c) Forces in the Range 0.5 MN to 10 MN

These forces are generated by an Amsler hydraulic testing machine. Unlike the dead-load machines this is not a fundamental machine as it uses a piston to generate force. It must therefore be calibrated in some way.

The calibration is obtained by measuring the diameter of the piston and cylinder and using a free-piston pressure gauge to measure the pressure in the ram. The machine has not been designed to enable the piston to be rotated and errors due to static friction will occur. These are allowed for by taking measurements with the piston both rising and falling. The accuracy achieved with this machine is approximately 0.2%.

The machine is designed to apply compression forces only. To enable it to be used in tension a cross-over rig has been constructed to produce forces in tension up to 5 MN.

Information on force calibration in general is given by Swindells and Debnam (1977).

7. Gas Flow

The gas-flow section of the Group is oriented to the development of the use of critical-flow techniques and their application to the calibration of gas-flow measuring instruments. The laboratory as such has no facilities for the calibration of flow devices and has no plans to expand into this field.

If the pressure drop across a nozzle is greater than a factor of 2 then sonic flow occurs and the nozzle is said to be a critical-flow nozzle. The pressure however is absolute pressure, not a pressure difference. The Mach number is always unity in the throat regardless of the gas properties and the mass flow through the device does not vary when the downstream pressure varies over a finite range. The downstream pressure must not rise sufficiently to cause the flow in the throat to become subsonic. It is this property of constant mass flow which has enabled critical-flow nozzles to be widely used for flow calibration and control. However the nozzles do require calibration.

It has been shown in our laboratory (Caw et al 1977) that it is possible to use critical flow nozzles in a fluid analogy of the Wheatstone-bridge network. Four critical-flow nozzles take the place of the resistors in the bridge, and a differential pressure gauge replaces the voltmeter. Such a bridge may be used to make comparisons between nearly equal nozzles at equal pressures over a required pressure or Reynolds-number range. It may also be used to compare a single large nozzle with two small ones connected in parallel to make a total conductance, equal to that of the single large one.

Thus a flow scale may be generated, 1, 2, 4, 8, for example, and in principle the calibration of one member of the set over a range of pressures should be sufficient to determine the coefficients of all the others over this pressure range. Where '1' is the nozzle of smallest area, '2' is a nozzle of double this area, and so on. The bridge system is tolerant of slowly fluctuating supply pressure or gas temperature.

Work is continuing on the use of pressure dividers (i.e. two nozzles in series, or half of the Wheatstone-bridge arrangement) to relate the coefficient of a critical-flow nozzle to the Reynolds number without recourse to absolute calibration. Bridges will not give information on the conductance at one pressure relative to the conductance at another pressure.

By using bridge methods it is possible to compare critical-flow nozzles with an uncertainty of 1 to 2 parts in 10^4.

Conclusion

A brief description outline has been given of the standards and calibration activities, Mass and Related Quantities Group. Some of the research activities have only been touched upon as they relate to the specific work described.

The staff of the group consists of:

3 research scientists
3 experimental officers
7 technical officers
1 technical assistant.

References

Bell, G.A. and Patterson, J.B. (1982). Precision Measurements and Fundamental Constants II, Ed. by B.N. Taylor and W.D. Phillips, Nat. Bur. Stand., Spec. Publ. 617, in press.

Berman, A. (1979). *Vacuum*, 29, 417.

Bignell, N. (1982a). *J. Phys. E: Sci. Instrum.*, 15, 378.

Bignell, N. (1982b). These Proceedings.

Caw, W.A., Bryant, N.W., and Bell, G.A. (1977). 6th Australasian Hydraulics and Fluid Mechanics Conference, Adelaide.
Cook, A.H. (1961). *Phil Trans. Roy. Soc. Series A.*, **254**, 17.
Giacaomo, P. (1982). *Metrologia*, **18**, 33.
Harrison, E.R., Hatt, D.J., Prowse, D.B., and Wilburham, J. (1976). *Metrologia*, **12**, 115.
Humphries, J. (1979). The Calibration and Verification of Volumetric Measures. Division of Applied Physics Technical Paper No. 1.
Jones, F.E. (1978). *J. Res. Nat. Bur. Stand.*, **83**, 419.
Leck, J.H. (1964). Pressure Measurement in Vacuum Systems (Chapman & Hall).
NPL (1954). Balances, Weights and Precise Laboratory Weighing. Notes on Applied Science No. 7.
NPL (1955). Volumetric Glassware. Notes on Applied Science No. 6.
Patterson, J.B. (1982). These proceedings.
Prowse, D.B. and Caw, W.A. (1980). *J. Phys. E: Sci. Instrum.*, **13**, 278.
Prowse, D.B. (1982a). Precision Measurements and Fundamental Constants II, Ed. by B.N. Taylor and W.D. Phillips, Nat. Bur. Stand. Spec. Publ. 617, in press.
Prowse, D.B. (1982b). These proceedings.
Swindells, B. and Debnam, R.C. (1977). Measurement of Force by Elastic Devices. (HMSO, London).

Calibration of Masses

D.B. Prowse
CSIRO Division of Applied Physics, Sydney, Australia 2070

Abstract

The calibration of sets of masses starting with a 1 kg standard is described in detail. The principles of designing weighing schemes suitable for least-squares analysis are presented. The propagation of errors and calculations of uncertainties are analysed. Because buoyancy corrections are required for precision weighings, sections are included on mass values and the calculation of air density.

1. Introduction

This paper describes how to obtain accurate values for sets of masses from 1 kg to 1 mg and from 1 kg to 10 kg, starting with a 1 kg standard. Various schemes are discussed and it is shown how the uncertainties may be calculated. Because the observations are extremely time-consuming this analysis is mainly suitable for the calibration of reference masses of the highest precision. For comparisons at this level it is necessary to understand the mass basis of the comparison and the buoyancy corrections due to the density of air. These are dealt with in section 3. A knowledge of balances and weighing procedures is assumed.

2. Terminology

There is considerable confusion, and divergence of opinion, over the use of the words *weight* and *mass*. To help in understanding the way they are used in this paper the following definitions are given.

The noun *weight* is used to denote a force or the object producing the force.

The noun *mass* is used to denote an object which has a mass value. It is this property which is under consideration in this paper.

The verb *to weigh* denotes the process by which the mass value of a mass (or an object) is determined.

By these definitions the *weight* of an object is measured in *newtons*, whereas the *mass* is measured in *kilograms*.

In section 5 the words *weight, weights* etc. are used in the statistical sense of *weighting*. However this use will be clear from the context.

3. Mass and Mass Values

When an object is weighed in air it experiences an upthrust or buoyant force (loss of weight) equal to the mass of air displaced. It is this buoyant force which causes many problems and much confusion in weighing. It is not practical to weigh in vacuum because of the surface effects that would occur on the objects being weighed. However, what is termed the 'true mass' of an object is the mass that would be measured in a vacuum, providing everything else (surface layers, etc) was unchanged. Thus in a standards laboratory it is true mass values which are measured and all other values are calculated from these.

The following nomenclature is used in this section:

ρ - air density kg/m^3

V - volume of a mass m^3

d - density of a mass kg/m^3

α - coefficient of cubical expansion of a mass/°C

t - temperature °C

M - true mass kg.

If the mass is measured on a weighing system (e.g. a spring balance) it would be equal to M*, where

$$M^* = M - \rho V$$
$$= M\left(1 - \frac{\rho}{d}\right)$$

If two objects, denoted by subscripts 1 and 2, are weighed then the difference in mass is given by

$$\delta M = M_1 - M_2 - \rho_1 V_1 + \rho_2 V_2 \qquad (1)$$

Thus if M_1 is known then M_2 can be calculated provided that ρ_1, ρ_2, V_1 and V_2 are also known.

If the weighings are done at nearly the same time, and under the same conditions, then $\rho_1 = \rho_2$, and if the masses are made of the same material then $V_1 \simeq V_2$ and so (1) reduces to

$$M_2 = M_1 - \delta M.$$

This is the case which is normally encountered in the calibration of masses. When comparing objects made from differing materials then $V_1 \neq V_2$ and the effect of air buoyancy must be calculated or be sufficiently small that it can be neglected for the accuracy required.

When the latter is done the value obtained is said to be the *mass (weight) in air* of the object. This is the value of the masses (usually stainless steel) required to balance the object in air of nominal density 1.2 kg/m^3. Take as an example weighing out a quantity of water M_w that balances a stainless-steel mass M_{ss} of density 8000 kg/m^3.

If ρ = 1.2 kg/m^3 then

$$M_w\left(1 - \frac{1.2}{1000}\right) = M_{ss}\left(1 - \frac{1.2}{8000}\right).$$

Hence $\quad M_w = M_{ss}\ (1.00105)$ kg.

This means that if the stainless-steel mass balanced against the water is 1 kg then because of the upthrust due to the air the mass of water balancing the stainless steel is approximately 1.001 kg, i.e., a difference of 1 g in 1000 g.

In this case the true mass differs from the mass in air by 1 g.

Because the density of the material used for masses changes from one set to another, the idea of an *apparent mass value* has arisen. The apparent mass, M_a, is the amount of any specified material which will 'exactly' balance the unknown in a specified atmoshpere, $\rho_o = 1.2$ kg/m^3, at the specified temperature t = 20°C.

For an object with a true mass M, made from material of density d_M, the calculated apparent mass value is:

$$M_a = M[1 - (\rho_o/d_M)]/[1 - (\rho_o/d_a)] \qquad (2)$$

where d_a is the density of the appropriate 'ideal' material. The internationally accepted value for d_a is 8000 kg/m^3 or 8.0 g/cm^3, at 20°C. This is commonly called the *8.0 basis*.

Therefore to convert from a true mass basis to the 8.0 basis, equation (2) gives

$$M_{8.0} = M(1 - \frac{1.2}{d_M})/(1 - \frac{1.2}{8000}). \qquad (3)$$

For stainless steel of density 7800 kg/m^3, equation (3) gives

$$M_{8.0} = M(0.999\ 996\ 15).$$

Therefore when stainless steel reference masses are calibrated in terms of the platinum-iridium standard on a true mass basis it is necessary to apply a correction of almost 4 parts in a million to them so that they can be used for calibrations on the 8.0 basis.

When brass was used extensively for standard masses the mass basis then used was 8.4. To convert from 8.0 to 8.4 equation (2) gives:

$$M_{8.4} = M_{8.0}(1 - \frac{1.2}{8000})/(1 - \frac{1.2}{8390.9})$$

$$= M_{8.0}(0.999\ 993\ 01)$$

where: 8390.9 kg/m^3 is the density of brass at 20°C, the density at 0°C is 8400 kg/m^3.

For accurate work it may be necessary to allow for the change in density of the masses with temperature. This can be done by means of the formula:

$$V_t = V_o[1 + \alpha(t - t_o)].$$

For some masses with a long history the volume may have been measured in millilitres (ml). The relationship between the old litre and the cubic metre is

$$1L = 1.000\ 028 \times 10^{-3}m^3.$$

The change in buoyancy δM_b for a change $\delta\rho$ in air density when comparing masses that differ in density δd (or volume δV) is given approximately by

$$\delta M_b \simeq \delta\rho\ \delta V$$

$$= \frac{M}{d^2}\ \delta\rho\ \delta d \quad \text{kg}$$

(This formula only applies if δV is small, i.e., if the densities of the two masses are approximately equal.) Therefore comparing masses of different materials with a standard of density 8000 kg/m^3, will give the buoyancy force corrections shown in Table 1 for air density changes of 0.08 kg/m^3. This change in air density is the maximum that could be expected for most laboratories.

Table 1

Buoyancy corrections for a change in air density of 0.08 kg/m^3, relative to a standard of density 8000 kg/m^3.

Material	Density kg/m^3	δd kg/m^3	$\delta M_b/M$
Stainless steel	7 800	200	2.57×10^{-7}
Brass	8 400	400	4.78×10^{-7}
Aluminium	2 700	5 300	1.97×10^{-5}
Water	1 000	7 000	7.04×10^{-5}
Lead	11 300	3 300	3.1×10^{-6}
Tantalum	16 600	8 600	5.2×10^{-6}
Platinum Iridium	21 500	13 500	6.3×10^{-6}

From Table 1 it can be seen that except for the most precise weighings or under very adverse conditions there is no need to apply corrections for air density when comparing masses of different stainless steels, or when converting from a true mass basis to an 8.0 mass basis.

The subject of mass and mass values is discussed in detail by Pontius (1974).

4. Calculation of Air Density

(a) Air Density Equation

As shown in the previous section it is not necessary to measure the air density for the normal calibration of stainless-steel masses. For weighings of the highest accuracy, particularly where materials of different density are used, the air density must be measured. The usual way of measuring the air density is to record the pressure, temperature, relative humidity and the composition (although it is more usual to assume a fixed composition). The air density is then calculated by means of a formula. Most standards laboratories have their own formula, often significantly different from others.
A working party arranged by the BIPM has produced a formula (Giacomo 1982) which it is hoped will be used by all laboratories. It is this formula which is described in this section.

The air density in kgm^{-3} is given by

$$\rho = [3.483\ 53 + 1.44\ (x_{CO_2} - 0.0004) \times 10^{-3}\ \frac{P}{ZT}\ (1 - 0.3780\ x_v) \qquad (4)$$

where $x_v = h\, f(P,t)\, \dfrac{P_{sv}(t)}{P}$

$f(P,t) = 1.000\,62 + 3.14 \times 10^{-8}\, P + 5.6 \times 10^{-7}\, t^2$

$P_{sv}(t) = \exp\,[1.281\,180\,5 \times 10^{-5}\, T^2 - 1.950\,987\,4 \times 10^{-2}\, T + 34.049\,260\,34 - 6.353\,631\,1 \times 10^3/T]$

u = relative humidity in %

$h = u/100$

P = pressure in Pa

$T = 273.15 + t$ = air temperature in kelvin

t = air temperature in °C

x_{CO_2} is the molar fraction of carbon dioxide present which can, to sufficient accuracy, be set equal to 0.0004.

Z is the compressibility factor given by:

$$Z = 1 - \frac{P}{T}\,[1.624\,19 \times 10^{-6} - 2.8969 \times 10^{-8} t + 1.088 \times 10^{-10}\, t^2 + (5.757 \times 10^{-6} - 2.589 \times 10^{-8}\, t)x_v + (1.9297 \times 10^{-4} - 2.285 \times 10^{-6}\, t)\, x_v^{\,2}] + \frac{P^2}{T^2}\,[1.73 \times 10^{-11} - 1.034 \times 10^{-8}\, x_v^{\,2}]$$

(b) Relative Humidity

The measurement of relative humidity is probably the most difficult to perform. There are a number of electrical instruments available all of which require calibration, but probably the best and most accurate instrument is the wet-and-dry-bulb thermometer. If this is used then the relative humidity is calculated by means of equation (5) which is obtained from BS 1339 (1965).

$$U = 100e/e'' \qquad (5)$$

where $e = e' - 6.66 \times 10^{-6} P(t - t')$

$\log_{10} e'' = 7.5t\,(237.3 + t) + 0.785\,71$

$\log_{10} e' = 7.5t'(237.3 + t') + 0.785\,71$

t = dry bulb temperature in °C

t' = wet bulb temperature in °C

P = atmospheric pressure in Pa.

(c) Uncertainties in the Calculation of Air Density

Giacomo (1982) distinguishes between the uncertainty due to equation (4) itself and the uncertainty due to the quantities measured at the time of application of the equation. He gives separate estimates for each of these and the various components and leaves it to the user to decide how to combine them. Jones (1978) estimates that with state-of-the-art measurements of pressure, temperature, relative humidity and composition the uncertainty in the calculated air density is 3×10^{-4} kg/m^3. Prowse (1982) has made absolute measurements of air density and compared them with calculated values and obtained general agreement to 2×10^{-4} kg/m^3.

5. Calibration of Reference Standards of Mass

In this section the problem of determining the values of a set of reference masses starting with a known 1 kg mass is considered. In principle the problem is straightforward - compare masses in combinations so that the masses in each pan of the balance are nominally equal, and then calculate those that are unknown. The problem consists in determining the best combinations that will give the least uncertainty with the fewest weighings. There is no unique solution to this problem but there are a number of satisfactory solutions.

A number of papers have been written on this subject; see for example Hayford (1892), Hotelling (1944), Mood (1946), Bell (1949), NPL (1954), Pontius (1966), Prowse and Anderson (1974), Cameron et al (1977), Grabe (1978), Zuker et al (1980), Hunt (1981). In all of these papers the method of least squares is used to solve a set of linear equations. To aid in understanding the rest of the paper the following is an example of a simple calibration scheme with combinations of masses.

The mass values of a set, consisting of values 10, 5, 2, 2', 1, 1' may be determined in terms of the '10' by weighing the following combinations:

$$
\begin{aligned}
10 &= 5 + 2 + 2' + 1 + a_1 \\
5 &= 2 + 2' + 1' \quad + a_2 \\
2 &= 2' \quad + a_3 \\
2' &= 1 + 1' \quad + a_4 \\
1 &= 1' \quad + a_5
\end{aligned}
$$

The values a_1 .....a_5 are necessary because the masses will not be equal to their nominal values and the balance is not perfect. These equations may be rewritten in the following form

$$
\begin{aligned}
5 + 2 + 2' + 1 &= 10 - a_1 \\
5 - 2 - 2' - 1' &= a_2 \\
2 - 2' &= a_3 \\
2' - 1 - 1' &= a_4 \\
1 - 1' &= a_5
\end{aligned}
$$

These are called observational equations and in this case there are 5 equations for 5 unknowns (or unknown parameters). For a least-squares solution the number of equations must exceed the number of parameters. In matrix notation these equations are written

$$\begin{array}{c} \begin{array}{ccccc} 5 & 2 & 2' & 1 & 1' \end{array} \\ \left[\begin{array}{ccccc} 1 & 1 & 1 & 1 & \\ 1 & -1 & -1 & & -1 \\ & 1 & -1 & & \\ & & 1 & -1 & -1 \\ & & & 1 & -1 \end{array}\right] \end{array} \left[\begin{array}{c} 5 \\ 2 \\ 2' \\ 1 \\ 1' \end{array}\right] = \left[\begin{array}{c} 10-a_1 \\ a_2 \\ a_3 \\ a_4 \\ a_5 \end{array}\right]$$

All the blank spaces are zero.

The matrix on the left-hand side gives the design scheme. It is the design matrix that is discussed in the remainder of this paper, the column vectors are in general omitted.

A brief discussion of the method of least squares, using matrix theory, is given in appendix 1. Because most computers (even some desk-top models) have matrix packages, it is not necessary to understand the matrix theory in order to perform the least-squares calculations. It is however, necessary to realise that the variance - covariance matrix is obtained from the inverse of the normal-equations matrix. This is done by multiplying the inverse matrix by the sum of squares of residuals divided by the degrees of freedom. The degrees of freedom is calculated by subtracting the number of unknown parameters (k) from the number of observations (n), i.e. (n-k), in the notation used in appendix 1. The diagonal terms of the variance-covariance matrix are the variances of the unknowns and the off-diagonal terms are the covariances (or correlation) between any two unknowns.

In designing a weighing scheme the following principles need to be considered.

(i) The variances of the unknown parameters should be as small as possible.

(ii) The covariances should be zero or negligibly small compared with the variances.

(iii) The variances of the unknown parameters should be equal. This means that the masses are used an equal number of times in the weighing scheme.

(iv) The scheme should be efficient. This means that the desired variances should be obtained with the smallest number of weighings possible.

Although each of these principles needs to be considered, the most important is (ii) - particularly that the covariances should be zero. When this occurs the design matrix is said to be orthogonal. For a given number of observations the variances will be a minimum, or close to a minimum, if the matrix is orthogonal. If the matrix is not orthogonal then the unknown parameters are correlated in an artificial manner which depends upon the choice of the design matrix X. If the unknown parameters are needed in further calculations, as a result of using the masses for calibration or weighing, then their variances and covariances have to be inserted in a propagation-of-error formula implying correlations, see equation (8). The interpretation of the propagated errors will be doubtful, since in this case the formula depends upon the choice of the design matrix X, which to a large extent is arbitrary.

It is generally agreed (see e.g. Hotelling 1944) that comparing masses at the maximum load of the balance tends to minimize the variance of the unknown parameters. This is because the fixed variance associated with any given weighing is spread over the largest total mass possible. In addition, such weighings tend to maximize the total number of times each mass is used. This conflicts with the orthogonality condition as it has not been possible to find orthogonal schemes with all weighings at maximum load. Prowse and Anderson (1974) give a scheme with all the weighings at maximum load.

If sufficient masses are added to a set it is possible to design a scheme which gives small variances but is not very efficient, since a large number of observations are required. Another set with considerably less observations may give only marginally larger variances. It is therefore important to consider the efficiency of a scheme.

(a) Efficiency

Although there are a number of possible definitions of efficiency the one used by Prowse and Anderson (1974) will be used here.

$$E_i = (1/v_i)/n$$

$$i = 1,\ldots\ldots k \tag{6}$$

where v_i = diagonal element of the matrix $(X^{*\prime}X^*)^{-1}$, (i.e. the inverse of the normal equations matrix), and n = number of observations.

An overall efficiency can be defined by:

$$E = (\sum_i 1/v_i)/n \tag{7}$$

which may be looked upon as the effective number of single comparisons of unit weight per observation that would be required in order to obtain the same accuracy as the proposed scheme. Thus the larger the value of E (or E_i) the better the scheme. The maximum values of E_i and E are 1 and n, respectively.

(b) Orthogonal Design Schemes

Grabe (1978) produced the first orthogonal matrix in relation to the equations for the calibration of the submiltiples of a unit of mass of denominations 10, 5, 3, 2, 1, 1'. This scheme is given in appendix 4. Grabe made the important point of allowing a number of weighings to be repeated.

Subsequently Zuker *et al.* (1980) made a systematic search for orthogonal systems using a computer. For the system 10, 5, 2, 2', 1, 1' (which is the one which will be discussed in this paper) the two simplest systems are

(a)

5	2	2'	1	1'
1	1	1	1	
1	1	1		1
1	-1	-1	-1	
1	-1	-1		-1
	1	-1	1	-1
	1	-1	-1	1
	1	-1		
	1	-1		
	1		-1	-1
	1		-1	-1
		1	-1	-1
		1	-1	-1
			1	-1
			1	-1

(b)

5	2	2'	1	1'
1	1	1	1	
1	1	1		1
1	-1	-1	-1	
1	-1	-1		-1
	1	-1	1	-1
	1	-1	1	-1
	1	-1	-1	1
	1	-1	-1	1
	1		-1	-1
	1		-1	-1
		1	-1	-1
		1	-1	-1

These two schemes yield the same normal equations and inverse matrices, as follows:

Normal equations matrix

$$(X^{*\prime}X^{*}) = \begin{bmatrix} 4 & & & & \\ & 10 & & & \\ & & 10 & & \\ & & & 10 & \\ & & & & 10 \end{bmatrix}$$

Inverse matrix

$$(X^{*\prime}X^{*})^{-1} = \begin{bmatrix} 0.25 & & & & \\ & 0.1 & & & \\ & & 0.1 & & \\ & & & 0.1 & \\ & & & & 0.1 \end{bmatrix}$$

In these schemes the first two weighings are made against the standard '10'.

There is little to choose between the two systems: (b) is slightly more efficient however (a) has two extra equations which provide a cover of all possible combinations. Although orthogonal, these schemes have only two weighings at maximum load and the '5' is used 4 times while the other masses are each used 10 times.

Ignoring for the moment the uncertainty in the standard (i.e. the '10') then the variance of the unknown parameters (the values of the masses) are:

Mass	Variance
5	$0.25\ \sigma^2$
2, 2', 1, 1'	$0.1\ \sigma^2$

where σ^2 = (sum of squares of residuals)/(n-k).

The covariances are zero.

In calibrating a set of masses the first decade is calibrated in terms of the 1 kg as the standard. The next decade is then calibrated using one of the two 100 g masses as the standard and so on down to 1 mg.

When performing the weighings for such a calibration it would be usual to use a balance with a capacity of 1 kg for all the weighings in scheme (a) down to the last two which would be done on a 100 g balance. If that is the case then these two weighings are more accurate and should be appropriately 'weighted'. This is discussed in detail by Prowse and Anderson (1974) and the inclusion of weighting factors (γ_i) in the least-squares analysis is described in appendix 1. Thus the weighting factors for each equation in scheme (a) would be

$$1, 1, 1, 1, 1, 1, 1, 1, 1, 1, 1, 1, 100, 100$$

where it is assumed that the next balance is 10 times more accurate then the one for the heavier weighings giving a weighting factor of 100. In practice, for good standardising balances, it appears that a factor of 10 is more appropriate than 100.

Incorporation of these weighting factors does destroy the orthogonality of the design matrix, but because it only involves the two unit masses these alone have a covariance. Since most sets do not have two unit masses, the 1' is usually added from another set and as they are not used together in calibrations the correlation between them is unimportant.

It is apparent that weighting cannot be used with scheme (b). However there are a number of balances with a capacity of 2 (or even 3) instead of 1. If weighting is used for all the weighings that could be performed on these balances then the orthogonality would be destroyed. It is therefore appropriate to use these balances for whatever weighings are within their capacity (as they are more accurate) but to ignore the weighting factors.

For the weighting factors as given above the normal equations matrix and the inverse matrix for scheme (a) are:

$$(X^{*\prime}X^{*}) = \begin{bmatrix} 4 & & & & \\ & 10 & & & \\ & & 10 & & \\ & & & 208 & -198 \\ & & & -198 & 208 \end{bmatrix} \qquad (X^{*\prime}X^{*})^{-1} = \begin{bmatrix} 0.25 & & & & \\ & 0.1 & & & \\ & & 0.1 & & \\ & & & 0.0512 & 0.0488 \\ & & & 0.0488 & 0.0512 \end{bmatrix}$$

The weighting factors are obtained from ratio of the squares of standard deviations of the balances used. However these do not need to be known very accurately as a change in the factor from 25 to infinity gives only a 9.5% improvement in the value of the appropriate inverse matrix elements v_i. For $\gamma_i = \infty$, $v_i = 0.05$.

In appendix 2 the calibration of a set of masses is presented using scheme (a).

The use of such schemes is not warranted for the routine calibration of masses. However in scheme (b) if the four repeated observations are ignored then there are only eight weighings for five unknowns. In using the scheme the 'repeated' observations are only weighed once but entered twice in the calculations. In this case $n-k = 3$ should be used in calculating the variance - covariance matrix. If this is done then it gives a very satisfactory method for the routine calibration of masses with an assessment of the uncertainty of the process.

It was pointed out by Zuker *et al* (1980) that an orthogonal scheme cannot be found when the standard is not the largest value in the set. This means that an orthogonal scheme cannot be used in the above fashion for calibrating masses above 1 kg. This is discussed further in section 5(e).

(c) Calibration of Nominally Equal Masses

The calibration of a number of masses of the same denomination is a different problem; the number of possible weighings can be large and it is not obvious whether they should be compared singly, in pairs, etc. Banerjee (1975) discusses weighing problems in relation to the weighing of objects. Nearly all of the schemes in his book assume that large mass differences between the combinations of objects can be read from the balance, or obtained by means of standards. However he does give a description of Hadamard matrices, which can be used to calibrate nominally equal masses.

Hadamard showed that if the elements of a square matrix X (of dimension k) are equal to ±1 then the determinant has a maximum value of $k^{k/2}$ and the matrix is then orthogonal in the sense that (X'X) is diagonal. The non zero elements of (X'X) are all equal to k and if X is a weighing scheme the masses have a minimum variance of σ^2/k.

The only Hadamard matrices of real interest for the calibration of masses are those for which the masses on each side of the balance are equal. This appears to restrict then to those with $k = 2^{2m}$ (where m is a positive integer). Practical considerations restricts this to k = 2, 4, 8, 16. The Hadamard matrix for k = 16 is as follows

(k=2, k=4, k=8 and k=16 mark the nested upper-left blocks of the matrix)

$$X = \begin{bmatrix}
+1 & +1 & +1 & +1 & +1 & +1 & +1 & +1 & +1 & +1 & +1 & +1 & +1 & +1 & +1 & +1 \\
+1 & -1 & +1 & -1 & +1 & -1 & +1 & -1 & +1 & -1 & +1 & -1 & +1 & -1 & +1 & -1 \\
+1 & +1 & -1 & -1 & +1 & +1 & -1 & -1 & +1 & +1 & -1 & -1 & +1 & +1 & -1 & -1 \\
+1 & -1 & -1 & +1 & +1 & -1 & -1 & +1 & +1 & -1 & -1 & +1 & +1 & -1 & -1 & +1 \\
+1 & +1 & +1 & +1 & -1 & -1 & -1 & -1 & +1 & +1 & +1 & +1 & -1 & -1 & -1 & -1 \\
+1 & -1 & +1 & -1 & -1 & +1 & -1 & +1 & +1 & -1 & +1 & -1 & -1 & +1 & -1 & +1 \\
+1 & +1 & -1 & -1 & -1 & -1 & +1 & +1 & +1 & +1 & -1 & -1 & -1 & -1 & +1 & +1 \\
+1 & -1 & -1 & +1 & -1 & +1 & +1 & -1 & +1 & -1 & -1 & +1 & -1 & +1 & +1 & -1 \\
+1 & +1 & +1 & +1 & +1 & +1 & +1 & +1 & -1 & -1 & -1 & -1 & -1 & -1 & -1 & -1 \\
+1 & -1 & +1 & -1 & +1 & -1 & +1 & -1 & -1 & +1 & -1 & +1 & -1 & +1 & -1 & +1 \\
+1 & +1 & -1 & -1 & +1 & +1 & -1 & -1 & -1 & -1 & +1 & +1 & -1 & -1 & +1 & +1 \\
+1 & -1 & -1 & +1 & +1 & -1 & -1 & +1 & -1 & +1 & +1 & -1 & -1 & +1 & +1 & -1 \\
+1 & +1 & +1 & +1 & -1 & -1 & -1 & -1 & -1 & -1 & -1 & -1 & +1 & +1 & +1 & +1 \\
+1 & -1 & +1 & -1 & -1 & +1 & -1 & +1 & -1 & +1 & -1 & +1 & +1 & -1 & +1 & -1 \\
+1 & +1 & -1 & -1 & -1 & -1 & +1 & +1 & -1 & -1 & +1 & +1 & +1 & +1 & -1 & -1 \\
+1 & -1 & -1 & +1 & -1 & +1 & +1 & -1 & -1 & +1 & +1 & -1 & +1 & -1 & -1 & +1
\end{bmatrix}$$

It can be seen that the matrices for k = 2, 4, 8 etc are subsets of the next higher one.

Because in a weighing scheme it is desirable to have more observations than unknown parameters this is achieved by repeating the appropriate matrix. For example, for k = 2 we could have:

$$\begin{bmatrix} +1 & +1 \\ +1 & -1 \\ +1 & +1 \\ +1 & -1 \\ \cdots & \cdots \end{bmatrix}$$

The disadvantage of Hadamard matrices is that there must be 2, 4, 8, 16 etc. masses in each set. This raises the problem of the standard on which to base the calibration. It can be seen that the first observation consists of weighing all the masses together against a known standard. It is desirable to have a matrix for $k = 10$, but as one does not exist in this form the standard would have to be a combination of known masses.

Banerjee (1975) gives an orthogonal matrix for $k = 10$ which is as follows:

$$X = \begin{bmatrix} 0 & +1 & +1 & +1 & +1 & +1 & +1 & +1 & +1 & +1 \\ +1 & 0 & +1 & -1 & +1 & -1 & +1 & -1 & +1 & -1 \\ +1 & +1 & 0 & +1 & -1 & +1 & +1 & -1 & -1 & -1 \\ +1 & -1 & +1 & 0 & -1 & +1 & -1 & -1 & +1 & +1 \\ +1 & +1 & -1 & -1 & 0 & +1 & -1 & +1 & +1 & -1 \\ +1 & -1 & +1 & +1 & +1 & 0 & -1 & +1 & -1 & -1 \\ +1 & +1 & +1 & -1 & -1 & -1 & 0 & +1 & -1 & +1 \\ +1 & -1 & -1 & -1 & +1 & +1 & +1 & 0 & -1 & +1 \\ +1 & +1 & -1 & +1 & +1 & -1 & -1 & -1 & 0 & +1 \\ +1 & -1 & -1 & +1 & -1 & -1 & +1 & +1 & +1 & 0 \end{bmatrix}$$

Here the variance of each unknown parameter is $\sigma^2/9$.

There are however two problems with this scheme. First, only nine masses are ever used at one time. Secondly the masses in each balance pan cannot be equal. This scheme can be used provided other known auxillary masses are available.

Hadamard matrices give the maximum theoretical efficiency, $E_i = 1$ or $E = k$. However when it comes to calibrating four, say, 1 kg masses all the weighings must be done on a 10 kg balance with a resulting loss in accuracy. But for masses of 1 g or smaller, for which all the weighings can be made on a microbalance, then Hadamard matrices are ideal.

The best method for comparing nominally equal masses (say A, B, C & D) in pairs so as not to exceed the capacity of the balance is as follows:

A	B	C	D	
1				$S + a_1$
	1			$S + a_2$
		1		$S + a_3$
			1	$S + a_4$
1	-1			a_5
1		-1		a_6
1			-1	a_7
	1	-1		a_8
	1		-1	a_9
		1	-1	a_{10}

where S is the known standard.

The normal equations and inverse matrices are:

$$(X^{*\prime}X^{*}) = \begin{bmatrix} 4 & -1 & -1 & -1 \\ -1 & 4 & -1 & -1 \\ -1 & -1 & 4 & -1 \\ -1 & -1 & -1 & 4 \end{bmatrix} \qquad (X^{*\prime}X^{*})^{-1} = \begin{bmatrix} 0.4 & 0.2 & 0.2 & 0.2 \\ 0.2 & 0.4 & 0.2 & 0.2 \\ 0.2 & 0.2 & 0.4 & 0.2 \\ 0.2 & 0.2 & 0.2 & 0.4 \end{bmatrix}$$

The extension to any number of parameters is quite straightforward.

It should be noted that this scheme is numerically equivalent to a graphical square, which gives a simple method of performing the calculations without a computer (Glazebrook 1955), but provides no estimate of the variance-covariance matrix.

(d) Estimation of Uncertainty

(i) Law of Propagation of Errors. The variance of f(a, b, c, ...) is given by

$$\text{var}(f) = \left(\frac{\partial f}{\partial a}\right)^2 \text{var}(a) + \left(\frac{\partial f}{\partial b}\right)^2 \text{var}(b) + \ldots\ldots\ldots$$

$$+ 2\,\frac{\partial f}{\partial a} \cdot \frac{\partial f}{\partial b}\,\text{cov}(a, b) + 2\,\frac{\partial f}{\partial a} \cdot \frac{\partial f}{\partial c}\,\text{cov}(a, c) + \ldots\ldots \qquad (8)$$

where: f is any function of the parameters a, b, c, ...,
var(a) is the variance of the parameter a,
cov(a, b) is the covariance between a and b.

It is this equation which is used to calculate the variance of a group of masses which are to be used for further calibrations.

(ii) Uncertainty of the Calibration Scheme. Each decade consists of masses in the ratios 5:2:2':1:1' while the first decade will include the standard (1 kg, 100 g depending on the set). The head mass from the set (i.e. 1 kg or 100 g) will have to be calibrated by the method described in section 5(c) for nominally equal masses. For subsequent decades the role of the standard is taken by the "1" from the previous decade; thus the 100 g, 10 g, etc. masses become intermediate standards whose uncertainty is propagated directly to the masses in the decade they head and hence to those in subsequent decades.

The uncertainty in a decade arising from the previous decades is termed the propagated uncertainty. For clarity the uncertainty arising from the initial standard is treated as a systematic uncertainty (understood as an "absolute limit"). It could, if desired, be treated as a random uncertainty in the same way as for the intermediate standards.

The random uncertainty has a "local" component arising from measurements in the decade under consideration and, a "propagated" component arising from the random uncertainty in the intermediate standards. The local or observed variance of the unknown parameters β_i (masses) is σ_i, which is extracted from the variance-covariance matrix. The propagated variance is

$$\sum_h r_{ih}^2\,\sigma_h^2$$

where r_{ih} is the nominal value of the ratio β_i/β_h and h takes values corresponding to all the intermediate standards between β_o and β_i. The variance of β_i is then

$$\text{var}\,\beta_i = \sigma_i^2 + \sum_h r_{ih}^2\,\sigma_h^2 \qquad (9)$$

Example: For a 20 g mass in a set starting with a standard of 1 kg, there is no propagated variance for the first decade therefore σ_h^2 is the variance of the 100 g mass used as the intermediate standard for the next decade, and is equal to 0.1 σ_{100}^2 (section 5(b)). Thus:

$$\text{var } [20] = 0.1\sigma_{20}^2 + \left(\frac{20}{100}\right)^2 0.1\sigma_{100}^2.$$

Similarly let r_{ij} be the nominal value of the ratio β_i/β_j. If β_i is the mass in a decade whose intermediate standard is β_j, then the systematic uncertainty of β_i is r_{ij} times the systematic uncertainty of β_j. Thus the systematic uncertainty of β_i equals r_{io} u_o, where u_o is the uncertainty in the value of β_o, the initial standard, and r_{io} is the ratio of the mass under consideration to the mass of the initial standard.

For a further description see Prowse and Anderson (1974), and Appendix 2.

(e) Calibration of Masses larger than 1 kg

Although the standard mass cannot be the highest mass of a set, it was pointed out by Zuker *et al* (1980), if an orthogonal system can be formed then one of the normal equations will constitute the relation between the highest mass and the defining unit standard.

Consider the normal equations for the design scheme (a) given in section 5(b). Written out in full they are:

$$4\ [5] = 2\ [10] + a_1 + a_2 + a_3 + a_4 \tag{10}$$

$$10\ [2] = 2\ [10] + a_1 + a_2 - a_3 - a_4 + a_5 + a_6 + a_7 + a_8 + a_9 + a_{10} \tag{11}$$

$$10\ [2'] = 2\ [10] + a_1 + a_2 - a_3 - a_4 - a_5 - a_6 - a_7 - a_8 + a_{11} + a_{12} \tag{12}$$

$$10\ [1] = [10] + a_1 - a_3 + a_5 - a_6 - a_9 - a_{10} - a_{11} - a_{12} + a_{13} + a_{14} \tag{13}$$

$$10\ [1'] = [10] + a_2 - a_4 - a_5 + a_6 - a_9 - a_{10} - a_{11} - a_{12} - a_{13} - a_{14} \tag{14}$$

The numbers in brackets, e.g. [5], represent the masses. Assuming that [1] is the standard then equation (13) gives

$$[10] = 10[1] - a_1 + a_3 - a_5 + a_6 + a_9 + a_{10} + a_{11} + a_{12} - a_{13} - a_{14} \tag{15}$$

This enables the value of mass [10] to be calculated. Then by substituting equation (15) in the other equations the values of the other unknown masses may be calculated in terms of the standard [1].

The variance of each unknown is calculated as follows. The residuals and hence the sum of squares of residuals are calculated from the observational equations. Then, as previously,

$$\sigma^2 = \Sigma(\text{residuals})^2/(n - k).$$

By use of equation (8):

from (15) $\qquad \text{var}[10] = 10^2\ \text{var}[1] + 10\sigma^2$

from (10) and (15) $\text{var}\ [5] = \left(\frac{1}{4}\right)^2 \{2^2\ \text{var}[10] + 4\sigma^2\},$

and similarly for the other masses.

A numerical example is given in Appendix 3.

6. Conclusion

The main thrust of this paper has been to describe a somewhat unfamiliar method of calibrating standards of mass. It depends heavily on the matrix theory of least squares. This method has the advantage that the variance-covariance matrix is readily obtained enabling the variance of each unknown parameter to be calculated, thereby enabling a laboratory to precisely define its level of uncertainty. The general principles outlined allow weighing schemes to be designed for special situations.

References

Banerjee, K.S. (1975). Weighing Designs (Dekker, New York).
Bell, G.A. (1949). Calibration of a set of weights. Report MIR 178. Division of Metrology, CSIRO, Sydney.
B.S. 1339 (1965). Definitions, formulae and constants relating to the humidity of the air (British Standards Institution).
Cameron, J.M., Croarkin, M.C., and Raybold, R.C. (1977). *Nat. Bur. Stand.*, Tech Note 952.
Giacomo, P. (1982). *Metrologia*, 18, 33.
Glazebrook, R. (1950). A Dictionary of Applied Physics. Vol III. 239 (Peter Smith, New York).
Grabe, M. (1978). *Metrologia*, 14, 143.
Hayford, J.F. (1892). On the least square adjustment of weighings. U.S. Coast and Geodetic Survey, Report 1892, Part 2, 515. Washington: Government Printing Office.
Hotelling, H. (1944). *Ann. Math. Statist.* 15, 297.
Hunt, R.A. (1981). NPL Report MOM 50.
Jones, F.E. (1978). *J. Res. Nat. Bur. Stand.* 83, 419.
Mood, A.M. (1946). *Ann. Math. Statist.* 17, 432.
NPL. (1954). Balance, Weights and Precise Laboratory Weighing. Notes on Applied Science No. 7.
Pontius, P.E. (1966). *Nat. Bur. Stand.* Tech Note 288.
Pontius, P.E. (1974). Mass and Mass Values. NBS Monograph 133.
Prowse, D.B., and Anderson, A.R. (1974). *Metrologia*, 10, 123.
Prowse, D.B. (1982). Precision Measurements and Fundamental Constants II, Ed. by B.N. Taylor and W.D. Phillips, Nat. Bur. Stand. (U.S.), Spec. Publ. 617 in press.
Rao, G.R. (1965). Linear Statistical Inference and its Applications. (New York, Wiley).
Zuker, M., Mihailov, G. and Romanowski, M. (1980). *Metrologia*, 16, 51.

Appendix 1

Method of Least Squares

The treatment given by Prowse and Anderson (1974) is reproduced here.

Consider the problem of finding a scheme for calibrating a given set of masses. Such a scheme of observations constitutes a linear model in which the parameters to be estimated are "fixed" (i.e. non-stochastic). The standard statistical procedure applied is the method of least squares. See, for example, Rao (1965) whose notation is used in this appendix.

As a standard notation, write $X = (x_{ij})$ for the $n \times k$ design matrix of a scheme; $\beta = (\beta_1,\ldots.\beta_k)$ for the vector of unknown parameters (masses); and Y for the observation vector. Y is a vector of random variables, with $E(Y_i) = x_{i1}\beta_1 +\ldots+ x_{ik}\beta_k$. In matrix notation

$$E(Y) = X\beta$$

and

$$D(Y) = \sigma^2 G$$

where D(Y) is the variance-covariance (or dispersion) matrix and G is assumed to be known. The model reduces to its simplest form by means of a transformation such as

$$Y^* = G^{-\frac{1}{2}}Y, \quad X^* = G^{-\frac{1}{2}}X,$$

which leads to

$$E(Y^*) = X^*\beta, \ V(Y^*) = \sigma^2 I,$$

where I is the $n \times n$ unit matrix.

Since the observations from a practical weighing scheme can be assumed to be uncorrelated, it follows that in practice G is diagonal, the diagonal entries being, of course, positive. Then

$$G = \text{Diag.} \ (1/\gamma_1^2,\ldots.1/\gamma_n^2),$$

and $G^{-\frac{1}{2}} = \text{Diag.} \ (\gamma_1,\ldots\gamma_n)$ which is the weighting matrix. The equation $Y^* = G^{-\frac{1}{2}}Y$ may thus be interpreted as weighting the observations in inverse proportion to their estimated variances; the variances of the weighted observations are then all equal.

Application of the least-squares principle gives the "normal equation"

$$X^{*\prime}X^*\beta = X^{*\prime}Y^*,$$

with solution

$$\beta = (X^{*\prime}X^*)^{-1} \ X^{*\prime}Y^*.$$

Provided $X^{*\prime}X^*$ is nonsingular, each entry of β has minimum variance among the class of linear unbiased estimators of the corresponding entry in β. This nonsingularity condition holds if and only if the rank of X^* is k, and for $n \geqslant k$. The variance-covariance matrix is

$$D(\beta) = (X^{*\prime}X^*)^{-1}\sigma^2,$$

where $(n - k)\sigma^2$ has the unbiased estimator

$$(Y^* - X^*\beta)'(Y^* - X^*\beta) = Y^{*\prime}Y^* - Y^{*\prime}X^*\beta.$$

Appendix 2

Calibration of a Set of Masses from 500 g to 1 g.

Consider the calibration of a set of masses containing the following masses

500, 200, 200', 100, 50, 20, 20', 10, 5, 2, 2', 1

The masses are calibrated using the scheme (a) in section 5(b). The balances used have capacities of 1 kg, 100 g, 20 g, 2 g. In this case to make the calibration clearer the weighting factor used is 1 not 100 as suggested in the main text.

The calculations are set out in the following tables using the notation of Appendix 1.

$$
\underset{X}{\begin{bmatrix}
1 & 1 & 1 & 1 & 0 \\
1 & 1 & 1 & 0 & 1 \\
1 & -1 & -1 & -1 & 0 \\
1 & -1 & -1 & 0 & -1 \\
0 & 1 & -1 & 1 & -1 \\
0 & 1 & -1 & -1 & 1 \\
0 & 1 & -1 & 0 & 0 \\
0 & 1 & -1 & 0 & 0 \\
0 & 1 & 0 & -1 & -1 \\
0 & 1 & 0 & -1 & -1 \\
0 & 0 & 1 & -1 & -1 \\
0 & 0 & 1 & -1 & -1 \\
0 & 0 & 0 & 1 & -1 \\
0 & 0 & 0 & 1 & -1
\end{bmatrix}}
\quad
\underset{Y(1\ kg)}{\begin{bmatrix}
1000.006\ 632\ 263 \\
1000.006\ 006\ 5 \\
.005\ 241\ 4 \\
.005\ 530\ 2 \\
.001\ 130\ 7 \\
.000\ 449\ 9 \\
.000\ 775 \\
.000\ 792\ 2 \\
-.001\ 16 \\
-.001\ 143\ 6 \\
-.001\ 929 \\
-.001\ 920\ 5 \\
.000\ 341\ 9 \\
.000\ 331\ 4
\end{bmatrix}}
\quad
\underset{Y(100\ g)}{\begin{bmatrix}
100.000\ 729\ 5 \\
100.000\ 691 \\
.000\ 145\ 2 \\
.000\ 176\ 6 \\
-.000\ 061\ 3 \\
-.000\ 110\ 2 \\
-.000\ 087\ 6 \\
-.000\ 090\ 9 \\
.000\ 071\ 8 \\
.000\ 063\ 8 \\
.000\ 164\ 9 \\
.000\ 168\ 8 \\
.000\ 029\ 2 \\
.000\ 026
\end{bmatrix}}
\quad
\underset{Y(10\ g)}{\begin{bmatrix}
10.000\ 092 \\
10.000\ 109\ 2 \\
-.000\ 006 \\
-.000\ 018 \\
-.000\ 009\ 4 \\
.000\ 018\ 67 \\
.000\ 003\ 73 \\
.000\ 005\ 03 \\
.000\ 009\ 1 \\
.000\ 012\ 1 \\
.000\ 006\ 63 \\
.000\ 006\ 63 \\
-.000\ 012\ 57 \\
-.000\ 012\ 12
\end{bmatrix}}
$$

$$
X^{*\prime}Y^{*} = \underset{(1\ kg)}{\begin{bmatrix}
2000.023\ 104\ 4 \\
2000.002\ 405\ 4 \\
1999.994\ 563\ 9 \\
1000.008\ 592\ 1 \\
1000.005\ 275\ 3
\end{bmatrix}}
\quad
\underset{(100\ g)}{\begin{bmatrix}
200.001\ 742\ 3 \\
200.000\ 884\ 3 \\
200.001\ 782\ 4 \\
100.000\ 219\ 1 \\
99.999\ 941\ 0
\end{bmatrix}}
\quad
\underset{(10\ g)}{\begin{bmatrix}
20.000\ 177\ 20 \\
20.000\ 264\ 50 \\
20.000\ 220\ 44 \\
10.000\ 010\ 70 \\
10.000\ 145\ 44
\end{bmatrix}}
$$

The rest of the calculations are given in detail only for the 1 kg decade.

$$
X^{*\prime}X^{*} = X'\ G^{-1}\ X = \begin{bmatrix}
4 & & & & \\
 & 10 & & & \\
 & & 10 & & \\
 & & & 10 & \\
 & & & & 10
\end{bmatrix}
$$

$$
(X^{*\prime}X^{*})^{-1} = \begin{bmatrix}
0.25 & & & & \\
 & 0.1 & & & \\
 & & 0.1 & & \\
 & & & 0.1 & \\
 & & & & 0.1
\end{bmatrix}
$$

The solution is given by :

$$
\beta = (X^{*\prime}X^{*})^{-1}\ X^{*\prime}Y^{*}
$$

i.e.

$$
\begin{bmatrix}
500 \\
200 \\
200' \\
100 \\
100'
\end{bmatrix}
=
\begin{bmatrix}
0.25 & & & & \\
 & 0.1 & & & \\
 & & 0.1 & & \\
 & & & 0.1 & \\
 & & & & 0.1
\end{bmatrix}
\begin{bmatrix}
2000.023\ 104\ 4 \\
2000.002\ 405\ 4 \\
1999.994\ 563\ 9 \\
1000.008\ 592\ 1 \\
1000.005\ 275\ 3
\end{bmatrix}
=
\begin{bmatrix}
500.005\ 776 \\
200.000\ 241 \\
199.999\ 456 \\
100.000\ 869 \\
100.000\ 528
\end{bmatrix}
$$

Substituting these values in each of the original equations gives the residuals

0.000 005 9
-0.000 005 9
-0.000 021 4
0.000 021 4
-0.000 014 9
0.000 002 6
0.000 009 2
-0.000 008 1
0.000 013 8
-0.000 002 6
-0.000 001 4
-0.000 009 9
-0.000 010 2
0.000 000 3

Squaring each of these values and adding gives the sum of squares of residuals = 1.767×10^{-9}.

The variance-covariance matrix is

$$\frac{1.767 \times 10^{-9}}{9}\begin{bmatrix} 0.25 & & & & \\ & 0.1 & & & \\ & & 0.1 & & \\ & & & 0.1 & \\ & & & & 0.1 \end{bmatrix} = \begin{bmatrix} 4.91 \times 10^{-11} & & & & \\ & 1.96 \times 10^{-11} & & & \\ & & 1.96 \times 10^{-11} & & \\ & & & 1.96 \times 10^{-11} & \\ & & & & 1.96 \times 10^{-11} \end{bmatrix}$$

Taking the square roots of the diagonal elements gives the standard deviation of the unknowns

500	7.0×10^{-6}
200	4.5×10^{-6}
200'	4.5×10^{-6}
100	4.5×10^{-6}
100'	4.5×10^{-6}

The value of the 100 mass is then used as the standard for the next decade, and this gives the observational matrix Y shown above. This is then repeated for the next decade.

Estimation of Uncertainty:

Two cases are considered in turn.

(a) The uncertainty of the 1 kg standard is a random variable. The standard deviation of the 1 kg standard is 5.0 μg. The variance of the observed variables is calculated from equation (9). For convenience the total uncertainty is calculated as a 3σ limit. These are shown in Table 2.

(b) The uncertainty of the 1 kg standard is a systematic uncertainty. The uncertainty of the 1 kg standard is considered to be a total systematic uncertainty of 15 μg. The systematic component is combined with the 3σ limit of the random uncertainty to give the total uncertainty of each mass as shown in Table 3.

Table 2

The Uncertainty of the 1 kg Standard is a Random Variable

Nominal Mass g	Observed Variance (μg^2)	Propagated Variance (μg^2)	Total Variance (μg^2)	Standard Deviation σ (μg)	Total Uncertainty (3σ) (μg)
1000	-	25	25	5	15
500	49.1	25/4 = 6.25	55.4	7.44	22.3
200	19.6	25/25 = 1.0	20.6	4.54	13.6
100	19.6	25/100 = .25	19.85	4.46	13.4
50	7.09	19.85/4 = 4.96	12.15	3.49	10.47
20	2.84	19.85/25 = 0.79	3.63	1.91	5.73
10	2.84	19.85/100 = 0.20	3.04	1.74	5.22
5	0.464	0.760	1.22	1.11	3.33
2	0.186	0.122	0.308	0.56	1.68
1	0.186	0.030	0.216	0.47	1.41

Table 3

The Uncertainty of the 1 kg Standard is a Systematic Uncertainty

Nominal Mass (g)	Observed Variance (μg^2)	Propagated Variance (μg^2)	Total Variance (μg^2)	Standard Deviation (μg)	Systematic Uncertainty (μg)	Total Uncertainty (μg)
1000	-	-	-	-	15	15
500	49.1	-	49.1	7.51	7.5	28.5
200	19.6	-	19.6	4.43	3.0	16.3
100	19.6	-	19.6	4.43	1.5	14.8
50	7.09	19.6/4 = 4.9	12.0	3.46	0.75	11.33
20	2.84	0.78	3.62	1.90	0.30	6.00
10	2.84	0.20	3.04	1.74	0.15	5.37
5	0.464	2.84/4 = 0.71	1.174	1.08	0.075	3.32
2	0.186	0.114	0.300	0.55	0.030	1.68
1	0.186	0.028	0.214	0.46	0.015	1.40

Appendix 3

Calibration of a 1 - 10 kg set of masses

Consider scheme (a) from section 5(a). For the masses in kg the observational equations are

[5]	[2]	[2']	[1]	[1']	
1	1	1	1		[10] + 0.0195
1	1	1		1	[10] + 0.0159
1	-1	-1	-1		-0.0203
1	-1	-1		-1	-0.0160
	1	-1	1	-1	-0.0067
	1	-1	-1	-1	-0.0158
	1	-1			-0.0110
	1	-1			-0.0107
	1		-1	-1	-0.0040
	1		-1	-1	-0.0045
		1	-1	-1	0.0071
		1	-1	-1	0.0068
			1	-1	0.0044
			1	-1	0.0046

Equations (10) to (14) give

$$[5] = \frac{1}{4}\{2[10] - 0.0009\} \qquad (10')$$

$$[2] = \frac{1}{10}\{2[10] + 0.0190\} \qquad (11')$$

$$[2'] = \frac{1}{10}\{2[10] + 0.1298\} \qquad (12')$$

$$[1] = \frac{1}{10}\{[10] + 0.0525\} \qquad (13')$$

$$[1'] = \frac{1}{10}\{[10] + 0.0084\} \qquad (14')$$

As the value of the standard [1] is 1000.00621 g then equation (13') gives

$$[10] = 10000.0621 - 0.0525 = 10000.0096$$

Substituting this value in equations (10'), (11'), (12') and (14') gives

$$[5] = 5000.0046$$
$$[2] = 2000.0038$$
$$[2'] = 2000.0149$$
$$[1'] = 1000.0018$$

By substituting these values back in the observational equations the residuals are obtained.

$$\sigma^2 = \Sigma(\text{residuals})^2/(n - k) = 81 \times 10^{-8}/9 = 9 \times 10^{-8}\ g^2$$

The standard deviation of the standard [1] is 5 µg. Hence:

$$\text{var}\ [1] = 25 \times 10^{-12}\ g$$

$$\text{var }[10] = 10^2 \times 25 \times 10^{-12} + 10 \times 9 \times 10^{-8} = 9.025 \times 10^{-7}$$

$$\sigma[10] = 9.5 \times 10^{-4}\text{ g}$$

$$\text{var }[5] = \frac{1}{16}(4 \times 9.025 \times 10^{-7} + 4 \times 9 \times 10^{-8}) = 2.48 \times 10^{-7}$$

$$\sigma[5] = 5.0 \times 10^{-4}\text{ g}$$

$$\text{var }[2] = \frac{1}{100}(4 \times 9.025 \times 10^{-7} + 10 \times 9 \times 10^{-8}) = 4.5 \times 10^{-8}$$

$$\sigma[2] = 2.2 \times 10^{-4}\text{ g}$$

$$\text{var }[1'] = \frac{1}{100}(0.025 \times 10^{-7} + 10 \times 9 \times 10^{-8}) = 1.8 \times 10^{-8}$$

$$\sigma[1'] = 1.4 \times 10^{-4}\text{ g}$$

If the uncertainty in the standard is considered to be a systematic uncertainty of magnitude 15 μg then the uncertainties are calculated as above with the random uncertianty of the standard being zero. The masses of magnitude 1, 2, 5 and 10 kg have systematic uncertainties of 15, 30, 75 and 150 μg respectively.

Appendix 4

Orthogonal weighing scheme for masses of denomination 5, 3, 2, 1, 1'.

This scheme was suggested by Grabe (1978). The design matrix is:

$$X = \begin{bmatrix} 1 & 1 & 1 & 1 & -1 \\ 1 & 1 & 1 & -1 & 1 \\ 1 & 1 & & 1 & 1 \\ 1 & -1 & -1 & 1 & -1 \\ 1 & -1 & -1 & -1 & 1 \\ 1 & -1 & & -1 & -1 \\ & 1 & -1 & -1 & \\ & 1 & -1 & -1 & \\ & 1 & -1 & & -1 \\ & 1 & -1 & & -1 \\ & & 1 & -1 & -1 \\ & & 1 & -1 & -1 \end{bmatrix}$$

The normal equations and inverse matrices are:

$$(X'X) = \begin{bmatrix} 6 & & & & \\ & 10 & & & \\ & & 10 & & \\ & & & 10 & \\ & & & & 10 \end{bmatrix}$$

$$(X'X)^{-1} = \begin{bmatrix} 1/6 & & & & \\ & 0.1 & & & \\ & & 0.1 & & \\ & & & 0.1 & \\ & & & & 0.1 \end{bmatrix}$$

Laboratory Pressure-Measuring Instruments and Their Calibration

N. Bignell
CSIRO Division of Applied Physics, Sydney, Australia 2070

Abstract

An outline of some high quality instruments used for laboratory pressure measurements from 0.1 - 10^8 Pa is given. Some details of their calibration are discussed.

1. Introduction

A fairly well-established pressure scale exists over some 14 orders of magnitude, but the equipment and techniques used to realise the scale and the accuracy of realisation vary greatly over this range. It is possible to generate pressures beyond the ends of this pressure scale but these cannot be measured satisfactorily with present techniques. The accuracy of measurement is highest around atmospheric pressure, where an accuracy of a few parts in 10^6 is achieved using mercury manometers. Piston gauges are mostly used above the atmospheric pressure range.

While S.I. units will be used in this paper, instruments frequently require calibration in other units. Piston gauges have weights which correspond to round numbers in some pressure unit, and, whilst it is then convenient to calibrate in that unit calibration in any other unit with that particular machine can be accommodated by using conversion factors. Mercury manometers are read in the unit of the scale, usually a length unit, and this may be converted to any required pressure unit.

A primary pressure standard is one which may be calibrated using established standards in quantities other than pressure, such as for example length and mass standards, the value of gravitational acceleration etc. A secondary standard is calibrated by reference to primary standards. Liquid columns and particularly mercury columns are regarded as primary standards, while piston gauges are usually regarded as secondary standards. This is, to a certain extent, a value judgement based on the accuracy with which the pressure may be determined. Piston gauges may be operated as primary standards and many mercury columns are operated as secondary standards.

A tertiary standard is, of course, one calibrated from a secondary standard, though it may be calibrated directly from a primary standard. Tertiary standards are generally less accurate and less stable than secondary standards and depend on more factors for their stability. Two high quality tertiary standards use the force produced by a pressure as a measure of the pressure. These instruments are fairly complicated and will be described later. In a tertiary standard stability depends on many factors and it is not always evident if something has malfunctioned or drifted in value slightly. For a secondary standard such as a piston gauge, however, the operation is sufficiently simple and the construction so uncomplicated that if something were to go wrong it would usually be obvious to a trained operator. Some tertiary standards are simple enough in operation but are just not very accurate, for example, Bourdon tube gauges. These will not be discussed.

2. Liquid Columns

If the pressure at the top of a liquid column is P_1 and at the bottom is P_2, then

$$P_2 - P_1 = \rho gh, \quad (1)$$

where ρ is the density of the liquid, g the value of gravitational acceleration and h the vertical height of the column. The formula also applies to columns of gas, which are sometimes used for the generation of very small pressures (e.g. Bignell, 1976); however, liquids are more convenient for most manometric purposes. Mercury is, of course, much used. Its low vapour pressure, high and well-known density and the fact that it does not wet glass make it the best manometric liquid at atmospheric pressures. The world's largest mercury column has a height of 40 m and $P_2 - P_1$ is about 5 MPa, by equation (1).

Short columns of mercury using micrometers to measure the height are also used, but, here, there is sometimes an advantage to be gained in using liquids of lower density, usually low vapour pressure oils. These wet the walls of the glass and sometimes dissolve gases.

To obtain greater amplification of the reading the tube may be inclined, since, then, the slant height may be 10 times or more the vertical height. This is by no means the extent of the ingenuity displayed in the design of micromanometers, as instruments in this range are called. A survey is found in Brombacher, 1970.

Even further amplification is required for vacuum work, and the design of the McLeod gauge allows mercury columns to work in the range 0.1 to 7 Pa. A volume of gas at the pressure to be measured is compressed by a rising mercury column; its pressure is thus increased by a known factor to a value which can be measured more easily.

(a) Mercury columns

Because the density of mercury is well known (Cook, 1961), g can be accurately measured at the place required and h is a length which can also be accurately determined, mercury columns using equation (1) can be operated as primary standards. In practice, the simplicity of equation (1) is deceptive and many corrections and precautions are necessary. Chief among the corrections is that due to the effect of temperature on the density of the mercury.

The density of mercury, ρ_{Hg}, is given as a function of its temperature T°C by the relation

$$\rho_{Hg}(T) = 13\,545.867\ (1 + 18\,115 \times 10^{-8}(T - 20) + 0.8 \times 10^{-8}(T - 20)^2)^{-1}. \quad (2)$$

It is also available from tables. The centre of the mercury meniscus is lower, due to surface tension, than it would be for an infinite surface. This depression depends on the diameter of the tube and is negligible for tubes with a diameter greater than about 22 mm. Correction tables are found in Brombacher (1960). For U tubes with arms of equal diameter the corrections cancel.

For work of high accuracy, the effect of pressure on the density of mercury should be considered. Thus for barometric work, where the average pressure of the mercury is half the atmospheric pressure, or about 50 kPa, the figure 13 545.840 should be used in front of the main bracket in equation (2). A list of other factors which must be taken into account for work of the highest accuracy is found in Harrison *et al* (1976).

(b) Standard Haas barometer

The Haas barometer is an instrument of simple design and of a quality often used in calibration work in meteorology. The instrument is shown in Fig. 1 and consists of a vertical scale mounted on a base plate which is used to measure the height of the mercury column. The level of mercury at the other surface is measured by a micrometer also firmly attached to the same base plate. The temperature of the scale is measured by a thermometer fixed to an upright, and the mercury temperature is measured with a thermometer immersed in it. The pressure at the top of the column is very small and is measured with a McLeod gauge.

This instrument can achieve repeatabilities of about 33 ppm. To achieve this accuracy, the mercury density must be corrected to the temperature of use, and a correction must also be made to the scale length using the relation

$$h = h(T)[1 + \lambda(T - T_o)]$$

where h is the height required and h(T) is the measured height at temperature T. T_o is the temperature at which the scale reads true, or at which the calibration is valid.

(c) Interference manometer

This is a mercury column instrument of the highest quality, capable of an accuracy of a few parts in 10^6. It is in essence, a U-tube manometer, and it owes its high accuracy to careful stabilization and measurement of the temperature and to the method of measurement of the height changes. This is done by interference of light reflected from the surface of the mercury, using a specially designed protective float to reduce the effect of vibrations on the surface.

The condition of zero pressure is established initially by evacuating both arms of the U. The pressure to be measured is then slowly admitted to one arm and as the surfaces move apart, interference between the beams of light reflected from the floats results in a succession of fringes which may be counted. Each fringe represents a difference in height of one half wavelength between the surfaces. Full details may be found in Harrison *et al* (1976).

(d) Commercial barometers

Descriptions of several types of these, such as the Fortin, Kew pattern and other fixed cistern types are found in Brombacher, Johnson & Cross (1960) and also in Daborn Hatley and Ruffell (1974). These types are usually operated as secondary standards i.e. they are calibrated by comparison with a primary reference barometer.

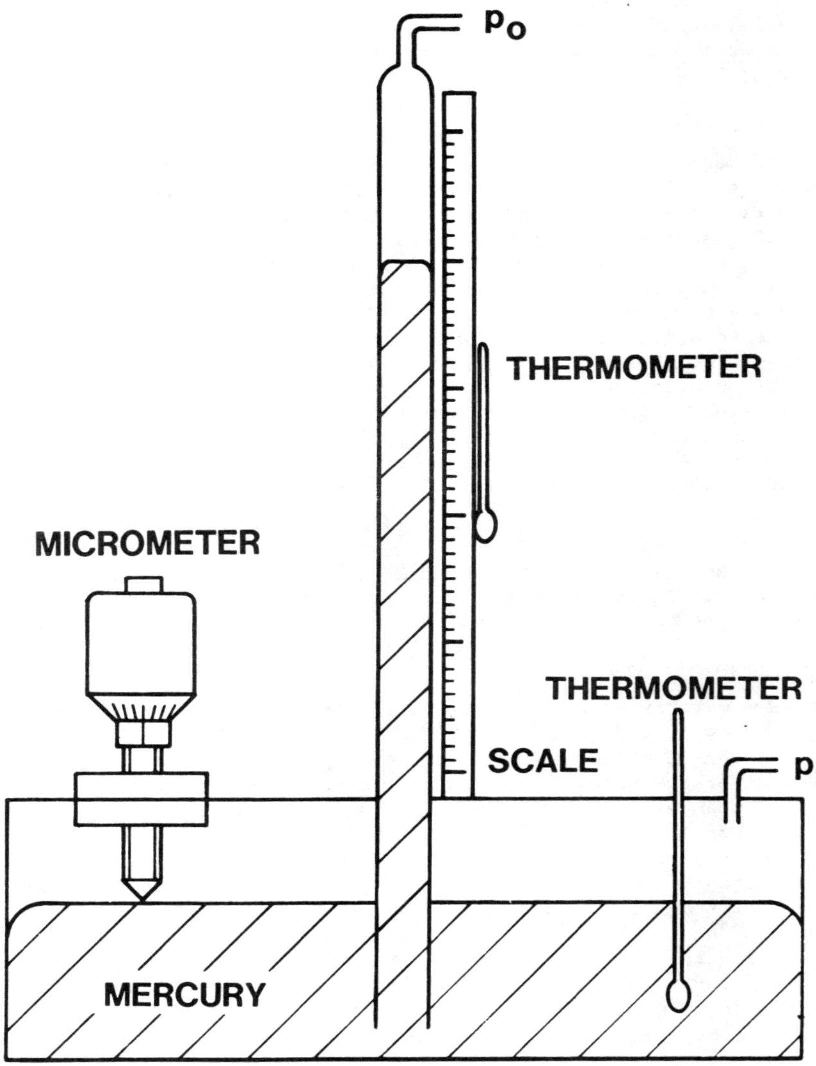

Fig. 1. Diagrammatic layout of the Standard Haas barometer, a quality primary standard barometer.

3. The Piston Gauge

(a) The pressure equation

In the piston gauge we can see most clearly the definition of pressure as force per unit area. The absolute pressure at the base of the piston produced by a piston gauge such as that shown in Fig. 2 is given by

$$P = F/A_{eff} + P_{Atm}. \quad (3)$$

Here F is the total downward force on the piston and A_{eff} is the effective area of the piston/cylinder combination; P_{Atm} is the atmospheric pressure at the level of the base of the piston.

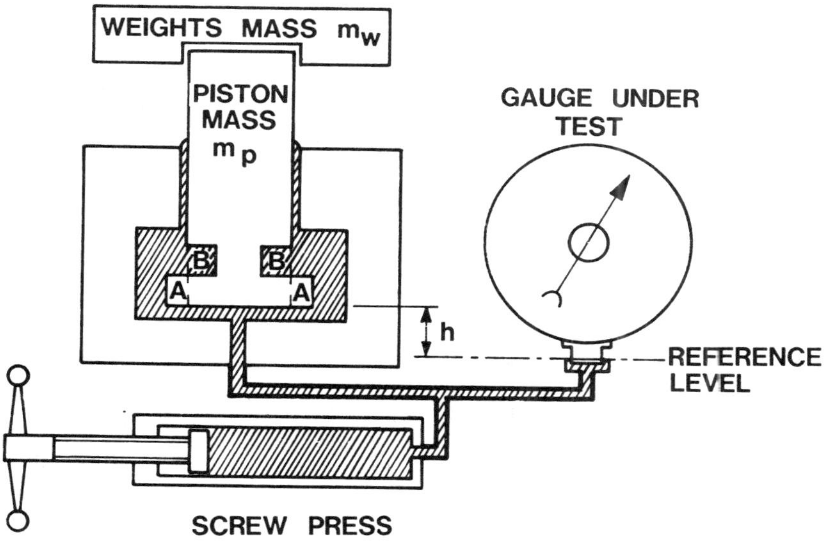

Fig. 2. A typical piston gauge shown with a screw press for pressure generation (mostly found in oil operated gauges) and a gauge being calibrated.

Though equation (3) is simple, the quantities F and A_{eff} are not easy to measure. The effective area of the piston/cylinder combination changes with pressure and can often be represented as

$$A_{eff} = A_o(1 + aP), \quad (4)$$

where A_o and a are constants. How these constants are determined for our standard will not be discussed here; for an account of this see Dadson *et al* (1965) and Johnson and Newhall (1953). The effective area also changes with temperature, and so we have

$$A_{eff} = A_o(1 + aP)(1 + \lambda(T - T_{ref})), \quad (5)$$

where T_{ref} is the reference temperature at which the area was measured.

The force F acting on the piston (mass m_p) due to its weight and that of the loading weights (mass m_w), corrected for their buoyancy in air, is

$$F = (m_w + m_p)g(1 - \rho_a/\rho_w), \tag{6}$$

where ρ_a is the density of the air and ρ_w is the density of the weights, or of the standards against which they were weighed when apparent mass values are quoted.

Piston gauges using liquids as the pressure medium also have a correction to the load due to the surface tension of the liquid, which exerts a downward pull $C\tau$ on the piston, as it leaves the cylinder, where C is the circumference of the piston at that place and τ is the surface tension of the liquid.

If some parts of the piston extend beyond the main diameter, shown as A in Fig. 2, then there will be an upthrust, due to the fluid of density ρ_f, acting on the piston. This buoyancy correction can also give a downward force if there are unfilled parts inside the main diameter such as the region B in Fig. 2. A carefully designed piston can thus have a zero buoyancy correction if the volume V_A of region A, equals the volume V_B of region B. Accordingly, the downward force is given by $(V_B - V_A)(\rho_f - \rho_a)g$, where ρ_a is subtracted from ρ_f because the buoyancy correction in air for the piston has already been made in equation (6).

If the pressure is required at a level other than at the base of the piston then there will be a correction due to the fluid head. A frequently used reference level is the base of the gauge outlet. In this case a correction pressure $\rho_f gh$ will have to be added, where h is the distance of the piston base above (positive) or below (negative) the reference level. In addition, the atmospheric pressure will be changed at that level by $\rho_a gh$.

Gathering these terms together, the absolute pressure at the reference level is

$$P = \frac{(m_w + m_p)g(1 - \rho_a/\rho_w) + (V_B - V_A)g(\rho_f - \rho_a) + C\tau}{A_o(1 + aP)(1 + \lambda(T - T_{ref}))} + \rho_f gh + P_{atm} \tag{7}$$

The gauge pressure (also called the applied pressure) at the reference level P_G = P - atmospheric pressure at the reference level

$$= P - (P_{Atm} + \rho_a gh). \tag{8}$$

Substituting for P using equation (7)

$$P_G = \frac{(m_w + m_p)g(1 - \rho_a/\rho_w) + (V_B - V_A)g(\rho_f - \rho_a) + C\tau}{A_o(1 + aP)(1 + \lambda(T - T_{ref}))} + (\rho_f - \rho_a)gh \tag{9}$$

(b) Comparison of piston gauges

Piston gauges are best calibrated by comparison with a standard gauge for which A_o and a are known. This is done by a "cross float" technique in which the gauges are connected together as shown in Fig. 3. Sometimes the differential pressure gauge is omitted and the gauges are simply connected together. In both cases, the aim is to make the pressure produced by each gauge the same by adding small box weights to the loading weights.

If a differential pressure gauge is used, then the technique is simply to add weights to achieve a null reading. Since the zero

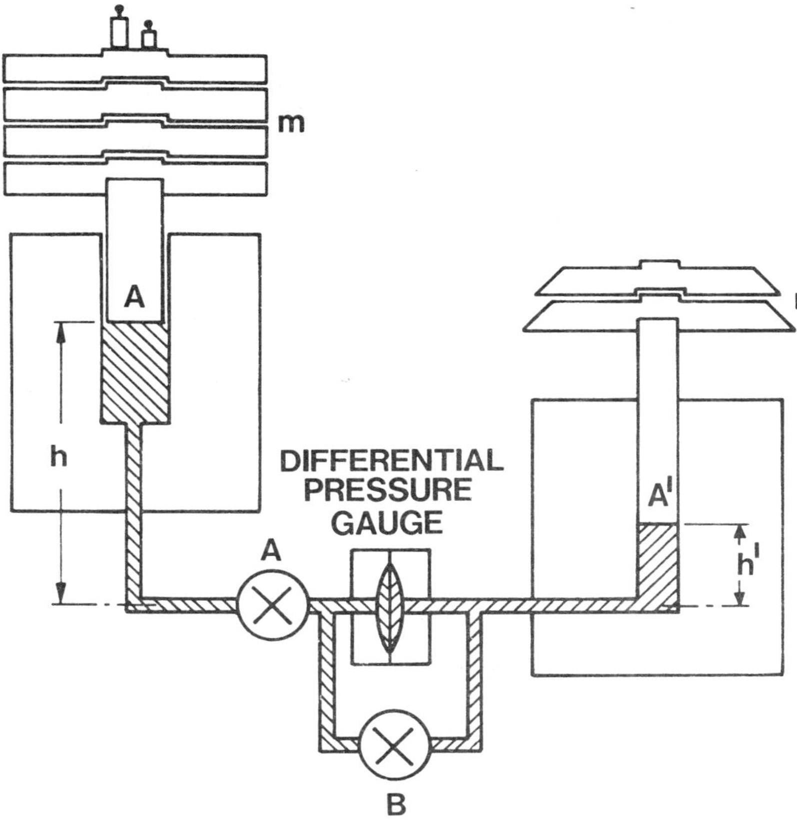

Fig. 3. Two piston gauges connected together in the cross-float configuration for comparison of effective areas. In this case, a differential pressure gauge and two constant volume valves are shown connecting the two piston gauges.

points of sensitive differential pressure gauges are liable to drift, a means of checking the zero must be included. This is the purpose of the constant volume valves shown in Fig. 3. They are called constant volume valves because they do not alter the volume of the system, and hence the pressure, when they operate. To zero the differential pressure gauge, valve B is opened and valve A closed. To balance the pressures produced by the piston gauges valve A is opened and valve B closed.

At the balance condition

$$P_G = P_G' \tag{10}$$

where the superscript is used to denote one of the gauges. If we take the head correction into the top line of equation (9), then

$$P_G = \frac{(m_w + m_p)g(1 - \rho_a/\rho_w) + (V_B - V_a)g(\rho_f - \rho_a) + C\tau + A_o(\rho_f - \rho_a)gh}{A_o(1 + aP)(1 + \lambda(T - T_{ref})} \tag{11}$$

From (10) and (11)

$$\frac{A_o(1 + aP)(1 + \lambda(T - T_{ref}))}{A'_o\ (1 + a'P)(1 + \lambda'(T - T'_{ref}))} = \frac{L}{L'} \tag{12}$$

where L and L' are the total loads (including some assigned loads) or the top lines of equation (11). The temperature correction terms of equation (12) may be assumed to have been absorbed into A_o and A'_o, so that (12) becomes

$$\frac{A_o(1 + aP)}{A'_o\ (1 + a'P)} = \frac{L}{L'}$$

$$= \frac{A_o}{A'_o}\ (1 + (a - a')P) = L/L' \tag{13}$$

using the algebra of small quantities and remembering that a and a' are small.

In practice, about 8 or more balance points are taken, each giving different values of L and L'. Then L/L' plotted against pressure should give the straight line of equation (13), with slope $(A_o/A'_o)(a-a')$ and intercept on the L/L' axis equal to A_o/A'_o. However there are several circumstances in which a curve such as that shown in Fig. 4 is obtained instead. These generally involve an error in the estimation of L or L', or both, due frequently to the small corrections to the load caused by fluid buoyancy and head. A method to overcome this is to add a small correction term to one load in the calculations, and to allow this term to assume the value which gives the best fit to equation (13). If we use the algebra of small quantities,

$$\frac{L + \Delta L}{L' + \Delta L'}$$

$$\approx (L + \Delta)/L', \text{ where } \Delta = \Delta L - \Delta L'/L').$$

Accordingly, it does not matter which load has the addition (or subtraction) of a small correction; one is sufficient to account for errors in both loads. The calculations to determine A_o/A'_o, a - a' and Δ are carried out using a computer program copies of which may be obtained from the Division of Applied Physics. The straight line obtained by making this correction to the data used to obtain the curve is also shown in Fig. 4.

(c) Uncertainties

There are some factors involved in the operation of piston gauges which make the error larger than expected from a straightforward analysis of equation (9). These are:

(i) oil may collect on the floating element and change the load;

(ii) if the piston is not operated in the mid-point of its range its effective area may be different from the calibration value;

(iii) air drafts and other aerodynamic forces can change the load;

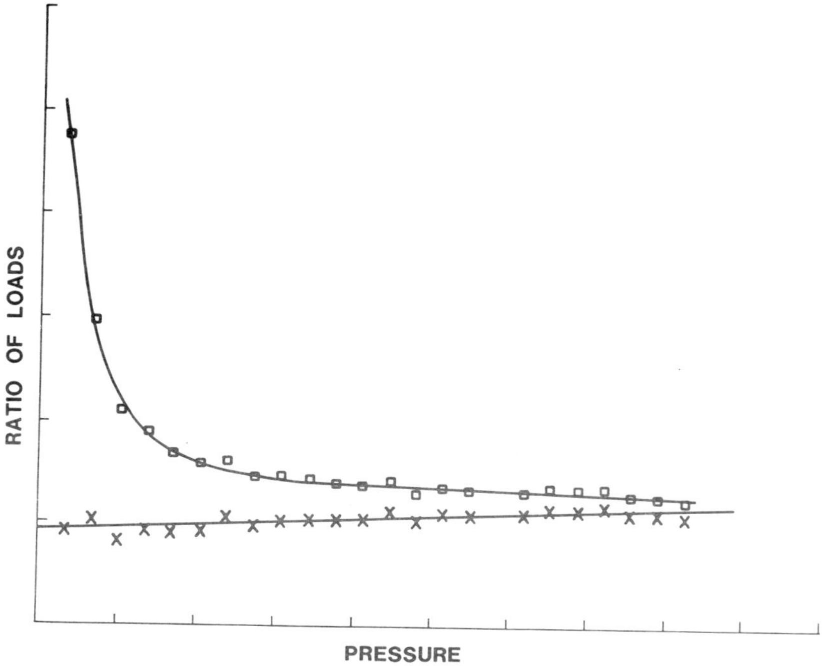

Fig. 4. The ratio of loads for two piston gauges plotted against pressure. The curve shows the uncorrected loads; the straight line is the ratio of the loads corrected as discussed in the text.

(iv) stray magnetic forces can change the load;

(v) air bubbles can reduce the density of the oil or alter the value of h in the head correction term; air tends to dissolve in the oil at higher pressures so the effect may not be constant;

(vi) if oil of a different viscosity is used A_0 will be different;

(vii) the speed and the direction of rotation of the piston can alter the pressure produced.

To be sure that none of these sources of error is operating and to make accurate measurements of V_A, V_B and h requires a great deal of effort.

When two piston gauges are compared by the cross-float method most of the error sources are compounded though error in the value of g cancels. The point of balance can be measured more accurately than it can be calculated from the various terms in equation (12). This allows Δ to be left as a factor to be determined as indicated in the preceding section. A number of tedious measurements can be avoided by this method when piston gauges of industrial quality (0.03%) are being calibrated. It is easier to take a few extra balance points to improve the accuracy of the fitting process than to make the measurements needed to determine L and L' accurately.

The variance of an effective area measured by this method is taken as

$$s_r^2 + s_A^2$$

where s_r^2 is the estimated variance of the intercept of the straight line and s_A^2 is the variance of the area of the standard. For the expansion coefficient the variance is

$$s_a^2 + (A_o/A_o')^2 \, s_M^2$$

Where s_a^2 is the variance of the expansion coefficient of the standard piston gauge and s_M^2 is the estimated variance of the slope of the line.

4. Tertiary Instruments

Two instruments mentioned in the introduction will be described for their interest and for their ability to achieve good accuracy conveniently. These instruments and others of similar design are much used in laboratory work where pressures must be measured with high precision.

(a) Oscillating-quartz force transducer

When pressure is applied to a flexible bellows it expands or, if confined, the bellows exerts a force. In this way, a bellows may be thought of as similar to a piston in a cylinder and may be characterised by an effective area. Unfortunately, hysteresis will occur as the bellows expands, however if this is not allowed to happen then hystersis can be reduced considerably. What is needed then is a method of measuring the force with small displacement, that is to say, a high-stiffness force transducer.

Such a transducer has been made using the change in resonant frequency of a specially shaped quartz oscillator when the force acts on it. This change can be measured easily to very good accuracy.

These devices are calibrated using piston gauges or mercury manometers depending on the range required. As the relationship between the pressure P and the period τ is given by the quadratic

$$P = A + B\tau + C\tau^2,$$

at least three different pressures must be used, but it is better to use rather more than this and to use a least-squares criterion to give a quadratic of best fit. These instruments are capable of measuring pressures with an uncertainty of 1 in 10^4.

A differential pressure gauge having similar accuracy has been made using a diaphragm as the elastic element on which the pressure acts rather than a pair of bellows acting in opposition. There is a change in the effective area of the bellows with pressure so that the sensitivity of a differential gauge employing them varies with line pressure.

(b) Force feedback

Elastic distortion gauges such as the Bourdon tube gauge are not strictly linear in their stress/strain relation and, when stressed, tend to exhibit hysteresis. Quartz has good elastic properties and can go a long way towards eliminating this problem. A further technique is to eliminate the strain by applying a force sufficiently large to prevent movement of the element. This force is provided electromagnetically, and the current needed to do this is taken as a measure of the pressure applied. For this, the current flows through a stable resistor, and the EMF across it is read with a digital voltmeter which may be conveniently calibrated to read in units of pressure. The manufacturer claims a 90 day stability of 0.0035%.

Because the stability of these instruments depends on a quartz elastic element, a force transducer a resistor and a digital voltmeter they are generally regarded as tertiary instruments. However they are capable of good stability and are very convenient to use.

References

Bignell, N. (1976). *J. Phys. E.: Sci. Instru.* 9, 730.

Brombacher, W.G. (1970). *Nat. Bur. Stand. (U.S.)*, Monogr. 114.

Brombacher, W.G., Johnson, D.P., Cross, J.L. (1960). *Nat. Bur. Stands. (U.S.)*, Monogr. 8.

Cook, A.H. (1961), *Phil. Trans. Roy.Soc. London, Ser. A*, 254, 125.

Daborn, J.E., Hatley, M.B., Ruffell, D.E. (1974). *Nat. Phys. Lab.* Report MOM 7.

Dadson, R.S., Greig, R.G.P., Horner, A. (1965). *Metrologia*, 1, 55.

Harrison, E.R., Hatt, D.J., Prowse, D.B., Wilbur-Ham, J. (1976). *Metrologia*, 12, 115.

Johnson, D.P., Newhall, D.H. (1953). *Trans. Am. Soc. Mech. Engrs.*, 75, 301.

Density Measurement of Solids and Liquids

J.B. Patterson
CSIRO Division of Applied Physics, Sydney, Australia 2070

Abstract

This report describes the facilities, equipment and basic techniques required for the measurement of the density of solids and liquids, as well as some unusual techniques for special density measurement.

1. Introduction

The basic methods of density measurement rely on an accurate determination of the mass of a solid or liquid, of known volume. Special instruments are also available for the direct measurement of liquid density and the major requirements are listed below:

(a) a balance for determining the mass of solids and liquids to be used;

(b) a solid or liquid of known density, to be used as a standard;

(c) pycnometers (density bottles) for the measurement of liquid and powder densities;

(d) a set of hydrometers; and

(e) a temperature-controlled bath, for maintaining the solids and liquids at constant standard temperatures.

2. Basic Equipment

(a) Balances

The balance to be used will depend on the methods of measurement and its limit of uncertainty. Top-loading, equi-arm or single-pan balances may all be used for particular purposes. Equi-arm or single-pan balances must have the facility for suspending objects from below the balance pan. Some equi-arm balances have heavy spheres attached to the ends of the balance beam in order to provide extra momentum to the swinging beam in cases where the weighed object is immersed in a high-viscosity liquid.

All loads applied to the balances must be smaller than the balance capacity, and the sensitivity must be such that the accuracy of measurement is satisfied.

(b) Standards

The basic liquid for use as a density standard is pure water. With the introduction of the metric system, the kilogram was designed to be equivalent in mass to $1 \times 10^{-3}\ m^3$ of pure water at its temperature of maximum density (approx. 4°C). However, in later experiments conducted by Guillaume and others (1910), the maximum density of water was re-measured using the kilogram and metre as the basic standards, and instead of being 1000 kg/m^3 as intended, was found to be 999.972 kg/m^3.

Water density tables were established by Chappuis (1907) and Thiessen and others (1900) and although the tables differ, it is remarkable that no fewer than five sets of tables have been published between 1937 and 1975, based on this work.

The absolute value of the density of water is dependent on the amount of air dissolved in the water, as well as any change in the ratio of the different isotopes of hydrogen and oxygen present in the sample. The Division of Applied Physics undertook a measurement of the dilatation of air-free water up to 40°C, and the results have yielded a density $\rho(t)$ given by

$$\rho(t) = \rho_{max} - \frac{(t - 3.989)^2 (t + 335.1596)}{558.7666\ (t + 72.2553)}\ kg/m^3 \qquad (1)$$

where ρ_{max} (kg/m^3) is the maximum density of the water sample at 3.989°C (Bell and Patterson 1981).

It was found by Steckel and Szapiro (1963) that any difference in the isotopic ratios of hydrogen or oxygen in any one sample of water changes the absolute density value but not the dilatation. Equation (1) will thus hold true for any sample of pure, air-free water if its maximum density, in kg/m^3, is substituted for ρ_{max}.

The effect of dissolved air on the density of water over the temperature range 4°C to 20°C, has also been measured in the Division (Bignell 1982). The density difference between air-free samples and those fully saturated with air at different temperatures (t°C) is given by

$$\Delta\rho = (4.38 - 0.103t) \times 10^{-3}\ kg/m^3$$

where the air-free sample is greater in density.

Table 1 gives a comparison of the various density tables in common use, assuming the maximum density of pure, air-free water at 4°C to be 999.9720 kg/m^3

Table 1
Water density kg/m^3

t°C	Chappuis	Thiessen	Tilton & Taylor	CSIRO
4	999.9720	999.9720	999.9720	999.9720
10	999.7002	999.6991	999.6998	999.7007
15	999.0989	999.0985	999.0998	999.1012
20	998.2063	998.2024	998.2042	998.2053
25	997.0490	997.0429	997.0450	997.0462
30	995.6501	995.6452	995.6472	995.6481
35	994.0333	994.0299	994.0314	994.0322
40	992.2215	992.2133	992.2142	992.2158

For some work, water may be unsuitable because of its large surface tension, and the fact that some irregular shaped objects are liable to retain air bubbles on their surfaces when immersed. Other liquids of high purity may be used as density standards, if their composition remains stable when exposed to air. White spirit and xylene are two suitable standard liquids, but their density must be measured and monitored regularly, either by calibration against water, or by using a calibrated sensitive hydrometer.

Solids may be used as standards if the density of the material is consistent for different samples. For example, single-crystal silicon is considered to be a suitable material, but long-term stability and reproduceability have not been established as yet.

Any stable composite object, such as a glass plummet filled with mercury, can also be used as a transfer standard between water and unknown liquids.

(c) Pycnometers

Pycnometers or density bottles are fixed-volume devices which can be used to determine the density of liquids or powders. There are many design variations for this type of device but the essential requirements are that the liquid must fill the vessel to a fixed mark without air bubbles, and that the loss of liquid due to evaporation must be minimal.

A full description of the methods of calibration and use of the most common pycnometers, together with their specifications, are given in British Standard 733.1965.

By use of a differential weighing technique on pycnometers, reported in Washburn and Smith (1934), the density difference between two liquids can be found with a high degree of accuracy.

(d) Hydrometers

Hydrometers are direct-reading instruments for the measurement of liquid density and are very quick and easy to use. They are commercially available in sets to cover the density range of most liquids.

Specifications and the method of reading hydrometers are given in Australian Standard 2026-1977, as is the theory for the design of special hydrometers. Special hydrometers have been designed to measure opaque liquids; the specifications for milk hydrometers are given in Australian Standard 2148-1978.

Although the official SI unit of density is the kilogram per cubic metre, most density hydrometer scales are graduated in grams per millilitre or grams per cubic centimetre. The relationship between them is given by

$$1 \text{ gm/ml} = 1000 \text{ kg/m}^3 = 1 \text{ gm/cc}$$

since the litre is now defined as equivalent to 1000 cubic centimetres.

Other density scales (specific gravity, Baume, Twaddell) have been introduced for special liquids, and a summary of the definitions and relationship with the normal density scale are given by Rands and Bigg (1961).

(e) Temperature Control

For work of high accuracy, the temperature of the solid or liquid must be maintained within certain limits while the measurement is in progress.

The vessel containing the liquid is generally held in a temperature-controlled bath for a suitable period before the measurements commence, and preferably during the measurements. The range of temperature within which the bath should be maintained depends on the dilatation coefficients of the liquid and solids to be measured, and the overall uncertainty limits of the measurement.

3. Measurement Technique

The techniques for the use of pycnometers and hydrometers are already described in the Australian and British Standards quoted above. For the calibration of hydrometers, a hydrostatic method has been devised by Cuckow (1949), and is the method used in the Division of Applied Physics.

The derivation of the formula for the calibration of pycnometers is given in Appendix 1 and for hydrometers in Appendix 2.

The basic technique for the measurement of solid density is to shape the solid into a convenient geometrical form whose volume can be determined by direct length measurement and then weighed in air. The absolute measurement on the density of water by Guillaume (1910) depended on accurate length comparator measurements on a cylindrical block. Other methods used at the same time depended on interferometric length measurements on low-expansion glass cubes.

The solid chosen by the Division of Applied Physics was an evacuated hollow sphere made from ultra low-expansion glass, the diameter of which was measured in a special interferometer designed by Saunders (1972). The measurements were conducted over the temperature range 4°C to 40°C and, following an accurate determination of the mass, values of the density of the sphere were calculated for this temperature range.

Another technique for solid density measurement is to apply Archimedes' Principle, and determine the loss in weight experienced by the solid when immersed in a liquid of known density. The derivation of the formula is given in Appendix 3. It is to be noted that for accurate work, the correction for air buoyancy on the weights is to be applied.

For a quick result, the object may be hung by a thin suspension wire from a moveable support above a top-loading balance, the pan of which supports a container holding a liquid of known density.

The object is in a cage or cradle, and is lowered slowly into the liquid so that no air bubbles form on the object or cage. The object and cage must be totally submerged and that part of the suspension passing through the liquid surface must be of small crossection to reduce surface tension effects. The change in weight reading of the balance from a zero reading is noted.

The object is removed from the cage, which is then relowered into the liquid and the change in weight reading is again noted. It must be pointed out that the zero reading of the balance may change, due to loss of fluid when removing the object and cage from the liquid.

The difference between the two noted readings is a direct measurement of the loss in weight experienced by the object, and when equated against the density of the liquid, gives the volume of the object. The mass of the object is then measured separately, and the density calculated. The accuracy of this method is limited by the sensitivity of the large capacity balance required to counterpoise the liquid contained in the vessel.

For greater accuracy, either the equi-arm or the single-pan balance is recommended. The vessel, containing the standard liquid under temperature control, is positioned under the balance such that objects may be suspended from the pans and submerged in the liquid.

For some equi-arm balances, it is advantageous to have objects suspended from both pans of the balance and immersed in a liquid contained in an interconnecting vessel. This form of apparatus is very convenient when measuring liquid densities by a two-plummet method, described as follows.

A long thin mercury filled glass plummet is made such that its volume differs from a more stout plummet of equal or near equal mass, by approximately 1×10^{-4} cubic metres. The difference in mass between each plummet, when immersed in water and the unknown liquid in turn, is measured and the density of the unknown liquid is calculated as shown in Appendix 4.

However, most measurements of density by hydrostatic methods can be carried out on a single-pan balance. For accurate work, the internal weights of the balance have to be calibrated.

Liquid density may be measured by using the same principle, if an object of known mass and volume is weighed in the liquid. The formula for this procedure is derived in Appendix 5.

A special balance, called the Westphal Balance, is designed to read liquid density directly. It consists of a pivoted beam graduated on one arm, from which a calibrated sinker is suspended in the unknown liquid and counter-weighted on the other arm. The beam is brought to the balanced condition by means of rider weights on the graduated arm from which the density can be read directly.

4. Measurement Errors using Hydrostatic Methods

The errors associated with using this technique, and the methods of reducing such errors, are as follows.

a) The loss in weight experienced by an object suspended in a liquid is affected by bubbles on the surface of the object. These may be removed by using a small-bore vacuum line, or a special vessel may be designed so that the liquid fills the vessel around the object under vacuum conditions. Further, it helps to have a mechanical means of loading and unloading the objects from the suspension cage, without removing either from the liquid. The zero drift of the balance can be monitored by using this technique.

b) To reduce the effect of surface tension on that part of the suspension passing through the liquid surface, the thinnest wire capable of supporting the load is used. It is best if the liquid "wets" the wire surface and so a clean rough surface is preferred. Some suspension wires may be coated with different materials to produce such surfaces.

   Unless the wire is uniform in diameter, it is best to have the liquid surface cut the suspension wire at the same point when weighing the cage in the "loaded" or "unloaded" condition. This will also eliminate the buoyancy error caused by different lengths of wire below the surface.

The author has found that when using swinging equi-arm balances the position of the turning points of the swing are affected by changes of surface tension on the wire. In this case, a minimum amplitude of swing is preferred.

c) The temperature of the liquid must be controlled to maintain its density sufficiently constant to satisfy the accuracy requirements. A normal laboratory recirculating temperature controller is suitable for maintaining temperatures to within 0.01°C, but for high accuracy work, a better control system that maintains the temperature to within 0.001°C may be necessary.

5. Special Techniques of Density Measurement

(a) Vibration of a Fixed Volume

This type of apparatus was developed by Kratky and others (1967) and is now available commercially. It consists of a temperature controlled glass U-tube, which is vibrated electronically at a high frequency which varies according to the density of the liquid or gas contained in the tube. The apparatus requires calibration using two substances of known density (usually air and water), and several models of different resolution are available.

Important advantages are that the sample volume of the liquid or gas is small and that measurements can be carried out while the gas or liquids are flowing through the U-tube.

(b) Flotation Techniques

The types of floats used in this technique and the methods of controlling them are many and varied. The position of a float freely suspended in an unknown liquid is used to measure the liquid density, and the following methods may be used to control the float movement:

(1) adding incremental weights to the float;

(2) changing the liquid temperature;

(3) changing the pressure above the liquid; or

(4) changing the current in a coil winding to provide a magnetic pull on a magnet in the float.

Most float methods require a calibration using liquids of known density and all have systematic errors of some kind. The reader is referred to papers by Bowman and Schoonover (1965), Beams and Clarke (1962), Clarke (1967), and Hales (1970).

An extension of the float method is one that is used to measure the density of small crystals. A mixture of liquids is made up to a density, such that the unknown crystals will just float. The density of the liquid is then measured by normal methods.

(c) Special Techniques

Methods based on:

(1) falling drops;

(2) balanced liquid columns;

(3) Boyle's law; and

(4) elastic helix

are explained by Hidnert and Peffer (1950).

Another method is to drop small solids into liquid columns with a graded density value from top to bottom, either by varying the temperature along the column or by mixing liquids of different density. The point at which the solid becomes stationary is a measure of its density. References may be obtained from Moret (1961) or Spaepen (1958).

6. Conclusion

This report describes the equipment and methods used to determine solid and liquid densities. Because of space limitations, the description of some methods are necessarily brief and lack details of technique.

Further information may be obtained from some of the references, particularly Rands and Bigg (1961).

References

Beams, J.W., and Clarke, A.M., (1962). *Rev. Sci. Instrum.*, **33**, 750.

Bell, G.A., and Patterson, J.B. (in press) in Precision Measurement and Fundamental Constants II Ed. by B.N. Taylor and W.D. Phillips Nat. Bur. Stand., Spec. Publ. 617.

Bignell, N., (1982). *J. Phys. E.: Sci. Instrum.*, **15**, 378.

Bowman, H.A., and Schoonover, R.M., (1965). *J. Res. Nat. Bur. Stand.* **69C**, 217.

Chappuis, P., (1907). *Bur. Int. des Poids et Mesure. Trav. et Mêm.* **13**, D1.

Clarke, A.L., (1967). *J. Sci. Instrum.*, **44**, 382.

Cuckow, F.W., (1949). *J. Soc. Chem. Ind.*, **68**, 44.

Guillaume, CH-ED et al., (1910). *Bur. Int. des Poids et Mesure. Trav. et Mêm.* **14**, 1.

Hales, J.L., (1970). *J. Phys. E.*, **3**, 855.

Hidnert, P., and Peffer, E., (1950). *Nat. Bur. Stand. Circ.* **487**.

Kratky, O., Leopold, H., and Stabinger, H., (1969). *Zeitschr. Angew. Physik.* **27**, 273.

Moret, H., (1961). *Rev. Sci. Instrum.*, **32**, 1157.

Rands, J.B., and Bigg, P.H., (1961). Notes on Applied Science No. 25 NPL.

Saunders, J.B., (1972). *J. Res. Nat. Bur. Stand.* **76C**, 11.

Spaepen, J., (1957). *Mededel. Koninkl, Vlaam. Adak. Wetenschap. Belg.* **19**, 5.

Steckel, F., and Szapiro, S., (1963). *Trans. Farad. Soc.*, **54**, 331.

Thiessen, M., (1900) et al. *Physik Tech. Reichs, Wiss. Abhandl* 3.1; 4.1.

Tilton and Taylor (1937). *J. Res. Nat. Bur. Stand.* **18**, 205.

Washburn, E.W., and Smith, E.R., (1934). *B.S. J. Res.* **12**, 305.

Appendix 1

Measurement of Liquid Density using a Pycnometer

When the term "weight" is used in these appendices it means the mass of the weight used to counterpoise the weighed object.

Symbols

Mass of pycnometer	$= M_g$	Density of glass	$= g$
Mass of water to fill pycnometer	$= M_d$	Density of water at temperature t_d	$= d$
Mass of liquid to fill pycnometer	$= M_\rho$	Density of liquid at temperature t_ρ	$= \rho$
Internal volume of pycnometer at temperature t_ρ	$= V$	Density of air	$= \sigma$
Weight of pycnometer	$= W_G$	Density of weights	$= \Delta$
Weight of pycnometer and water	$= W_d$	Dilatation of glass	$= \alpha$
Weight of pycnometer and liquid	$= W_\rho$	Temperature of liquid	$= t_\rho$
		Temperature of water	$= t_d$

Equations of Equilibrium

Weighing empty pycnometer in air of density σ_1:

$$M_g(1 - \sigma_1/g) = W_g(1 - \sigma_1/\Delta). \tag{1.1}$$

Weighing water-filled pycnometer in air of density σ_2:

$$M_g(1 - \sigma_2/g) + M_d(1 - \sigma_2/d) = W_d(1 - \sigma_2/\Delta). \tag{1.2}$$

Weighing liquid-filled pycnometer in air of density σ_3:

$$M_g(1 - \sigma_3/g) + M_\rho(1 - \sigma_3/\rho) = W_\rho(1 - \sigma_3/\Delta). \tag{1.3}$$

Using equations 1.1, 1.2 and 1.3

$$\rho = \frac{M_\rho}{V} = \frac{M_\rho . d}{M_d[1 + \alpha(t_\rho - t_d)]}$$

$$= \frac{[W_\rho(1 - \sigma_3/\Delta) - W_G \dfrac{1 - \sigma_1/\Delta}{1 - \sigma_1/g} . (1 - \sigma_3/g)](d - \sigma_2)}{[W_d(1 - \sigma_2/\Delta) - W_G \dfrac{1 - \sigma_1/\Delta}{1 - \sigma_1/g} (1 - \sigma_2/g)][1 + \alpha(t_\rho - t_d)]} + \sigma_3.$$

If

$$\frac{1 - \sigma_3/g}{1 - \sigma_1/g} = \frac{1 - \sigma_2/g}{1 - \sigma_1/g} = 1$$

then

$$\rho = \frac{\{W_\rho(1 - \sigma_3/\Delta) - W_g(1 - \sigma_1/\Delta)\}(d-\sigma_2)}{\{W_d(1 - \sigma_2/\Delta) - W_g(1 - \sigma_1/\Delta)\}\{1 + \alpha(t_\rho - t_d)\}} + \sigma_3.$$

Assuming the air density is constant for all weighings

$$\rho = \frac{(W_\rho - W_g)(d - \sigma)}{(W_d - W_g)\{1 + \alpha(t_\rho - t_d)\}} + \sigma.$$

Appendix 2

Calibration of Hydrometers

Mass of the hydrometer	$= M$
Weight of hydrometer in liquid at calibration point	$= W_d$
Weight of hydrometer in air	$= W_\sigma$
Volume of hydrometer below liquid surface, reading ρ	$= V$
Volume of hydrometer above liquid surface, reading ρ	$= v$
Diameter of hydrometer stem at liquid surface	$= D$
Acceleration due to gravity	$= g$
Density of weights	$= \Delta$

It can be shown for normal hydrometers that corrections for the expansion of the glass over a temperature range of ±10°C is negligible, and that changes in air density between weighings can be ignored.

Thus

a) For a hydrometer weighed in air of density σ:

$$M - (V\sigma + v\sigma) = W_\sigma(1 - \frac{\sigma}{\Delta}). \tag{2.1}$$

b) For a hydrometer floating freely in a liquid of density ρ, surface tension T_1:

$$M + \frac{\pi D T_1}{g} = V\rho + v\sigma \tag{2.2}$$

c) For a hydrometer weighed in a liquid of density d, surface tension T_2, to the same depth as case (b):

$$M + \frac{\pi D T_2}{g} = Vd + v\sigma + W_d(1 - \frac{\sigma}{\Delta}). \tag{2.3}$$

By rearranging equations 2.1, 2.2 and 2.3 and eliminating M, V & v we obtain

$$\rho = (d - \sigma) . \frac{W_\sigma + \dfrac{\pi D T_1}{g}}{(W_\sigma - W_d) + \dfrac{\pi D T_2}{g}} + \sigma$$

Appendix 3

Measurement of Solid Density by Hydrostatic Weighing

Symbols

Mass of specimen	$= M$	Density of air	$= \sigma$
Volume of specimen	$= V$	Density of liquid	$= \rho$
Weight of specimen in air	$= W_\sigma$	Density of weights	$= \Delta$
Weight of specimen in liquid	$= W_\rho$	Density of specimen	$= D$

Equations of Equilibrium

Weighing specimen in air of density σ_1:

$$M - V\sigma_1 = W_\sigma (1 - \sigma_1/\Delta). \qquad (3.1)$$

Weighing specimen in liquid, air density σ_2:

$$M - V.\rho = W_\rho (1 - \sigma_2/\Delta). \qquad (3.2)$$

Subtracting equation 3.2 from 3.1:

$$V(\rho - \sigma_1) = W_\sigma(1 - \sigma_1/\Delta) - W_\rho(1 - \sigma_2/\Delta).$$

From equation 3.1

$$M = W_\sigma(1 - \sigma_1/\Delta) + V.\sigma_1.$$

Therefore

$$D = \frac{M}{V} = \frac{W_\sigma\{1 - (\sigma_1/\Delta)\}.(\rho - \sigma_1)}{W_\sigma\{1 - (\sigma_1/\Delta)\} - W_\rho\{1 - (\sigma_2/\Delta)\}} + \sigma_1.$$

For a constant value of air density

$$D = \frac{W_\sigma.(\rho - \sigma)}{W_\sigma - W_\rho} + \sigma.$$

NOTE: W_ρ is the weight difference between the loaded and unloaded condition of the suspension cage holding the specimen.

Appendix 4

Measurement of Liquid Density by Hydrostatic Weighing using Two Plummets

Symbols

Mass of lean plummet	$= M_L$	Volume of lean plummet	$= V_L$
Mass of stout plummet	$= M_S$	Volume of stout plummet	$= V_S$
Weight difference in air	$= W_\sigma$	Density of air	$= \sigma$
Weight difference in water	$= W_d$	Density of water	$= d$
Weight difference in liquid	$= W_\rho$	Density of liquid	$= \rho$
		Density of weights	$= \Delta$

Equations of Equilibrium

Weighing in air of density σ_1:

$$M_L - M_S - (V_L - V_S)\,\sigma_1 = W_\sigma(1 - \sigma_1/\Delta) \tag{4.1}$$

Weighing in water of density d, air density σ_2:

$$M_L - M_S - (V_L - V_S).\,d = W_d(1 - \sigma_2/\Delta) \tag{4.2}$$

Weighing in liquid of density ρ, air density σ_3:

$$M_L - M_S - (V_L - V_S).\,\rho = W_\rho(1 - \sigma_3/\Delta) \tag{4.3}$$

Subtracting equation 4.1 from 4.2

$$-(V_L - V_S)(d - \sigma_1) = W_d(1 - \sigma_2/\Delta) - W_\sigma(1 - \sigma_1/\Delta)$$

From equation 4.3

$$-(V_L - V_S)\,\rho = W_\rho(1 - \sigma_3/\Delta) - (M_L - M_S)$$

$$\rho = \frac{W_\rho(1 - \sigma_3/\Delta)}{-(V_L - V_S)} - \frac{M_L - M_S}{-(V_L - V_S)}\,.$$

If $M_L - M_S$ is small and $V_L - V_S$ is sufficiently large

$$\rho = \frac{W_\rho(1 - \sigma_3/\Delta)}{-(V_L - V_S)}$$

$$= \frac{W_\rho(1 - \sigma_3/\Delta).(d - \sigma_1)}{W_d(1 - \sigma_2/\Delta) - W_\sigma(1 - \sigma_1/\Delta)}\,.$$

For a constant value of air density

$$\rho = \frac{W_\rho (d - \sigma)}{W_d - W_\sigma} .$$

Appendix 5

Measurement of Liquid Density by Hydrostatic Weighing using a Standard Object

Symbols

Mass of standard object	$= M$	Density of air	$= \sigma$
Volume of standard object at temperature t	$= V$	Density of liquid at temperature t	$= \rho$
Weight of object in air	$= W_\sigma$	Density of weights	$= \Delta$
Weight of object in liquid	$= W_\rho$		

Equations of Equilibrium

Weighing standard object in air of density σ_1:

$$M - V.\sigma_1 = W_\sigma (1 - \sigma_1/\Delta). \quad (5.1)$$

Weighing standard object in liquid, air density σ_2:

$$M - V.\rho = W_\rho (1 - \sigma_2/\Delta). \quad (5.2)$$

Subtracting equation 5.2 from 5.1

$$V(\rho - \sigma_1) = W_\sigma (1 - \sigma_1/\Delta) - W_\rho (1 - \sigma_2/\Delta)$$

$$\rho - \sigma_1 = \frac{W_\sigma (1 - \sigma_1/\Delta) - W_\rho (1 - \sigma_2/\Delta)}{V} .$$

Therefore

$$\rho = \frac{W_\sigma (1 - \sigma_1/\Delta) - W_\rho (1 - \sigma_2/\Delta)}{V} + \sigma_1 .$$

For a constant value of air density

$$\rho = \frac{(W_\sigma - W_\rho) . (1 - \sigma/\Delta)}{V} + \sigma .$$

NOTE: W_ρ is the weight difference between the loaded and unloaded condition of the suspension cage holding the object.

Calibration of Standards of Length

Maurice J. Puttock
CSIRO Division of Applied Physics, Sydney, Australia 2070

Abstract

The fundamental standard of length is currently the wavelength of a particular visible radiation of the rare-gas isotope krypton-86. The practical standards of length are secondary wavelength standards, gauge blocks, length bars, line scales and tapes or wires. The purpose of this paper is to review the procedures and techniques necessary to ensure that these standards are calibrated with the accuracy appropriate to their position in a hierarchical system for the dissemination of length. Particular attention is given to the calibration of gauge blocks.

1. Introduction

The field of length measurement covers a range of dimensions from nanometres to kilometres and in relation to the fundamental standard, the radiation (in vacuum) of krypton-86 at 605.8 nm, this represents ratios from 10^{10} to 10^{-2}.

The physical standards of length which are of concern in the context of this paper range in value from nanometres for secondary wavelength standards (secondary in the sense of being second to the fundamental standard) to metres for high-precision survey tapes or wires. The physical forms of these standards are related to their function (e.g. tapes for survey, gauge blocks for precision engineering, etc.) but in every case their calibration must be in terms of the fundamental unit, namely the *metre* which is currently defined as "being the length equal to 1 650 763.73 wavelengths in vacuum of the radiation corresponding to the transition between the levels $^2p_{10}$ and $^5d_5$ of the krypton-86 atom".

There is a strong possibility that the metre will be redefined within the next few years, maybe in 1983, in terms of the speed of light (in vacuum) and the second of time (itself defined from the atomic standard of frequency). The reasons for this change in definition will not be discussed here; it should be noted, however, that there will be no resulting change in the length of the metre in a practical sense - any actual change will be so small as to be totally undetectable by any known techniques. The practical realization of the metre at the highest accuracy will continue to be through wavelength standards and in this regard there will be an advantage to be gained by using stabilized lasers which are considerably more precise than the Kr-86 primary standard.

2. Wavelength Standards

The actual wavelength of a particular radiation emitted by a lamp containing an isotope is dependent on many factors such as the physical construction of the lamp, the type and amount of energy used to excite the isotope, the temperature of the lamp, etc. It follows therefore that to obtain a *standard* wavelength from a given source these factors must be carefully specified. In the case of Kr-86 primary standards the relevant specification has been promulgated by the Committee International des Poids et Mesures, CIPM (1960), and if this specification is adhered to, within the tolerances stated therein, the wavelength emitted will be its standard value within 1 in 10^8.

The calibration of other (secondary) wavelength standards in terms of the primary standard is a somewhat involved process requiring specialised equipment and techniques using the phenomenon of interference of light waves (generally referred to as interferometry). Only a few of the national laboratories undertake this task and it is not appropriate, or necessary, to discuss the details here. In particular it is not necessary because the CIPM has promulgated the values for a number of such secondary wavelength standards (together with specification of lamp and operating conditions as appropriate); 4 additional secondary standards from Kr-86 and 4 each from Hg 198 and Cd 114 were promulgated in 1963, CIPM (1963). These secondary standards have associated uncertainties of from 2 to 7 in 10^8.

In addition, 2 secondary wavelength standards, one in the visible at 633 nm and one in the infra-red at 3390 nm, from He-Ne lasers have been promulgated, CIPM (1973). These lasers are stabilised with respect to absorption lines of I_{127} and of methane respectively and have a precision of 1 in 10^{11} or better and hence have uncertainties closely the same as the Kr-86 primary standard in terms of which they have been standardised.

It should be noted that there are various ways of stabilising the wavelength of the radiation emitted by the He-Ne laser (this laser is currently the only one readily available to the metrologist as a length standard) which give uncertainties in the range 1 to 5 in 10^7. Stabilised lasers in this category are readily available commercially and provide excellent secondary standards for the metrologist.

3. Interferometry

Before proceeding to the calibration of the other forms of length standards it is essential that the basis of the technique, whereby optical wavelengths of 400 to 650 nm can be used to measure lengths up to metres, should be understood; it should be remembered that the number of such wavelengths in a metre range from 1.7 to 2.5×10^6. respectively.

The technique is known as interferometry and is basically extremely simple in concept. If we consider visible radiation to be electric field fluctuations in the form of sine waves then the wavelength standards already discussed are sine waves where the wavelength, i.e. distance from crest to crest, is constant. Now if a beam of such radiation is somehow split into two similar beams and subsequently recombined then, depending on how far each 'half-beam' has travelled,when they are recombined they will be in or out of phase.

If they are in phase the resultant beam is a replica of the original and hence is visible as light; if they are exactly (i.e. 180°) out of phase then they cancel each other out and the resultant beam has no amplitude, hence no light is to be seen. (See Fig.1).

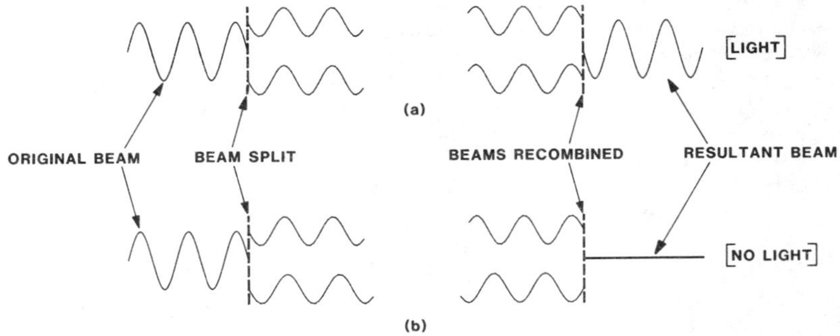

Fig.1 Principle of optical interference
(a) Recombined beams in phase
(b) Recombined beams 180° out of phase.

The simplest form in which this phenomenon of interference of light waves is used in metrology is in the measurement of flatness of a lapped surface by means of an optical flat.

An optical flat is a piece of optical glass or quartz in the form of a thick disc (thickness = 1/4 to 1/5 of diameter) both faces of which are polished and with one worked to a very high degree of flatness (usually 0.02 to 0.05 µm for a 50 mm diameter flat). If this optical flat is placed with its high-accuracy surface in contact with the lapped surface to be tested the surface of the flat will act as a beam splitter, i.e., it will reflect some of the light and will allow some to go through and be reflected at the lapped surface (Fig.2). When that part of the ray reflected at the lapped surface re-enters the optical flat it will, depending on how far it has travelled, be in phase or out of phase with the other part reflected at the surface of the flat. Hence, depending on the excess distance travelled, the rays will recombine to give light or darkness. If the optical flat is inclined at a very small angle the fringes will appear as dark and light bands across the lapped surface. The degree of straightness of these bands will be a measure of the flatness of the lapped surface. Since from one dark band, usually called a "fringe", to the next represents a total path difference of one wavelength, λ, this corresponds to a change in separation of optical flat and lapped surface of $\lambda/2$.

NOTE: Another phenomenon is that which causes the ray reflected at the glass surface to be put automatically out of phase with the ray reflected from the lapped surface. The effect of this is to produce a dark fringe at contact between flat and lapped surfaces instead of light, as you might otherwise expect.

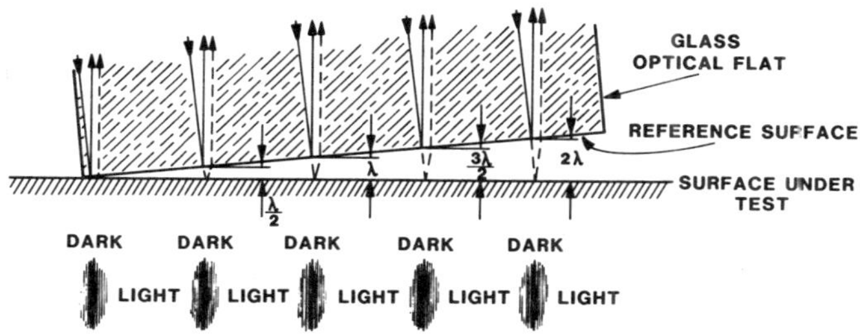

Fig.2 Formation of interference fringes with optical flat.

White light produces rainbow-coloured fringes. Measurement of fringe straightness, etc., is made considerably easier if light waves of only one wavelength (monochromatic) are used. These produce dark fringes on a uniform background. Suitable sources are:

(i) Sodium lamp. Yellow light, $\lambda/2 \triangleq 0.3$ μm usually used when optical flat is in contact with test surface (as in Fig.2).

(ii) Mercury lamp. Green filter, $\lambda/2 \triangleq 0.25$ μm usually used when optical flat separated from test surface (e.g., in a flatness interferometer).

In various ways this technique of interferometry can be used to determine not only flatness but also other features. It is convenient to divide these various ways of using interferometry into two broad groups, namely:

Direct interferometry where the process is static and the length is determined from the fringe pattern.

Indirect interferometry where the interferometry is used to measure a displacement, i.e., a fringe counting or similar technique.

Applications to the calibration of length standards, particularly those of the highest quality, are of considerable importance and will be considered more fully in the appropriate sections which follow.

4. End Standards

There are three types of end standard, as defined by their cross-section, in use in the world.

The most well known is, of course, that of rectangular cross-section known as 'gauge blocks' (in the U.K. they have been called for many years 'slip gauges'). By far the majority of gauge blocks in use are of nominal length 100 mm and less. It is possible, however, to obtain gauge blocks up to 3 m nominal length; the international standard ISO 3650 Gauge Blocks is, however, restricted to a maximum length of 1 m.

The other types are those of square cross-section made in the U.S.A. and often referred to as 'Hoke gauges' (see ANSI B89.1.9) and those of circular cross-section made in the U.S.A. (see ANSI B89.1.9) and in the U.K. (see BS 5317) and often referred to as 'Length Bars' or 'End Bars'.

It is convenient to consider the calibration of end standards as falling into two general categories dependent on length - those up to 100 or 150 mm and those of greater lengths. Since more than 95% of all gauge blocks in use come within the first of these categories, namely <150 mm, prime attention will be given to them.

In general end standards are available in a number of quality grades; in the ISO standard for gauge blocks ISO 3650 Grades 00, Calibration, 0, 1 and 2 in descending order of quality. In this case, it is generally recommended that Grade 00 be calibrated by interferometry, Calibration and Grade 0 either by interferometry or by comparison with Grade 00, and Grades 1 and 2 by comparison with a higher grade. This is fairly logical but it should be modified in practice by considering the use to which a particular set of gauges is to be put. For example, a well organized laboratory will progressively downgrade the use of a set of gauges so that a Grade 00 set could be used ultimately for the tasks usually assigned to Grades 1 or 2; obviously it would be foolish to calibrate such a set by interferometry just because it was marked Grade 00 (few, if any, organizations change the grade marking when downgrading a set of gauges).

The dimensional features of direct concern in the calibration of end standards are:

flatness of end faces;
parallelism of end faces, i.e. variation in length;
squareness of end faces to side faces or axis;
length.

The hardness of the end faces is normally not measured because of the damage such measurement entails; reliance is placed on the manufacturer to provide adequately hard surfaces. It is not usual either to measure the surface roughness unless there is good reason to suspect that it is not adequate; 'wringing property' is however assessed (not measured), the assessment being based on experience of what a 'good wring' should be.

One further point should be discussed here and that is 'what do we understand the length of an end standard to be?'. A perusal of the various standard specifications, both national and international, reveals subtle differences, in terminology and definitions, which are confused and confusing. The interpretation which this laboratory has extracted from the confusion, and which is used in the rest of this discussion, is as follows:

(i) The gauge length includes one wringing contact.

(ii) For end standards used to calibrate other end standards or high precision measuring equipment (in gauge blocks this would be normally Grades 00 and Calibration) the gauge length is the central length (Fig.3).

(iii) For other standards the gauge length is the average length obtained from a number of measurements distributed over the end faces.

It should be noted that, since calibration is by direct interferometry for the top grades of end standards, the measured length, as will be seen later, includes the thickness of one wringing contact. When such a standard is used to calibrate another standard by comparison this wringing contact thickness is automatically included in the end result for the length of that other standard.

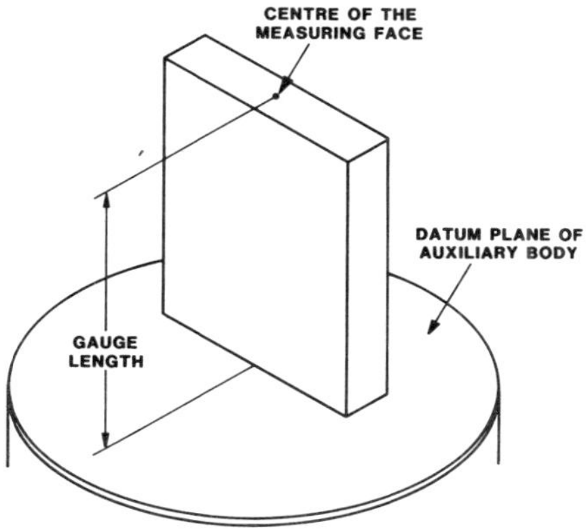

Fig.3 Gauge length.

(a) Calibration of Gauge Block Lengths by Direct Interferometry

There are two types of interferometer which are used in the measurement of gauge blocks. They can be classified as Michelson and Fizeau types. The optical schemes of these two types are shown in Fig.4 and 5; the essential difference is that in the Michelson the beam splitter is separate from the reference surface which is a fully reflecting surface, whilst in the Fizeau the reference surface and beam splitter are the same semi-reflecting surface. This leads to the fact that for the same difference in path length between beam splitter to reference surface and beam splitter to gauge and/or base-plate surface the Michelson type has twice the length capacity of the Fizeau.

In both interferometers the gauge block is wrung to a lapped steel baseplate or platen and the objective is to determine the central length of the gauge block referred to a temperature of 20°C - the international reference temperature (see ISO Standard ISO-1).

The measurements and calculations are virtually identical for both interferometers and it is therefore proposed to consider in detail the calibration of a gauge block using the Fizeau type as the example. A description of such an interferometer is given by Barnes and Puttock (1953).

The formation of the fringe pattern seen in the viewing aperture when a gauge block is wrung in position on the baseplate is illustrated in Fig.6. In this figure if the extra distance, ABC, travelled by the second part of the light is an exact number of wavelengths (λ) then when it recombines with the first part the two recombining ray, R_1, from the reference mirror and R_2 from the gauge block or baseplate, will, because of the automatic phase change at the reference mirror (see note in Section 3), be out of phase and no light will be seen.

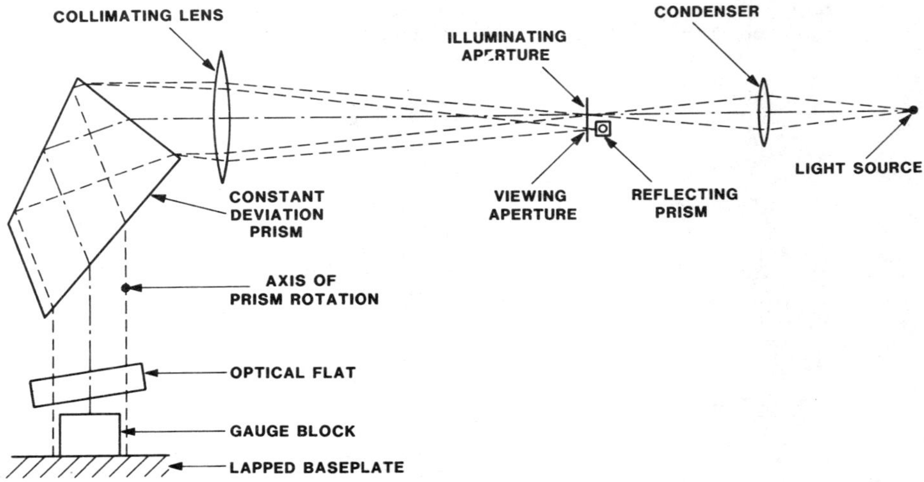

Fig.4 Fizeau-type gauge interferometer.

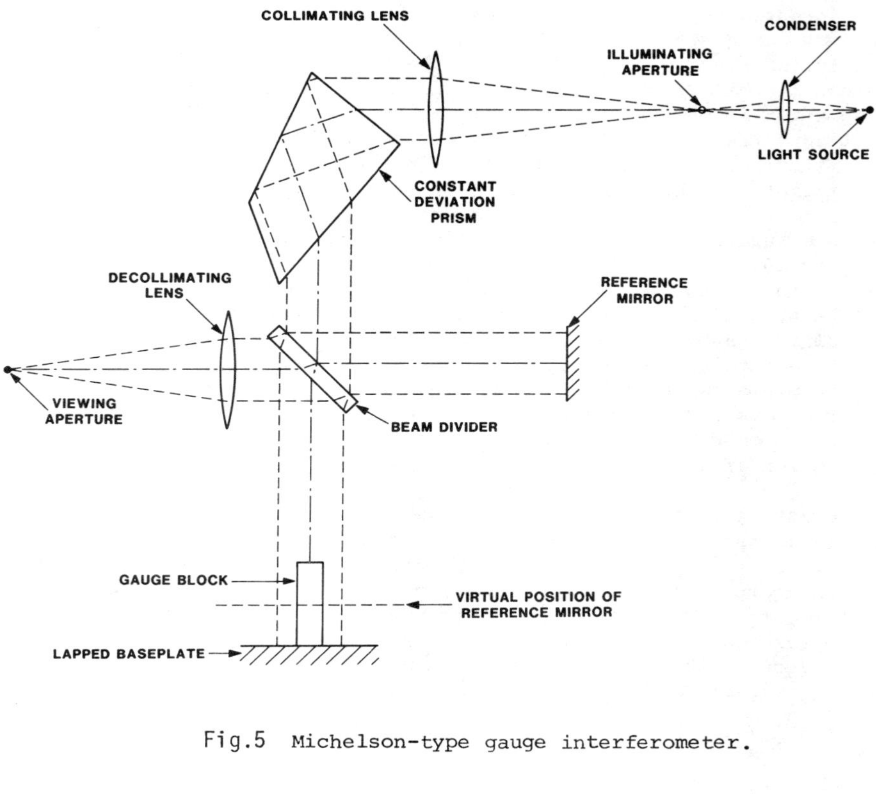

Fig.5 Michelson-type gauge interferometer.

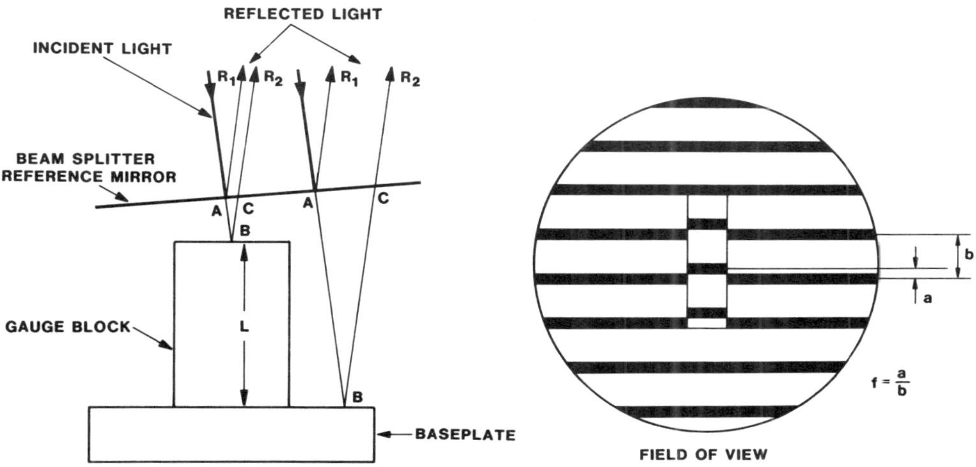

Fig.6 Formation of fringes in Fizeau-type gauge interferometer.

Similarly, if ABC is a whole number of wavelengths plus half a wavelength R_1 and R_2 will be in phase and light will be seen. With the beam splitter/reference mirror slightly tilted a series of fringes will be seen across the field of view.

If the fringes on the gauge block and baseplate are displaced, as shown in Fig.6, it means that the length of the gauge block is, if the beam splitter/reference mirror is tilted in the right direction, an exact (but unknown) number of $\lambda/2$ plus a fraction $f \times \lambda/2$. In practice the tilt of the beam splitter/reference mirror is adjusted so that there are approximately three fringes spaced along the end face of the gauge block with one through the centre of that face; this ensures that f is directly related to the central length as required by specification. By measuring f in a number of wavelengths (obtained by rotation of the constant deviation prism), in the case of Cd 114 red, green, blue and violet the four fractions f_R, f_G, f_B and f_V enable the length of the gauge block to be determined without knowing the whole numbers of $\lambda/2$ involved.

The information obtained so far is, in effect, that the length of the gauge block under the ambient conditions is given by

$$L_{\text{ambient}} = (N_R + f_R)\frac{\lambda_R}{2} = (N_G + f_G)\frac{\lambda_G}{2} = (N_B + f_B)\frac{\lambda_B}{2} = (N_V + f_V)\frac{\lambda_V}{2}$$

where N is the whole number of $\lambda/2$ in each case.

Now if we divide the nominal length of the gauge block by the nominal wavelengths λ' at "normal" air conditions (temperature 20°C, pressure 760 mm Hg, a water vapour pressure of 10 mm Hg and 0.03% by volume of CO_2), we get a series of nominal half wavelengths and fractions (N' and f' respectively) as follows:

$$L_{\text{nominal}} = (N'_R + f'_R)\frac{\lambda'_R}{2} = (N'_G + f'_G)\frac{\lambda'_G}{2} = (N'_B + f'_B)\frac{\lambda'_B}{2} = (N'_V + f'_V)\frac{\lambda'_V}{2}$$

If we now subtract these two equations we are left with the residual difference between the measured length and the nominal length which can be written

$$\delta L = [(N''_R + f''_R)\frac{\lambda'_R}{2} = \quad \ldots \quad \ldots \quad \ldots \quad \ldots] + kL$$

where

$\delta L = L_{ambient} - L_{nominal}$

$N'' = (N - N')$

$f'' = (f - f')$

and kL is a correction required because of the difference between the wavelengths at the conditions of the measurement (λ) and the wavelengths at "normal" air conditions (λ') plus a correction for the expansion of the gauge block derived from the difference between the temperature of the gauge block at the time of measurement and the reference temperature of 20°C.

The different wavelengths (λ_R, λ_G, λ_B, λ_V) can be considered as 4 linear scales having a common zero, corresponding to $\delta L = 0$, and for only one measured value of δL will all the fractions coincide. In fact the four scales, in units of $\lambda/2$, and a length scale, in units of micrometres, can be engraved on a slide rule and a cursor used to get a 'best fit' of the observed residual differences (see Fig.7). This process can also be carried out automatically on a computer or programmable calculator. The method is often referred to as the method of "exact fractions".

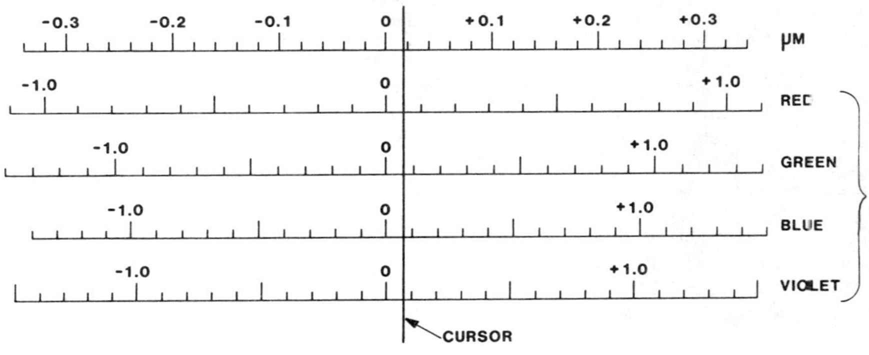

Fig.7 Length and wavelength scales - fringe fraction coincidence.

Having determined the length equivalent of the residual fringe fractions it is necessary to determine the correction kL. The basic data required for this is the temperature, pressure and humidity of the air and the temperature of the gauge block, all at the time of measurement.

The effect of these parameters on the length measurement is as follows:

Air pressure	1 mm Hg	$\triangleq$ 0.36 $\times 10^{-6}$L
Air temperature	1°C	$\triangleq$ 0.93 $\times 10^{-6}$L
Air humidity	1 mm Hg	$\triangleq$ 0.056 $\times 10^{-6}$L
Gauge block temperature (steel)	1°C	$\triangleq$ 12 $\times 10^{-6}$L

The actual corrections to be applied per unit length for the ambient air conditions are derived from the refractivity-density relationship established by Barrell and Sears (1939). In practice these corrections can be calculated with sufficient accuracy from the following simplified equations:

Correction for temperature and pressure:

$$c_1 = [A - \frac{Bh'}{(1+\alpha t)}] \times 10^{-6}$$

where A and B are constants dependent on the mean of the wavelengths used

$$h' = h(1 + \beta_t h)$$
$$h = \text{barometric pressure in mm Hg}$$
$$\beta_t = (1.049 - 0.0157t) \times 10^{-6}$$
$$\alpha = 0.003\,661$$
$$t = \text{temperature °C.}$$

Correction for humidity:

$$c_2 = C(f-10) \times 10^{-6}$$

where C is a constant dependent on the mean of the wavelengths used,

$$f = \text{vapour pressure of water in mm Hg.}$$

For the four cadmium wavelengths the values of the constants used above are:

$$A = 273.561$$
$$B = 0.386\,088$$
$$C = 0.058\,8.$$

In addition to the corrections c_1, c_2 and that for the gauge block temperature (expansion), two further corrections must be applied.

The first arises from the position of the viewing aperture (obliquity correction). In the case of the Michelson-type interferometer this position is on axis (i.e. on a normal to the gauge and platen surfaces) and the correction is zero. In the Fizeau-type interferometer described by Barnes and Puttock (1953) the illuminating and viewing apertures are off axis and the rays are not normal. The correction is a cosine one and is given by the expression $L\theta^2/2$ where θ is the angle subtended in radians; if the aperture is S mm off axis (i.e. S is half the separation of the centres of the illuminating and viewing apertures) and the focal length of the collimating lens is F then the correction is

$$\frac{1}{2}\left(\frac{S}{F}\right)^2.L.$$

The second correction arises from the size of the aperture (collimation effect) and is more complex; see Bruce (1955), Thornton (1955).

For the NPL-Hilger gauge interferometer in this Division, which has a viewing aperture of slit form 0.76 mm long, the collimation correction is closely linear up to the 100 mm capacity of the instrument and is $+0.09L \times 10^{-6}$. It should be noted that the linear relationship ultimately becomes an oscillating one at longer lengths. Moreover, it is wavelength dependent and for highest accuracy work should be calculated separately for each wavelength as a correction to be applied to each observed fringe fraction prior to obtaining the best fit coincidence.

Application of the above corrections to the measured value δL gives a value of the error, δL_O, from the nominal *optical length* L_O of the gauge block. Because the surfaces of the gauge block and of the base plate to which the gauge block is wrung have structure (surface roughness) the effective planes of reflection are below the plane touching the asperities, the "mechanical" surface. Since, in practice, it is the mechanical length L_M (see Fig.8) that is required, and since the surface effects "g" and "p" at the gauge and platen respectively are not not necessarily identical a correction equal to g-p must be applied to δL_O to give δL_M (the error in the mechanical length).

Note that this correction, which is usually referred to as the phase correction, includes the phase change on reflection at the platen and gauge surfaces arising from the complex refractive index of the material. When both gauge and platen are of steel this phase change contribution can be considered to be the same at each surface.

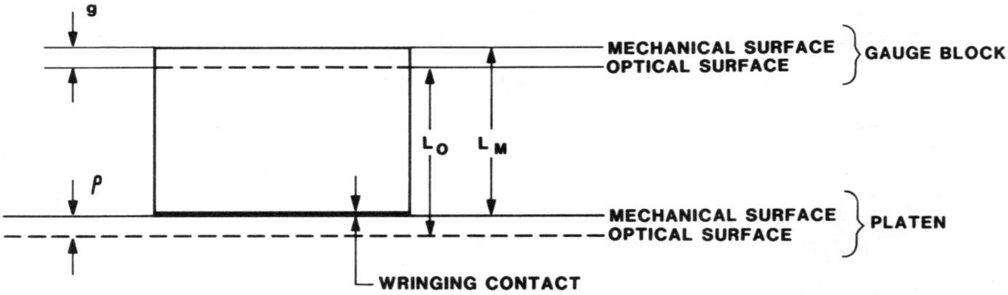

Fig.8 Relationship of mechanical and optical surfaces.

This phase correction (g-p) is usually of the order of a few nanometres (it can be, however, as much as 20 to 30 nm) and can be determined with an uncertainty of ±3 nm using a very high resolution measurement of the surface roughnesses, see Thwaite (1978). A less accurate method, which assumes that all gauges in a set have about the same surface roughness, is to take say 5 gauge blocks for which the individual optical lengths have been determined on a given platen and wring them one on top of the other to form a stack on the same platen and then to measure the overall optical length H_O of the stack. In this case the stack is made up of 4 mechanical lengths plus one optical length and the mean correction is given by $(H_O - \Sigma L_O)/4$.

A typical set of observations and corrections for a gauge block is given in Fig.9. Such a set would be repeated with the gauge block reversed, i.e. unmarked face up, and a mean of the two results taken as the result for that block.

Date: 19.9.78.

Interferometer: Hilger-NPL Gauge Interferometer.

Platen: No.119.

Gauge: 2.32 mm (Set Y4).

Temperature - air and gauge - Hg-in-glass thermometer 4679:

Start	Mid.	Finish	Mean	Correction	Corrected temp.
20.181	20.249	20.412	20.281	-0.039	20.242 °C

Humidity:

Dry Bulb	Wet Bulb	V.P.
20.3	14.4	9.3 mm Hg

Pressure:

Barometer Rdg	Barometer Temp.	Correction	Corrected Pressure
757.15	20.0	-3.23	753.92 mg Hg

Observations:

Surface upward	Source Cd. Wavelength Fractions (λ/2)					Error μm
		R	G	B	V	
Marked	Observed	.70	.45	.82	.50	
	Repeat	.65	.42	.85	.55	
	Mean	.68	.44	.84	.52	
	Nominal	.64	.36	.80	.40	
	Difference	.04	.08	.04	.12	
	Slide Rule	+0.05	+0.06	+0.07	+0.07	+0.016

Correction c_1 = $+2.415\times10^{-6}\times L = +0.0056$ μm

Correction c_2 = $-0.041\times10^{-6}\times L = -0.0001$

Gauge temp.correction = $-0.242\times12\times10^{-6}\times L = -0.0067$

Obliquity correction = $+0.036\times10^{-6}\times L = +0.0001$

Collimation correction = $+0.09\times10^{-6}\times L = +0.0002$

Surface correction (g - p) = +0.012

Total correction .. .. +0.011

Error in length of gauge = +0.027 μm

Fig.9 Typical results of interferometric calibration of a gauge block.

Although the procedures just described refer to the Fizeau-type interferometer they are directly applicable to the Michelson type, the only difference being that in the latter no obliquity correction is necessary since the apertures are on axis.

(b) Calibration of Gauge Block Lengths by Comparison

In contrast to calibration by direct interferometry, calibration by comparison is relatively straightforward. In addition to the need for a set(s) of reference gauge blocks which have already been calibrated at the appropriate level all that is required is a high quality high-magnification comparator.

Such a comparator should have a magnification of at least 20 000 X and preferably be of the type having two sensors (contacting opposite faces of a gauge block) to eliminate errors arising from imperfect contact on a base plate. The comparator should be adequately shielded from any heat sources (particularly the operator) during the measurement process.

The method of amplification embodied in the comparator can be either mechanical, inductive, capacitive, optical (optical levers or fringe-counting interferometer) or pneumatic. In the last case it is possible to use two sensors in the form of air jets which do not physically contact the gauge-block surfaces, Sim (1959).

In the majority of comparators the sensors are in the form of spherically ended probes with a specific force applied to ensure adequate contact. Care should be taken when comparing gauge blocks of dissimilar materials to ensure that the different properties do not give rise to different compressive effects and hence give an erroneous result. *If the spherical contacts are true spheres* then the compressive effects can be calculated and due allowance made. However, over a period of time small flats may develop (through wear) on the contacts and in some cases the manufacturer deliberately produces such flats on new instruments; in either case the differential effects become negligible and no allowance should be made.

(c) Determination of other geometric parameters of Gauge Blocks

(i) Flatness of end faces. The flatness of the wringing faces of gauge blocks is measured by interferometry. The departure from flatness of each face is determined from the curvature of the interference fringes on that face - if this is done using the green radiation of either Cd or Hg then 0.1 fringe curvature is equivalent to 0.025 μm departure from flatness.

When gauge blocks are calibrated by direct interferometry it is convenient to determine flatness errors during the process of determining the fringe fractions for length.

When gauge blocks are calibrated by comparison it is necessary to determine flatness errors by interferometry as a separate exercise. Relatively simple interferometers are available for this purpose and an optical scheme typical of such a flatness interferometer is shown in Fig.10.

(ii) Parallelism of end faces. When gauge blocks are calibrated by direct interferometry it is possible to determine the errors in parallelism from the slope of the fringes on the face of the gauge block relative to the fringes on the platen; as with the flatness measurement this is usually best done using the green radiation of Cd or Hg.

When gauge blocks are calibrated by comparison the parallelism errors are best determined from the *variations in length* obtained from comparator readings taken in a pattern over the surfaces - usually five such readings will suffice.

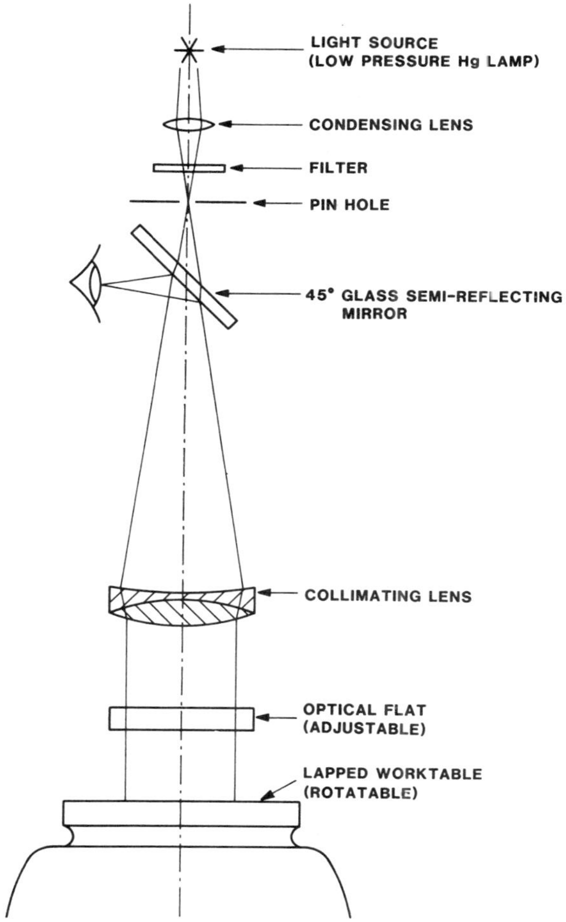

Fig.10 Schematic layout of flatness interferometer.

(iii) Squareness of side and faces. Where this is specified (not all gauge block standard specifications do so) the errors in squareness between the lapped end faces and the ground side faces can most readily be determined using an autocollimator and making use of the fact that the end faces can be considered, to the accuracy required, as being perfectly parallel.

For example, a gauge block is placed on a side face on a flat surface and an autocollimator set to read on one end face. The gauge block is then rotated through 180° on the side face and a second reading of the autocollimator taken on the opposite end face. Half the difference in reading is then the squareness error relative to that side face.

(d) Uncertainties in the Measurement of Gauge Blocks

(i) When measured by direct interferometry. The sources of uncertainties in the measurement of gauge blocks by direct interferometry arise in or from:

the estimation of fringe fraction,

the measurement of the gauge block temperature (including calibration of the temperature measuring equipment),

the uncertainty of knowledge of the coefficient of expansion of the gauge block material,

the measurement of air temperature, pressure and humidity,

the obliquity and collimation corrections,

the surface reflection corrections,

the wringing contact (possible variations of),

the effect of the interferometer optics (quality of wavefront),

the flatness and parallelism of the gauge block faces.

Some of these are random and some systematic in nature. It is not proposed to examine these in detail in this paper; for a thorough treatment reference should be made to Birch (1979).

In this Division we estimate that our overall uncertainties U, at the 99% confidence level are:

Blocks up to and including 25 mm: $U = \pm(0.0174 + 0.000\ 726L)$ μm

" over 25 mm up to and including 100 mm: $U = \pm(0.0174 + 0.000\ 599L)$ μm

where L is the nominal length in mm.

(ii) When measured by intercomparison. The sources of uncertainties in the measurement of gauge blocks by comparison arise in or from:

the uncertainty of the standard (reference) block,

errors (and uncertainty of those errors) in the comparator - these include instrument error, reading error, lack of repeatability, contact errors,

"observer's" error,

differential expansion of material of test and standard blocks,

temperature of test and standard blocks,

the flatness and parallelism of the gauge block faces.

The contributions from each of these sources must be estimated and then all combined (in a similar manner to that described by Birch (1979)) to give an overall uncertainty.

In this Division we estimate that our overall uncertainty U, at the 99% confidence level, for the comparison of Grade 0 blocks against Grade 00 standards is

$$U = \pm[0.02 + (x^2 + 0.032^2)^{1/2} + 0.000\ 075L]\ \mu m$$

where x is the uncertainty in the values for the Grade 00 standards
L is the nominal length in mm.

(e) Calibration of Length Bars and Long Gauge Blocks

There is one additional criterion over those required for gauge blocks as already discussed, namely that for end standards over 100 mm the length is referred to the horizontal position of the standard when it is supported at two positions 0.211 L from the ends.

The correction for the change in length from vertical to horizontal position is positive and is proportional to the square of the length of the standard. For a 1 m standard it is approximately -0.19 μm and for 500 mm approximately $+0.05$ μm.

It should also be noted that any squareness requirement for end face to side faces must relate to lines or axes defined by these support positions.

(i) Calibration by direct interferometry. Equipment and techniques are available for the calibration, by direct interferometry, of the longer end standards up to a maximum length of 1 metre. Only a few national laboratories have this facility, NPL in England, PTB in West Germany and NRLM in Japan, are examples, Kinder (1961), Sakurai and Yamamoto (1963). Actually, because of a number of factors inherent in discharge light sources (including Kr 86, Hg 198 and Cd 114) the practical difficulties of observing fringes and measuring fringe fractions in several wavelengths limit measurement by this form of interferometry to 500 mm. The equipment referred to above first measures a 500 mm standard and then compares that with a 1 m standard, i.e. it measures the 500 mm 'gap' by direct interferometry.

In this Division our capability is limited to 500 mm; this is the maximum capacity of a Michelson-type interferometer designed and constructed in the Division, Bruce (1956), and calibrations are made directly in terms of the radiations from Kr 86. The calibration of end standards longer than 500 mm is achieved by comparison against two shorter standards wrung together using equipment and techniques described in 4(e)(iii).

In general the procedures are closely the same as those described in 4(a) for gauge blocks. However, considerably more care is required in ensuring stable temperature conditions and in measuring the temperature of the standard; temperature-associated effects have the potential for the greatest sources of uncertainty in the measurement of long end standards.

(ii) Calibration by conversion to a line standard. It is possible to convert an end standard into a line standard by wringing on to its end faces small blocks, like gauge blocks, one side of which is polished and has a suitable line engraved at its centre (Fig.11), and then to measure the combination as a line standard (see Sect.5(b)). If each block is turned end for end and another line measurement made the mean of the two line measurements minus the mean of the length of the two blocks, each of which is measured by direct interferometry, gives the length of the end standard, i.e. referring to Fig.11,

$$L = \frac{(L_1 + L_2) - (a + b)}{2} .$$

This technique can be used alone or in conjunction with direct interferometry, up to the full 1 m, using a stabilised laser as the light source. In this case, because the laser emits only one wavelength (an adequate "spread" of laser wavelength is not yet a practical reality) it is necessary to know the fringe "order", i.e. the length of the standard should be known to the nearest half-wavelength, approx. 0.25 μm, and this can be obtained from the "line standard" measurement. This dual approach gives an improved accuracy.

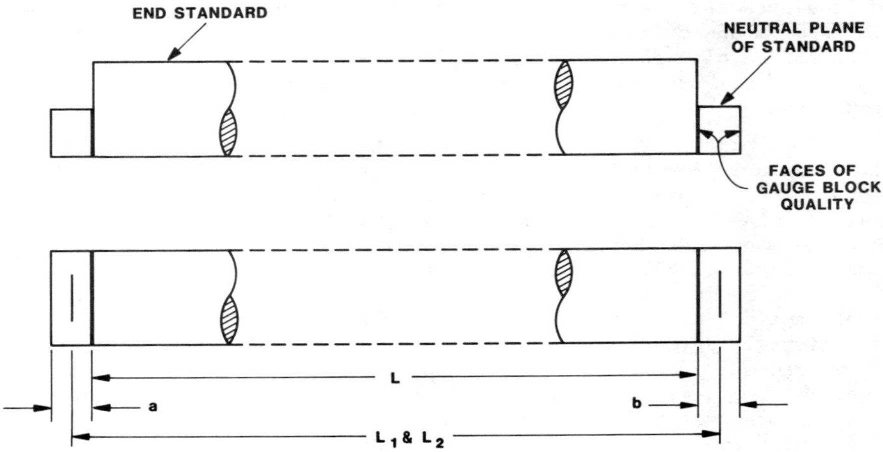

Fig.11 Conversion of an end standard to a line standard.

(iii) Calibration by Comparison. The comparison of length standards, either of standards of nominally equal length or of one standard against the combination of two standards wrung together, using high accuracy comparators is a straightforward technique similar to that for gauge blocks (see Sect.4(b). As in the case of interferometric calibrations temperature effects have considerable significance and care is required to minimise them.

The comparators used for this work are many and varied. Some are commercially available but many have been developed and constructed by individual laboratories. They can be considered in three classifications.

The first is "differential comparators", where the length standards (nominally equal) are placed side by side and the difference in their length is determined directly. Examples are the Brookes Level Comparator (Rolt 1929) and the Vertical End Bar Comparator (Holmes and Harrison 1963).

The second is "substitution comparators", where the two standards being compared sequentially occupy the space between the measuring probes. These are essentially the type of comparator with which the metrologist is most familiar and there is a very wide variety in design and in the type of sensor used. Most instruments are of the horizontal type.

The third is "displacement comparators" where the length of the standard is determined by the displacement of a probe or probes, the standard being stationary, or vice versa. The displacement, usually of a carriage running on very precise ways, is measured by means of a screw, a line standard (scale), an electromagnetic or electro-conductive device, or by laser interferometry. The development of fast, accurate, and simple laser interferometers has given a tremendous boost to length measurement. In this case it is relatively easy to convert any length-measuring machine, having sufficiently precise carriage movements, into a high-precision length-standard calibrating instrument.

A discussion on the use of laser interferometry in the measurement of length is given in a paper by Stock (1982).

5. Line Standards

Line standards of length are material standards characterized, as implied in the name, by the fact that their length is defined as the distance between lines. They may be considered as comprising of two main groups: flexible standards in the form of tapes and wires used principally in the field of surveying, and rigid standards often referred to as linear scales - the International Prototype metre, the old fundamental standard of length, falls into this group. The rigid standards are generally limited to 1 m, but occasionally lengths up to 3 m are used, particularly in measuring machines. The flexible standards range from about 5 m up to 100 m with probably the majority having lengths of 25, 30 and 50 m.

The form and materials of line standards vary according to their intended function but in every case there is one common requirement, namely, that the lines must be precisely formed so as to provide a precise delineation of the "length". Associated with this common requirement is the need for a means of precisely locating (viewing, pointing on) these lines; this is accomplished by micrometer microscopes or photoelectric scanning microscopes.

(a) Microscopes

(i) Micrometer microscopes. The term "micrometer microscopes" is usually reserved for a visual microscope having a micrometer-controlled movement of the eye-piece graticule, the readout of this micrometer usually being directly in terms of virtual movement in the object plane. A schematic diagram is given in Fig.12. In some microscopes the readout is visible directly in the field of view (rather than externally as shown in the figure) and this arrangement tends to reduce eye fatigue. A particular point to note is that the scale line is vertically illuminated; this is essential for reproducible results particularly if the line standard is used with a variety of microscopes.

The most accurate method of setting on the image of a line is by means of a graticule having two parallel lines the spacing of which is just greater than the width of the line image. Correct setting is when the narrow gaps on either side of the image are equal - the eye is particularly sensitive to apparent inequalities if the gaps are relatively small. A number of microscopes have a series of parallel setting lines, of different spacing, so that different-width scale lines can be accommodated by setting lines of appropriate spacing.

(ii) Photoelectric scanning microscopes. One of the significant sources of error in the use of microscopes is the observer; despite the high sensitivity of the human eye to the apparent gap inequalities referred to in 5(a)(i), what appears to be a symmetric setting is very often asymmetric. The answer is to replace the human observer with a photoelectric device which is theoretically free from such bias. This has the additional advantage of having the capability of automatic recording.

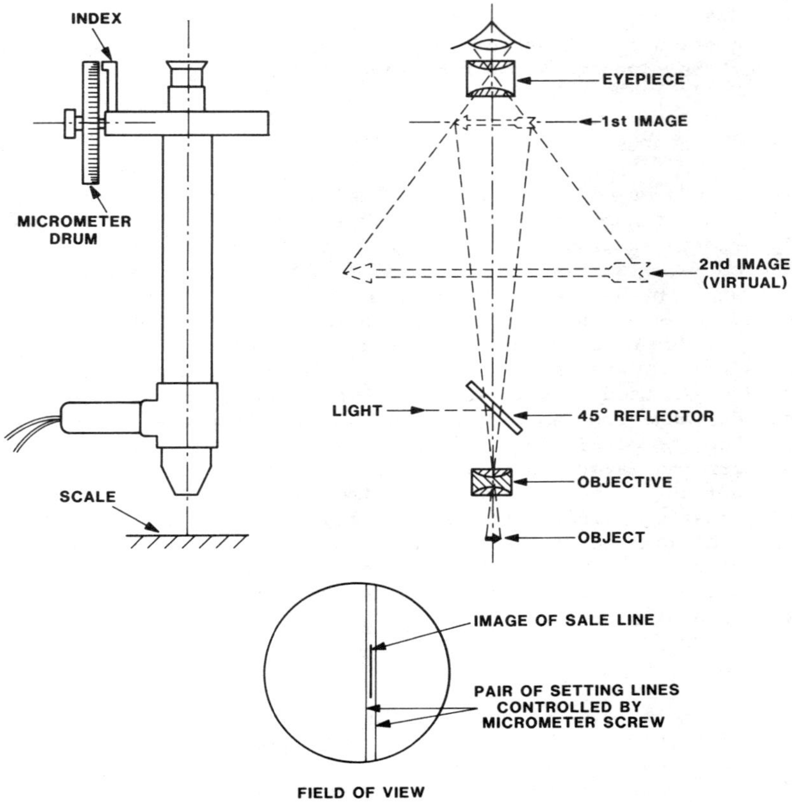

Fig.12 Micrometer microscope.

There is quite a variety of photoelectric scanning microscopes - some scan the image of the line, others illuminate the line with a scanning slit illuminator. In all cases the output signal is a measure of the amount that the scale line is off the optical axis of the microscope. In some applications a readout of the off-centre distance is obtained, in others, particularly when doing a dynamic calibration using laser interferometry, a trigger signal is obtained when the scale line is on axis.

A block diagram of the optical and electrical system of a photoelectric scanning microscope developed in this Division is shown in Fig.13. In this microscope the image of an illuminated slit is scanned across the scale line with a setting precision of 0.01 µm. A full description is given by Ritter and Goldberg (1964).

(b) Calibration of Rigid Line Standards

Rigid line standards range from the small standards, referred to as "stage micrometers" which are lines on a glass microscope slide typically with intervals of 100 µm, 50 µm or 20 µm and overall scale length of 10 mm, to the longer standards of 1 m (or more) of H section with very fine lines cut into the polished surface which is the neutral plane of the bar.

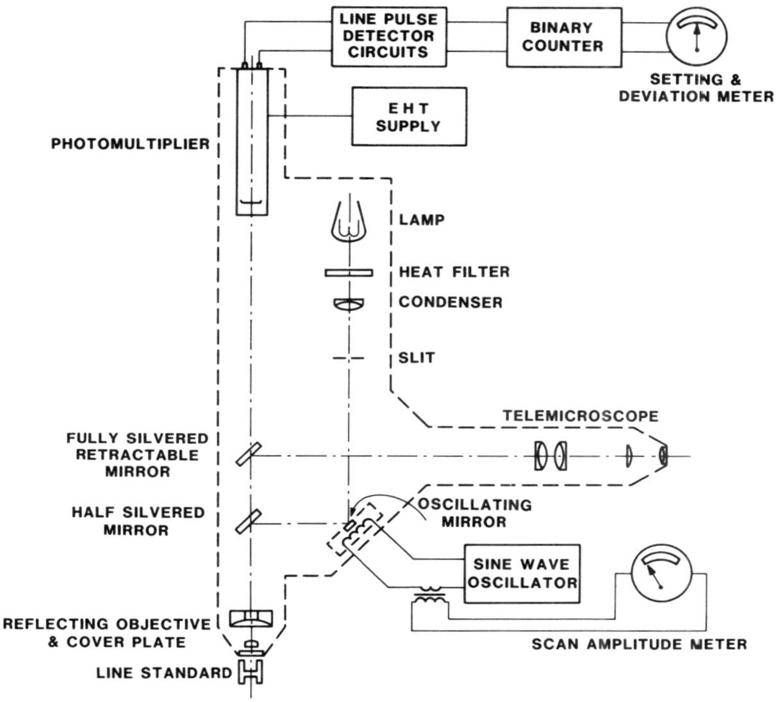

Fig.13 Photoelectric scanning microscope - block diagram.

Until recently the calibration of rigid line standards was done by intercomparison with other standards, themselves calibrated against the International Prototype Metre M held at BIPM in Sèvres, France. Subdivisional calibration of the 1 m intervals into decimetres, centimetres and millimetres was a very substantial exercise confined to national standards only. These then provided the basis for direct comparison with other standards. It is not proposed to discuss any of these procedures for the simple reason that they are obsolete technology; laser interferometry has provided the means for improved accuracy with reduced effort and cost.

Laser interferometers are commercially available and will measure linear displacement, of a corner reflector, with an uncertainty of 1 in 10^6 or better. Such interferometers can be readily associated with a moving carriage, a photoelectric (or visual) microscope and the scale to be measured to provide a high-accuracy calibration facility. Figure 14 is a schematic layout of such a line-standard interferometer constructed in this Division and which can calibrate a 1 m line standard, and all of its subdivisions if required, in terms of a calibrated He-Ne laser.

There are a number of points to be noted in setting up such a facility:

the base for the carriage and for the microscope mount must be very rigid and located in a vibration-free position;

the carriage supporting the line standard must move freely without stick/slip (ours is mounted on air bearings), must travel in a straight line, and must be capable of being positioned by means of sensitive, backlash-free controls;

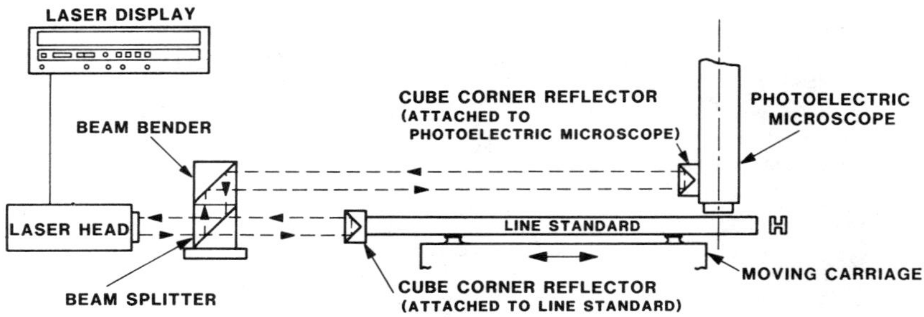

Fig.14 Schematic layout of line standard interferometer.

it is preferable to mount the corner reflector directly on the line standard if possible;

it is preferable also to locate the reference reflector on the microscope, thus giving automatic compensation for any small movements of the microscope during operation of the equipment;

the whole equipment, excluding the laser head and display unit, should be shielded to provide good temperature stability of the essential features.

Additionally, it should be noted that to obtain maximum accuracy the procedures should include reversal of the standard on the carriage and rotation of the microscope through 180°; the first to minimise the effect of any asymmetry in the grooves forming the lines on the standard, and the second to minimise any effects arising from the optics of the microscope. This procedure requires a minimum of four sets of measurement.

This type of line-standard interferometer can be adapted to automatic measurement by arranging that, as the line standard is continuously moved along, the photoelectric microscope provides a trigger signal when a line passes beneath it and a printout of the line position is obtained from the laser interferometer. In this way, for example, every millimetre interval in a 1 m scale can be calibrated relatively easily.

Similar line standard interferometers have been developed elsewhere, for example at NBS (Gillilland et al. 1966) and at NPL (Rowley 1964; Rowley and Stanley 1965).

Prior to the development of laser interferometers a number of the major national standards laboratories developed very sophisticated and highly accurate interferometers for the calibration of 1 m line standards in terms of the wavelength standard Kr 86. A general description of some of these is given by Baird (1963). These instruments, whilst in most cases having a higher precision, are not as versatile as the type just discussed.

The calibration of very short line standards, the stage micrometers, is carried out in a similar manner to that of the longer standards. Because of their length, the temperature effects are less troublesome, but because of the use to which they are put (calibration of microscopes), there is a need for a very small absolute uncertainty. In this Division, we have developed and constructed equipment for this purpose and this is described by Thwaite and King (1982). The equipment has a linear calibration range of 10 mm, and uncertainties of better than ±0.1 μm can be obtained provided the quality of the scale lines is extremely good. As with the longer standards the calibration procedure should at least include measurements with the scale reversed, to obtain improved accuracy.

(c) Calibration of Flexible Line Standards

The term "flexible line standards" in this context refers to tapes and wires as used in surveying and in many cases these have marked on them the major interval with or without a few subintervals. In the subsequent discussion consideration is given to the calibration of the higher quality standards where uncertainties of 1 in 10^6 to 1 in 10^5 are required.

The classical calibration procedure involved the use of a series of measuring microscopes mounted on pillars, on on a wall, such that all their focii were in one horizontal plane and their optical axes in one vertical plane. The microscopes were spaced at intervals of 4 m or 5 m and their positions, relative to the "zero" microscope at one end, determined by reference to a 4 m or 5 m line standard itself calibrated against a 1 m standard. Provision was made, by means of a carriage mounted on rails, to transfer the line standard from one pair of microscopes to the next. The tape (or wire) was then supported in single or multiple catenary under the microscopes and its interval(s) calibrated; suspension was over "friction-free" pulleys with the specified tension applied by free-hanging weights at each end. Provided that the environmental conditions were very good, the quality of the lines on the tape was also very good, and proper procedures were followed, an uncertainty in the measured interval, from 10 to 50 m, of 1 in 10^6 could be achieved; this accuracy required that the tape be made of invar (36% nickel steel) the low linear coefficient of which - usually about 1 in $10^6/°C$ - reduced the major source of uncertainty, namely the temperature effects.

This Division has developed, as have other national laboratories elsewhere, a calibration facility based on a laser interferometer, and in our case we have used a commercially available interferometer with special reflecting components of our own design and construction. A schematic lay-out showing the basic principle of the facility is shown in Fig.15.

The essential reflecting components are two corner-reflectors each built-up from three front-surface mirrors and designed to straddle a survey tape. A microscope is mounted above each mirror assembly with its focus at the virtual apex of the three mirrors; when the microscope is pointed on a scale mark on a tape then that scale mark is also at the virtual apex of the mirrors. This configuration minimises the effect of any rotational movement of the corner reflectors.

Each corner reflector and its associated microscope is mounted on a carriage running on an accurately aligned track on the top of a single free-standing reinforced-concrete beam approximately 2 m high, 0.6 m wide and 70 m long in a basement laboratory approximately 3.6 m square section by 80 m long. The track has been adjusted, using a laser alignment device also developed in the Division, so that the virtual apex of each corner reflector moves in a straight line over the full 60 m working length within 0.5 mm.

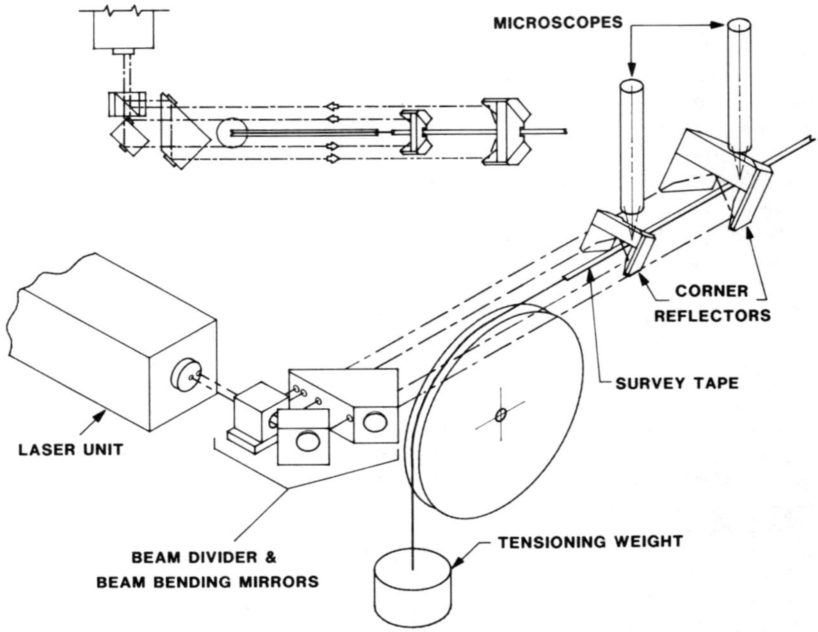

Fig.15 Optical scheme of the laser interferometer for calibration of survey tapes.

The interferometric concept is basically that of a Michelson interferometer in that a beam of light (in this case from a He-Ne laser) is split into two component beams which travel to and back from separate reflectors. These component beams are then recombined to give a measure of any displacement of one reflector relative to the other. By an appropriate sequence of settings of the microscopes on the tape lines and of carriage movements the separation of two lines on the tape can be determined.

The measured value then has to be corrected for departures from the standard atmospheric conditions at which the wavelength of the He-Ne laser in the interferometer has itself been calibrated. The appropriate corrections are determined by calculation from measurements of the temperature, pressure and humidity of the air taken during the tape calibration; it is proposed to develop a suitable refractometer so that the total correction can be determined directly. A correction must also be applied for the effect of thermal expansion of the tape and the actual temperature at which the measurements are made.

The normal resolution of the interferometer is 0.0001 mm which at 60 m is better than 1 in 10^8. The uncertainty at which the interval on a survey tape can be determined is, however, dependent on many factors such as the quality of the lines on the tape, uncertainties in the values of temperature, pressure and humidity, uncertainty in the knowledge of the laser wavelength (approximately 4 in 10^8), etc. On a good quality invar tape the uncertainty of measurement is 5 in 10^7 at 50 m. In addition to this improvement in accuracy, the actual time required to make the necessary observations has been reduced from that required by the old classical technique by 80 to 90%.

6. Conclusion

The calibration of end standards of length in the form of gauge blocks has been covered in some detail since it is this form of length standard which is most used in precision engineering and which constitutes the major demand on standardising laboratories. It is hoped that sufficient information has been given in relation to the calibration of the other standards of length to indicate the most appropriate equipment and techniques which should be considered in establishing any calibration facility.

References

ANSI B89.9.1.9.'Precision Inch Gage Blocks for Length Measurement'.

Baird, K.M. (1963). *Appl.Opt.* 2(5), 471.

Barnes, D.C. & Puttock, M.J. (1953). *Engineering* 176(4585), 757.

Barrell, H. & Sears, J.E. (1939). *Philos.Trans.* A238:1.

Birch, K.G. (1979). 'Uncertainties in the Measurement of Gauge Blocks by interferometry'. NPL Rep. MOM29.

Bruce, C.F. (1955). *Aust.J.Phys.* 8, 224.

Bruce, C.F. (1956). *J.Sci.Instrum.* 33, 478.

BS 5317. 'Metric Length Bars and their Accessories'.

CIPM (1960). Procés Verbaux 28, 71.

CIPM (1963). Procés Verbaux 31, 76.

CIPM (1973). Procés Verbaux 41, 112.

Gillilland, K.E., Cook, H.D., Mielenz, K.D., Stephens, R.E. (1966). *Metrologia* 2, 95.

Goldberg, J.L. & Ritter, H.J. (1964). *Microtecnic* 18, (2) 54 and (3) 117.

Holmes, P. & Harrison, E.R. (1963). *Microtecnic* 17(1), 8.

ISO-1.'Standard Reference Temperature for Industrial Length Measurements'.

ISO-3650. 'Gauge Blocks'.

Kinder, W. (1961). *Zeiss Werkzeitschrift* 43, 3.

Rolt, F.H. (1929). 'Gauges and Fine Measurements', Vol.1, (Macmillan).

Rowley, W.R.C. (1964). 'Lasers and their Applications', Proceedings of Conference, London, Inst.Electr.Eng. 4.

Rowley, W.R.C. & Stanley, V.W. (1965). *Instrum.Pract.* 19(12), 1106.

Sakurai, Y. & Yamamoto, H. (1963). Bulletin NRLM, Series 7, Nov.

Sim, P.J. (1959). *Machinery (NY)*, 94, 1009.

Thornton, B.S. (1958). *Aust.J.Phys.*, 8, 241.

Thwaite, E.G. (1978). *Metrologia*, 14, 53.

Thwaite, E.G. & King, H. (1982). *Precis.Eng.* 4(1), 15.

Stock, H.M.P. (1982). These proceedings.

Angle Standards and their Calibration

Peter J. Sim
CSIRO Division of Applied Physics, Sydney, Australia 2070

Abstract

This paper reviews some of the common forms of reference and working angle standards, some of the techniques available for their calibration, and some of the precise angle-measuring instruments associated with this work. The information should be of use to laboratories contemplating the establishment of calibration facilities for angle standards.

1. Introduction

For precise length measurement, it is essential to have traceability to a primary standard, which is currently defined in terms of the wavelength of a particular radiation from a krypton-86 lamp. In the case of angular measurement, no corresponding primary standard is required, since any angle can be established by appropriate subdivision of a circle. The ability to calibrate reference and working standards of angle is thus mainly dependent upon the ability to carry out accurate subdivision of a circle and not on the possession of, or ability to realize, a primary standard. In the case of small angles, however, it is usual, for practical reasons, to determine the angle in terms of standards of length.

This paper reviews some of the common forms of reference and working angle standards, some of the techniques available for their calibration, and some of the precise angle-measuring instruments associated with this work.

Specific references are given at appropriate places in the text but useful information may also be obtained from the relevant chapters of standard works on engineering metrology, including those by Hume (1970), Miller (1962), Scarr (1967), and Thomas (1974), as well as from Evans and Taylerson (1964).

2. Units of Angle

(a) Sexagesimal System

This system is used mainly for engineering purposes.

1 degree (°)	=	1/360 of a complete circle
1 minute (')	=	1/60 of a degree
1 second (")	=	1/60 of a minute.

(b) Radian System

This system is used mainly in the theoretical treatment of angle.

1 radian (rad)	=	the angle subtended at the centre of a circle by an arc of length equal to the radius
1 milliradian (mrad)	=	1/1000 of a radian.

It follows that

360 degrees	=	2π radians
1 degree	$\approx$	1.75×10^{-2} radians
1 minute	$\approx$	2.91×10^{-4} radians
1 second	$\approx$	4.85×10^{-6} radians.

(c) Centesimal System

This system has been used in some European countries. It is not part of the Système International d'Unités (SI) and no further reference will be made to it in this paper.

1 grade (^{g})	=	1/400 of a complete circle
1 minute (^{c})	=	1/100 of a grade
1 second (^{cc})	=	1/100 of a minute.

3. Basic Principles of Angle Calibration or Generation

(a) Fundamental Method - Subdivision of a Circle

Suppose we have a circle divided into 6 nominally equal angle intervals, a, b, c, d, e, f, each with a small error δa, δb, ... δf.

The six unknown errors can be determined by comparing the interval a with each of the other five intervals in turn and observing the five differences θ_{ab}, θ_{ac}, ... θ_{af}. Thus five equations are obtained:

$$\delta_a - \delta_b = \theta_{ab} \qquad (1)$$

$$\delta_a - \delta_c = \theta_{ac} \qquad (2)$$

......

$$\delta_a - \delta_f = \theta_{af} \qquad (5)$$

The fact that the sum of the six intervals is a complete circle, and hence the sum of the six errors must be zero, provides the sixth equation needed:

$$\delta_a + \delta_b + \delta_c + \delta_d + \delta_e + \delta_f = 0 \qquad (6)$$

These six equations are readily solved for the six unknowns to give:

$$\delta_a = (\theta_{ab} + \theta_{ac} + \theta_{ad} + \theta_{ae} + \theta_{af})/6$$

$$\delta_b = \delta_a - \theta_{ab}$$

....

$$\delta_f = \delta_a - \theta_{af}.$$

Errors in further subdivisions of these main intervals may be determined in a similar manner. In Equation (6) the sum of the errors is then not zero but is equal to the previously determined error in the main interval.

Thus, in the general case, if there are n subdivisions, then (n-1) equations are formed from the observed differences and the nth equation becomes

$$\delta_a + \delta_b + \delta_c + \ldots\ldots + \delta_n = \Delta$$

where Δ is the previously determined error in the main interval, and Δ=0 when the subdivisions form a complete circle.

In principle, the error in any angle may be determined by using such techniques.

(b) Method based on Trigonometrical Principles

It is frequently more convenient to calibrate or generate an angle by using a sine or tangent principle. An arm of fixed length forms the hypotenuse or adjacent side of a right-angled triangle and a measurement or setting is made of the length of the opposite side.

This is *not* a fundamental method but is dependent on the length standards used in measuring or setting the lengths of the sides of the triangle.

The method can be used, in principle, to generate any angle, but it is only recommended for angles up to 45 degrees; above this a rapid loss of precision is encountered. A frequent use is for generating small angles where, to the accuracy required,

$$\alpha = \sin \alpha = \tan \alpha \text{ (where } \alpha \text{ is in radians).}$$

4. Standards of Angle

(a) Precision Indexing Tables

Precision indexing tables are small rotary tables which, in the highest quality versions, are more accurate than any other readily available equipment used for generating precise angles over a full circle. One such table, the Moore 1440 Precision Index, has 1440 radial serrations on each of its upper and lower elements and can be indexed through any multiple of 0.25 degree with a total range of error of less than 0.2 second.

Precision indexing tables can be used, in conjunction with an autocollimator, for the calibration of precision polygons and some angle gauges and also, with the addition of a single reflector, for the calibration of rotary tables and optical dividing heads.

(b) Precision Polygons

Precision polygons are hardened steel or glass blocks in the form of regular polygons with flat reflecting faces. They may have up to 72 faces, but 6, 8 or 12 faces are more common.

Precision polygons can be used in conjunction with an autocollimator to calibrate angle-measuring devices such as rotary tables and optical dividing heads. Provided the faces are sufficiently flat the results are not subject to eccentricity error.

(c) Angle Gauges

An angle gauge is a hardened steel block having its two lapped working faces inclined at a nominal angle. Angle gauges are usually supplied in sets containing a series of gauges having different nominal angles, and they may be wrung together in combination such that their angles will either add or subtract.

One common set consists of 12 gauges:

41, 27, 9, 3, 1 degrees
27, 9, 3, 1 minutes
27, 9, 3 seconds.

With this set and a square block it is possible to reproduce any desired angle to within 1.5 seconds. The manufacture and adjustment of these gauges has been described by Knoyle (1950).

(d) Circular Scales

Circular scales are usually on glass and built into angle measuring instruments such as rotary tables, optical dividing heads and theodolites. They are rarely used as separate angle standards.

Such a scale is subject to error if eccentrically mounted, but this error can be almost completely eliminated by using two reading devices mounted 180 degrees apart.

5. Small-Angle Measuring Devices

(a) Precision Levels

Precision levels are used for accurate measurement of small angles of tilt in a vertical plane. They may be either of the conventional spirit type or of the more recent electronic type incorporating an analogue or digital readout unit connected to the measuring head by a flexible cable. A sensitivity of 10 seconds per division is useful for general precision work but higher sensitivities are available.

(b) Autocollimators

Autocollimators are optical instruments capable of measuring small angles of tilt of a reflecting face. There is, in principle, no restriction on direction, but many autocollimators are constructed for measurement in planes close to the horizontal and/or vertical.

Some autocollimators are visual types incorporating a micrometer eyepiece. Others are photo-electric types either with null setting of an electrical meter by means of a micrometer drum or with direct analogue or digital readout. High precision types have a resolution of 0.1 second of arc or better.

Autocollimators are typically used as reading devices in the calibration of precision polygons and angle gauges.

(c) Laser Interferometer Angle-Measuring Accessory

An accessory to the Hewlett Packard length measuring laser-interferometer system consists of two cube corner retroreflectors adjacent to each other in a single mount. The device uses the sine principle and is shown in Figure 1.

The distance H between the poles of the two retroreflectors represents the hypotenuse of a right-angled triangle, and a measured change in the path length differences of the reflected beams, i.e. $L_1 - L_2$, represents the side of the triangle opposite the angle being measured.

The resolution of the device is 0.1 second and, by applying appropriate sine corrections, it can be used for angles up to ±10 degrees.

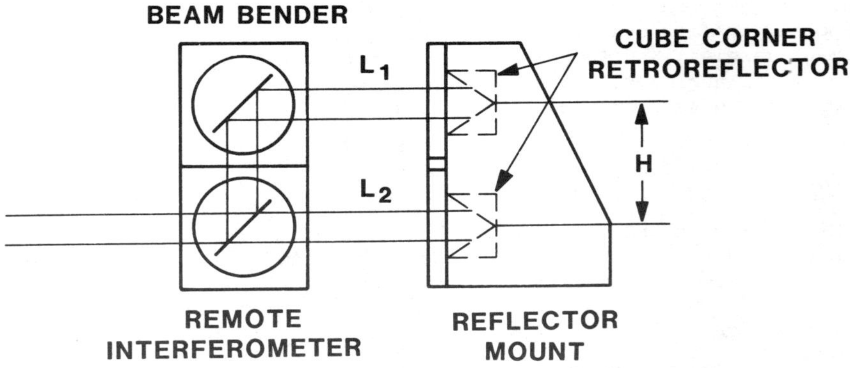

Fig.1. Laser-interferometer angle-measuring accessory.

(d) A Note on Straightness and Flatness Measurement

The three small-angle measuring devices described can all be used for examination of straightness and flatness. A generator of the surface under examination is stepped off in short spans equal to the spacing of a pair of feet attached to either the base of a level, the base of an autocollimator reflector mount or the base of a laser-interferometer twin-retroreflector mount.

In each case the successive angular tilts represent differences in height of the two feet. Accumulation of these values enables the straightness of the generator to be computed and the shape of a surface can be determined by combining the results from a number of generators.

6. Small-Angle Setting Devices

(a) Level Tester

One form of the equipment for the calibration of precision levels consists of a table which can be tilted about a pair of ball feet by means of a micrometer mounted on a long arm rigidly attached to the table. The micrometer has a ball-end on its spindle and provides the third point of support for the table. Such an instrument operates on the tangent principle. It is desirable that the span from the micrometer spindle to the line joining the two fixed ball feet be at least 750 mm. A level tester of this type can also be used to calibrate small-angle angle gauges.

(b) Small-Angle Generator

A small-angle generator is, in principle, similar to a level tester but is fitted with a plane mirror and used for the calibration of autocollimators.

In some versions, the micrometer head is replaced by a fixed ball foot and gauge blocks are used to set the angle of tilt; in this form the instrument operates on the sine principle (Evans & Taylerson 1964).

(c) Compound-Angle Reflecting Tangent Arm

A variation of the small-angle generator, in which the reflector is set at an angle to the axis of rotation, has been developed in this Division. The resulting compound-angle effect produces a proportionally smaller effective (i.e. in the desired plane) angle of tilt of the mirror for a given angle of tilt of the arm. In the instrument currently in use the ratio is 20:1. Figure 2 illustrates the device in a version designed for generation of small angles of rotation about a vertical axis. Note that the mirror surface is in a vertical plane and that the axis of rotation is inclined at 2.87 degrees to the horizontal.

Fig.2. Compound-angle reflecting tangent-arm

7. Calibration of Angle Standards

(a) General Information

Angle standards may be calibrated by *fundamental subdivisional* methods, by *trigonometric generating* methods or by *direct comparison* with appropriate reference standards. The method chosen will depend on the equipment available and the accuracy required; the techniques described here are those most often used.

When a subdivisional method is used it is common for a series of comparisons to be made which is more extensive than that leading to the equations in 3(a). *Each* interval to be calibrated is compared in turn with *every* other interval. A computational procedure is followed which leads to least squares estimates of the individual errors and enables a statistical analysis of the results to be made.

A calibration of this nature, in which the intercomparison is of intervals within a *single* standard or measuring device, is often known as a *simple* calibration. The observational and computational procedure for an example of a simple calibration is given in Appendix A(a).

In another technique the intervals of one device are compared with those of a second device, resulting in the determination of the errors in *both* without any prior knowledge of either. This is sometimes referred to as a *cross* calibration and an example is given in Appendix A(b).

When a further subdivision of previously calibrated main intervals is required, a cross-calibration technique is commonly used, with the sub-intervals of one main interval being compared with the sub-intervals of another main interval.

The subdivisional methods described here are essentially those of Johnson (1923), Gilet & Watson (1953), Evans & Taylerson (1964), and the instructions issued by NRLM, Japan in October 1980, for an international comparison of angle standards.

Other treatments include those of Cook (1954), Thwaite & Leslie (1963), Nock & Steiner (1966), Bouillon, Tremblay & Delisle (1970), Boguslavskii, Eliashberg, Sharova & Fedotova (1972), Brezina (1974,a) and Debler (1975). Some of these accounts refer specifically to linear measurement, but the techniques can be adapted to the angular case.

(b) Calibration of a Precision Indexing Table

(i) Cross calibration against another precision indexing table. This is the most satisfactory method if a second indexing table is available. One table is mounted on the other and a single high-quality mirror, mounted on the upper table, is viewed by a single high-quality autocollimator. A procedure similar to that of Appendix A(b) is followed.

(ii) Cross calibration against a high-quality precision polygon. This is the most practical method available to a laboratory possessing only one precision indexing table. The polygon is mounted on the table and is viewed by a single high-quality autocollimator as shown in Fig.3. The procedure is described in Appendix A(b).

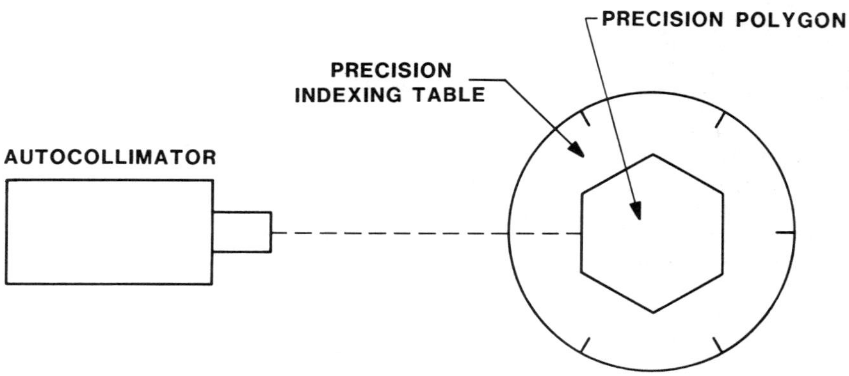

Fig.3. Cross calibration of a precision polygon and a precision indexing table.

(iii) Simple calibration using a single reflector and two high-quality autocollimators. This method is similar to that described later in 7(c)(i) and in Appendix A(a). The indexing table is rotated so that the reflector mounted on it is viewed by each of the autocollimators in turn. Rotation of the reflector on top of the table, or of the complete assembly in relation to the autocollimators, is required after each pair of observations. This is probably best achieved by mounting the precision index on a subsidiary rotary table with a good fine-setting facility.

NOTE: (1) No direct-comparison method of calibration, even against another calibrated high-precision indexing table, would be likely to yield results appropriate to the high performance of this type of instrument.

(2) The three methods described would only lead to meaningful results if the autocollimators were of very high-precision type, i.e. 0.01 second resolution and accuracy of 0.1 to 0.2 second over the small range used.

(c) Calibration of a Precision Polygon

(i) Simple calibration using two autocollimators. The polygon is mounted on a rotatable table and is viewed by two autocollimators set at an angle ϕ, corresponding to an integral number of polygon intervals, as shown in Fig.4. The number of intervals is varied as the calibration proceeds. The complete procedure is described in Appendix A(a).

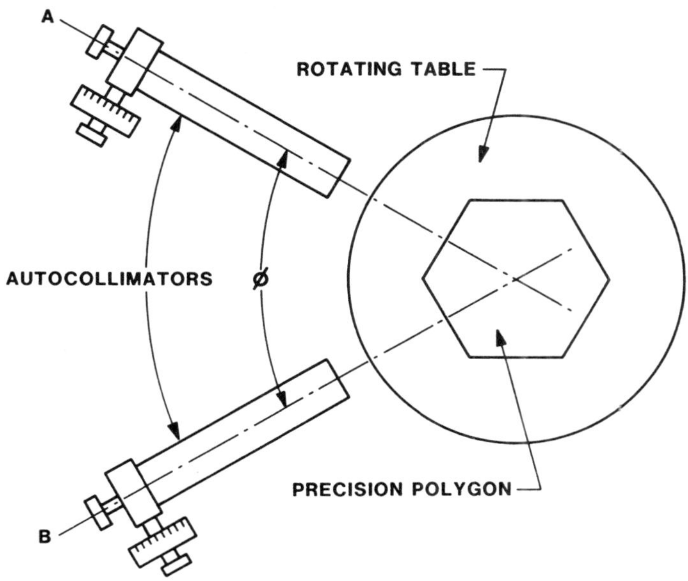

Fig.4. Simple calibration of a precision polygon

(ii) Cross calibration against a precision indexing table using a single autocollimator. This method is described in 7(b)(ii) and in Appendix A(b).

NOTE: Both method (i) and method (ii) are capable of yielding very precise results, depending on the quality of the polygon and of the autocollimators used.

(iii) Direct comparison with a precision indexing table. This method requires almost the same equipment to set up as (ii) but is a much shorter procedure; it would be used when the quality of the polygon under examination did not warrant the additional work necessary to obtain the improved precision available from the full cross calibration. In most cases the errors in the precision indexing table could be disregarded.

(iv) Methods involving cross calibration against, or direct comparison with, another precision polygon with the same number of faces. With one polygon being mounted directly on top of the other, a separate autocollimator may be used for viewing each polygon. Alternatively, a single autocollimator may be set so that part of a face of each polygon occupies half the field of view. In this latter case, individual observations are obtained by blocking off half the field at a time, and it is desirable to repeat the calibration with the polygons interchanged, in order to compensate for the asymmetrical viewing.

The cross calibration requires no previous knowledge of the errors of either polygon and the procedure would be similar to that described in Appendix A(b). In the direct comparison method the errors of one polygon (the reference polygon) must be known and the results would derive simply from the differences between observations on corresponding faces of the two polygons, corrected for the errors in the reference polygon.

(d) Calibration of Angle Gauges of One Degree and Greater

(i) Comparison with a precision indexing table. When the sizes correspond with the available intervals of a precision indexing table, direct comparisons may be made with that instrument using a single autocollimator.

(ii) Calibration from first principles. Three unknown sets of angle gauges may be calibrated from first principles by testing for equality of gauges of the same nominal size and testing for parallelism or squareness using identities such as 9 + 9 + 9 - 27 = 0 and 27 + 27 + 27 + 9 = 90 (Knoyle, 1950).

(iii) Comparison with a reference angle gauge. An angle gauge may be calibrated by direct comparison with a reference angle gauge of the same size using two autocollimators. The autocollimators are set to view the two faces of the reference gauge, which is then replaced by the unknown gauge. Evans & Taylerson (1964) describe the technique.

The comparison may also be made using a single autocollimator by mounting one gauge directly on top of the other on a rotary table, successive readings being made on corresponding faces of the two gauges by blanking off half the field. As with polygon comparisons by a similar method, it is desirable to interchange the gauges to compensate for asymmetrical effects.

(e) Calibration of Angle Gauges of less than One Degree

(i) Calibration using a small-angle setting device. The smaller angle gauges may be calibrated using one of the small-angle setting devices as described by Knoyle (1950) and Evans & Taylerson (1964).

In this Division it is the practice to use the compound angle reflecting tangent arm as an intermediate reflector between an autocollimator and the angle gauge mounted on a precision indexing table, as shown in Fig.5.

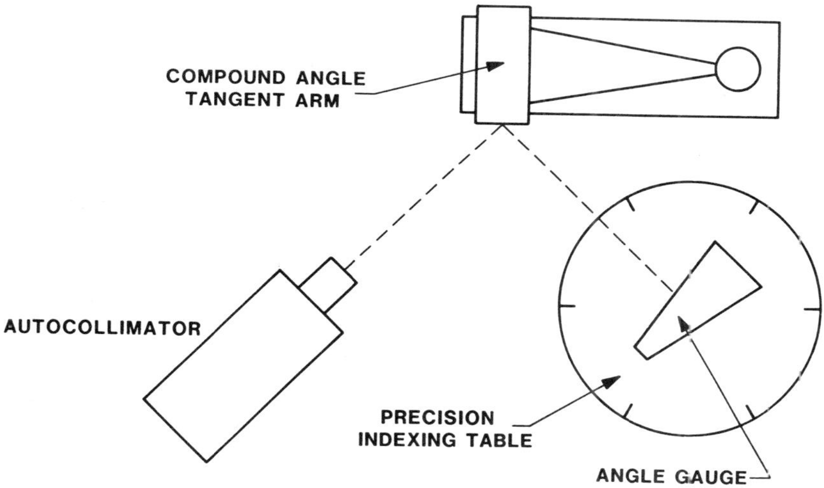

Fig.5. Calibration of an angle gauge.

(ii) Comparison with a reference angle gauge. The small angle gauges may also be calibrated by direct comparison with reference gauges as described in 7(d)(iii).

8. Conclusion

Some common methods for the calibration of angle standards have been described and references to published accounts of other techniques have been given.

It is hoped that this information will be of assistance to a Laboratory in assessing its angle calibration capability in relation to available equipment, in determining what equipment may need to be acquired and in establishing the proper techniques.

References

Boguslavskii, M.G., Eliashberg, V.M., Sharova, E.E. and Fedotova, L.I. (1972). *Meas.Tech.(USSR)* 15, 984.

Bouillon, G., Tremblay, G. and Delisle, C. (1970). *Can.J.Phys.* 48, 21.

Bouillon, G., Tremblay, G. and Delisle, C. (1973). *Can.J.Phys.* 51, 2516.

Brezina, I. (1974a). *Merova Technika* č.2, 23.

Brezina, I. (1974b). *Meas.Tech.(USSR)* 17, 666.

Cook, A.H. (1954). *Br.J.Appl.Phys.* 5, 367.

Debler, E. (1975). PTB-Bericht-ME-8. Physikalisch-Technische Bundesanstalt, Germany.

Evans, J.C. and Taylerson, C.O. (1964). 'Measurement of Angle in Engineering'. Notes on Applied Science No.26, National Physical Laboratory, London.

Gilet, P.M. and Watson, G.S. (1953). *Aust.J.Phys.* 6, 155.

Hume, K.J. (1970). 'Engineering Metrology', 3rd (metric) ed., MacDonald, London.

Johnson, W.H. (1923). 'Dictionary of Applied Physics', Vol. Ed. R. Glazebrook, MacMillan, London.

Knoyle, G.H. (1950). *Machinery (London)* 77, 275.

Miller, L. (1962). 'Engineering Dimensional Metrology', Arnold, London.

Noch, R. and Steiner, O. (1966). *Z. InstrumentenKde* 74, 307.

Scarr, A.J.T. (1967). 'Metrology and Precision Engineering', McGraw-Hill, London.

Thomas, G.G. (1974). 'Engineering Metrology', Butterworths, London.

Thwaite, E.G. and Leslie, R.T. (1963). *Aust.J.Phys.* 16, 82.

Appendix A

Procedures for Simple and Cross Calibrations

(a) Example of a Simple Calibration - Procedure for Simple Calibration of a Precision Polygon with n faces, using Two Autocollimators

(i) Set the polygon on a rotatable table and set the autocollimators at an angle of one polygon interval so that autocollimator A views the datum face and autocollimator B views an adjacent face.

(ii) Take the two autocollimator readings simultaneously. Record these readings and the calculated difference (B - A) as in Table 1.

(iii) Rotate the polygon through one interval in a direction to take the datum face away from autocollimator B. Repeat (ii).

(iv) Continue to repeat (ii) and (iii) until the polygon reaches its original position and the simultaneous observations on those faces have again been recorded.

(v) Repeat (ii) to (iv) with the direction of rotation reversed.

(vi) Calculate the mean of the two (B - A) values for each position of the polygon and calculate the difference between successive means as in the last column of Table 1.

(vii) With autocollimator A remaining fixed, reset autocollimator B so as to increase the angle between the autocollimators by one polygon interval.

(viii) Repeat (ii) to (vi), recording and calculating results as in Table 1.

(ix) Continue increasing the angle between the autocollimators by one polygon interval at a time, and repeating (ii) to (vi), until a set of observations has been recorded for all possible, i.e. (n - 1), spacings of the autocollimators.

(x) The differences in the final column of Table 1 now represent differences between single intervals of the polygon taken in all combinations.
Enter these differences into the appropriate places in a graphical square as in Table 2.
Ensure that the signs are such that each difference represents: (Column interval) - (Row interval). The sign will be affected by factors including direction of reading of the autocollimators and the direction of labelling of the polygon faces; the effect of all such factors must be checked in every calibration. Note that the differences along one diagonal of the square are zero as each represents the difference between an interval and itself.

(xi) Sum the rows and the columns. Calculate the estimated errors in the polygon intervals from:

$$\text{Estimated error in } r^{th} \text{ interval} = \frac{(\text{Sum of } r^{th} \text{ column}) - (\text{Sum of } r^{th} \text{ row})}{2n}$$

Determine the estimated accumulation errors from the datum face by taking progressive totals of the interval errors.

Table 1. Observations for a simple calibration of a polygon

Unit: Second of Arc

Polygon Faces observed by Autocollimators		Autocollimator Readings							
		Clockwise Rotation			Counter-Clockwise Rotation			Mean	Differ-ences
B	A	B	A	B-A	B	A	B-A	B-A	
60	0	10.55	9.14	+1.41	10.51	9.05	+1.46	+1.435	-
120	60	10.57	11.37	-0.80	10.52	11.32	-0.80	-0.80	-2.235
180	120	8.57	11.15	-2.58	8.76	11.37	-2.61	-2.595	-1.795
240	180	9.10	9.33	-0.23	9.23	9.46	-0.23	-0.23	+2.365
300	240	6.47	10.03	-3.56	6.47	10.07	-3.60	-3.58	-3.350
0	300	8.23	7.32	+0.91	8.16	7.27	+0.89	+0.90	+4.480
60	0	10.62	9.09	+1.53	10.49	8.98	+1.51	+1.52	+0.620
120	0	9.59	8.79	+0.80	9.60	8.90	+0.70	+0.75	-
180	60	7.65	10.91	-3.26	7.63	10.93	-3.30	-3.28	-4.030
240	120	8.28	11.00	-2.72	8.28	11.02	-2.74	-2.73	+0.550
300	180	5.60	9.22	-3.62	5.74	9.46	-3.72	-3.67	-0.940
0	240	7.24	9.83	-2.59	7.41	10.10	-2.69	-2.64	+1.030
60	300	9.49	7.08	+2.41	9.59	7.21	+2.38	+2.395	+5.035
120	0	9.58	8.90	+0.68	9.60	8.88	+0.72	+0.70	-1.695
180	0	7.63	5.70	+1.93	7.63	5.70	+1.93	+1.93	-
240	60	8.40	8.10	+0.30	8.31	8.02	+0.29	+0.295	-1.635
300	120	5.78	8.25	-2.47	5.64	8.12	-2.48	-2.475	-2.770
0	180	7.36	6.40	+0.96	7.36	6.41	+0.95	+0.955	+3.430
60	240	9.55	7.00	+2.55	9.48	6.96	+2.52	+2.535	+1.580
120	300	9.39	4.03	+5.36	9.38	4.03	+5.35	+5.355	+2.820
180	0	7.62	5.71	+1.91	7.65	5.67	+1.98	+1.945	-3.410
240	0	8.76	8.24	+0.52	8.73	8.18	+0.55	+0.535	-
300	60	5.98	10.41	-4.43	5.94	10.46	-4.52	-4.475	-5.010
0	120	7.55	10.39	-2.84	7.58	10.47	-2.89	-2.865	+1.610
60	180	9.69	8.53	+1.16	9.69	8.63	+1.06	+1.110	+3.975
120	240	9.79	9.27	+0.52	9.77	9.31	+0.46	+0.490	-0.620
180	300	8.21	6.66	+1.55	8.14	6.59	+1.55	+1.55	+1.060
240	0	8.77	8.28	+0.49	8.76	8.16	+0.60	+0.545	-1.005
300	0	3.01	5.52	-2.51	2.94	5.37	-2.43	-2.47	-
0	60	4.44	7.55	-3.11	4.40	7.46	-3.06	-3.085	-0.615
60	120	6.76	7.69	-0.93	6.74	7.64	-0.90	-0.915	+2.170
120	180	7.04	6.20	+0.84	6.91	6.03	+0.88	+0.86	+1.775
180	240	5.19	6.78	-1.59	5.13	6.61	-1.48	-1.535	-2.395
240	300	5.83	3.95	+1.88	5.74	3.81	+1.93	+1.905	+3.440
300	0	2.97	5.45	-2.48	2.86	5.34	-2.48	-2.480	-4.385

Table 2. Graphical square for a simple calibration of a polygon

Unit: Second of Arc

	0-60	60-120	120-180	180-240	240-300	300-0	Sum
0-60		(+2.219) +2.170 (-0.049)	(+4.007) +3.975 (-0.032)	(+1.619) +1.580 (-0.039)	(+5.016) +5.035 (+0.019)	(+0.591) +0.620 (+0.029)	+13.380
60-120	(-2.219) -2.235 (-0.016)		(+1.788) +1.775 (-0.013)	(-0.600) -0.620 (-0.020)	(+2.797) +2.820 (+0.023)	(-1.628) -1.695 (-0.067)	+0.045
120-180	(-4.007) -4.030 (-0.023)	(-1.788) -1.795 (-0.007)		(-2.388) -2.395 (-0.007)	(+1.009) +1.060 (+0.051)	(-3.416) -3.410 (+0.006)	-10.570
180-240	(-1.619) -1.635 (-0.016)	(+0.600) +0.550 (-0.050)	(+2.388) +2.365 (-0.023)		(+3.397) +3.440 (+0.043)	(-1.028) -1.005 (+0.023)	+3.715
240-300	(-5.016) -5.010 (+0.006)	(-2.797) -2.770 (+0.027)	(-1.009) -0.940 (+0.069)	(-3.397) -3.350 (+0.047)		(-4.425) -4.385 (+0.040)	-16.455
300-0	(-0.591) -0.615 (-0.024)	(+1.628) +1.610 (-0.018)	(+3.416) +3.430 (+0.014)	(+1.028) +1.030 (+0.002)	(+4.425) +4.480 (+0.055)		+9.935
Sum	-13.525	-0.235	+10.605	-3.755	+16.835	-9.875	+0.050 +0.050
Interval Error	-2.242	-0.023	+1.765	-0.623	+2.774	-1.651	
Accum. Error	-2.242	-2.265	-0.501	-1.123	+1.651	0.000	

$\Sigma v^2 \quad = \quad 0.034\ 378$

$s_o^2 \quad = \quad \Sigma v^2/25 \quad = \quad 0.001\ 375$

	1 interval	2 intervals	3 intervals
Variance	$5\ s_o^2/72$ $= 0.000\ 095$	$8\ s_o^2/72$ $= 0.000\ 153$	$9\ s_o^2/72$ $= 0.000\ 172$
Standard deviation	0.010	0.012	0.013

(b) Example of a Cross Calibration - Procedure for Cross Calibration of a Precision Polygon with m Faces and a Precision Indexing Table

(i) Set the polygon on the indexing table with the table reading zero and the autocollimator viewing the datum face of the polygon.

(ii) Take the autocollimator reading and record it as in Table 3.

(iii) Rotate the indexing table through one polygon interval in a direction to give increasing readings of the indexing table. Repeat (ii).

(iv) Continue to repeat (ii) and (iii) until the indexing table reaches its original zero position and the autocollimator reading has again been recorded.

(v) Repeat (ii) to (iv) with the direction of rotation reversed.

(vi) Calculate the differences between readings at successive positions and the mean of corresponding differences as in the last column of Table 3.

(vii) Reset the equipment so that with the indexing table still reading zero, the autocollimator views the next adjacent face of the polygon. This may be done by:

(1) rotating the polygon on the top face of the indexing table,
or (2) moving the autocollimator to a new position,
or (3) rotating the complete indexing table/polygon assembly.

Method (3) is perhaps most convenient if the indexing table can be mounted on a suitable rotatable table; with this method there is less handling or disturbance of the polygon during calibration.

(viii) Repeat (ii) to (vi), recording and calculating the results as in Table 3.

(ix) Continue resetting the equipment so that with the indexing table reading zero the autocollimator views the next adjacent face of the polygon, and repeating (ii) to (vi), until the number of sets of observations is equal to the number of faces of the polygon, i.e. until a set has been recorded with each polygon face in turn corresponding with the zero reading of the indexing table.

(x) The differences in the final column of Table 3 now represent differences between single intervals of the polygon and the chosen intervals of the indexing table, taken in all combinations.
Enter these differences into the appropriate places in a graphical square as in Table 4.
Ensure that the signs are such that each difference represents: (Column interval) - (Row interval). In the example given, each column contains the differences associated with a given polygon interval, and each row those associated with a given indexing table interval. The sign will be affected by factors such as the direction of reading of the autocollimator, the direction of labelling of the polygon faces and the direction of the scale of the indexing table. The effect of all such factors must be checked in every calibration.

(xi) Calculate the sum and the mean of each row and column, as well as the grand sum G of the rows and columns.
Calculate the estimated errors in the intervals of the polygon and of the indexing table from:

Table 3. Observations for a cross calibration - polygon/indexing table

Unit: Second of Arc

Index Table	Polygon Faces	Counter-clockwise Rotation		Clockwise Rotation		Mean Differences
		Autocol. Readings	Differences	Autocol. Readings	Differences	
0	0	5.22	-	5.25	-	-
60	300	3.44	-1.78	3.50	-1.75	-1.765
120	240	6.27	+2.83	6.30	+2.80	+2.815
180	180	5.84	-0.43	5.78	-0.52	-0.475
240	120	7.55	+1.71	7.53	+1.75	+1.730
300	60	7.50	-0.05	7.51	-0.02	-0.035
0	0	5.27	-2.23	5.29	-2.22	-2.225
0	60	9.28	-	9.36	-	-
60	0	6.97	-2.31	7.06	-2.30	-2.305
120	300	5.50	-1.47	5.49	-1.57	-1.520
180	240	8.35	+2.85	8.28	+2.79	+2.820
240	180	7.73	-0.62	7.76	-0.52	-0.570
300	120	9.45	+1.72	9.49	+1.73	+1.725
0	60	9.19	-0.26	9.38	-0.11	-0.185
0	120	9.34	-	9.30	-	-
60	60	9.25	-0.09	9.19	-0.11	-0.100
120	0	7.18	-2.07	7.09	-2.10	-2.085
180	300	5.54	-1.64	5.54	-1.55	-1.595
240	240	8.19	+2.65	8.17	+2.63	+2.640
300	180	7.58	-0.61	7.56	-0.61	-0.610
0	120	9.36	+1.78	9.24	+1.68	+1.730
0	180	6.22	-	6.24	-	-
60	120	7.90	+1.68	7.86	+1.62	+1.650
120	60	8.07	+0.17	7.88	+0.02	+0.095
180	0	5.77	-2.30	5.79	-2.09	-2.195
240	300	4.09	-1.68	4.10	-1.69	-1.685
300	240	6.82	+2.73	6.83	+2.73	+2.730
0	180	6.20	-0.62	6.27	-0.56	-0.590
0	240	8.33	-	8.26	-	-
60	180	7.61	-0.72	7.60	-0.66	-0.690
120	120	9.52	+1.91	9.52	+1.92	+1.915
180	60	9.51	-0.01	9.50	-0.02	-0.015
240	0	7.24	-2.27	7.23	-2.27	-2.270
300	300	5.61	-1.63	5.61	-1.62	-1.625
0	240	8.30	+2.69	8.30	+2.69	+2.690
0	300	8.95	-	9.01	-	-
60	240	11.64	+2.69	11.68	+2.67	+2.680
120	180	11.14	-0.50	11.19	-0.49	-0.495
180	120	12.94	+1.80	12.98	+1.79	+1.795
240	60	12.88	-0.06	12.88	-0.10	-0.080
300	0	10.62	-2.26	10.68	-2.20	-2.230
0	300	8.95	-1.67	9.00	-1.68	-1.675

Table 4. Graphical square for a cross calibration - polygon/indexing table

Unit: Second of Arc

Index Table \ Polygon	0-60	60-120	120-180	180-240	240-300	300-0	Sum	Mean M	Adjusted G/m^2-M	Accum.
0-60	(-2.306) -2.305 (+0.001)	(-0.141) -0.100 (+0.041)	(+1.670) +1.650 (-0.020)	(-0.660) -0.690 (-0.030)	(+2.641) +2.680 (+0.039)	(-1.732) -1.765 (-0.033)	-0.530	-0.088	+0.088	+0.088
60-120	(-2.097) -2.085 (+0.012)	(+0.068) +0.095 (+0.027)	(+1.879) +1.915 (+0.036)	(-0.451) -0.495 (-0.044)	(+2.850) +2.815 (-0.035)	(-1.523) -1.520 (+0.003)	+0.725	+0.121	-0.121	-0.033
120-180	(-2.162) -2.195 (-0.033)	(+0.003) -0.015 (-0.018)	(+1.814) +1.795 (-0.019)	(-0.516) -0.475 (+0.041)	(+2.785) +2.820 (+0.035)	(-1.588) -1.595 (-0.007)	+0.335	+0.056	-0.056	-0.089
180-240	(-2.257) -2.270 (-0.013)	(-0.092) -0.080 (+0.012)	(+1.719) +1.730 (+0.011)	(-0.611) -0.570 (+0.041)	(+2.690) +2.640 (-0.050)	(-1.683) -1.685 (-0.002)	-0.235	-0.039	+0.039	-0.050
240-300	(-2.225) -2.230 (-0.005)	(-0.060) -0.035 (+0.025)	(+1.751) +1.725 (-0.026)	(-0.579) -0.610 (-0.031)	(+2.722) +2.730 (+0.008)	(-1.651) -1.625 (+0.026)	-0.045	-0.008	+0.007	-0.042
300-0	(-2.260) -2.225 (+0.035)	(-0.095) -0.185 (-0.090)	(+1.716) +1.730 (+0.014)	(-0.614) -0.590 (+0.024)	(+2.687 +2.690 (+0.003)	(-1.686) -1.675 (+0.011)	-0.255	-0.043	+0.042	+0.000
Sum	-13.310	-0.320	+10.545	-3.430	+16.375	-9.865	G -0.005			
Mean M	-2.218	-0.053	+1.758	-0.572	+2.729	-1.644		G/m -0.001		
Adjusted M-G/m^2	-2.218	-0.053	+1.758	-0.572	+2.729	-1.644			G/m^2 0.000	
Accum.	-2.218	-2.271	-0.514	-1.085	+1.644	0.000				

$$\Sigma y^2 = 0.033\ 533$$

$$S_s^2 = \Sigma y^2/25 = 0.001\ 341$$

	1 interval	2 intervals	3 intervals
Variance	$5\ S_s^2/36$ = 0.000 186	$8\ S_s^2/36$ = 0.000 298	$9\ S_s^2/36$ = 0.000 335
Standard deviation	0.014	0.017	0.018

Estimated error in zth
interval of the polygon = (Mean of zth column - G/m^2)

Estimated error in zth
interval of the
indexing table = (G/m^2 - Mean of zth row)

These results are shown in the row and column labelled "Adjusted". Determine the estimated accumulation errors of the polygon from the datum face and of the indexing table from the zero position by taking progressive totals of the appropriate interval errors.

(c) Statistical Analysis of Graphical Squares

It can be shown that either of the types of graphical square described provides a least squares solution to the observational equations. Further statistical analysis of the results may be carried out by determining the improved estimate for each of the differences entered in the square. In the case of a simple calibration this is the difference between the two appropriate values in the row labelled "Interval Error" (see Table 2). For a cross calibration it is the difference between the appropriate value in the column labelled "Adjusted" and that in the row labelled "Adjusted" (see Table 4). These improved estimates of the differences are shown in the square in parentheses above the observed values, and the residuals between the observed values and the improved estimates are shown in parentheses below. The sum of the squares of these residuals is calculated in each case and the following analysis made:

(i) For a *simple calibration*, with n intervals:

Where sum of squares of residuals $= \Sigma v^2$

Estimated variance for observed differences $S_o^2 = \Sigma v^2/(n-1)^2$

Estimated variance for improved estimate of each interval error $= (n-1)S_o^2/2n^2$

Estimated variance for the accumulated error over r intervals $= r(n-r)S_o^2/2n^2$

(ii) For a *cross calibration* with m intervals in each device:

Where sum of squares of residuals $= \Sigma y^2$

Estimated variance for observed differences $S_s^2 = \Sigma y^2/(m-1)^2$

Estimated variance for improved estimate of each interval error $= (m-1)S_s^2/m^2$

Estimated variance for the accumulated error over z intervals $= z(m-z)S_s^2/m^2$

For each quantity the estimated standard deviation is the square root of the corresponding variance and the estimated uncertainty at the 99% confidence level is three times the standard deviation.

The analysis given here is based on that of Gilet and Watson (1953), modified, in the case of a simple calibration, to suit the extended procedure recommended in this paper and which requires twice the number of observations.

Gilet and Watson rightly draw attention to the fact that their analysis is strictly correct, for both types of square, only if all the observed differences are independent. With the procedures described this is *not* so, as each individual observation contributes to two successive differences. They point out that it is possible to repeat all the observations, then choose differences alternately from the two sets and combine them to form two squares in each of which the differences are wholly independent. In our experience, doing this rarely results in significant changes in the calculated variances but the procedure can always be followed for work of sufficient importance.

The uncertainties determined from a graphical square by these methods refer to random errors only. Systematic errors will also be present and must be assessed separately.

(d) Sources of Error

Precautions which must be taken to reduce both random and systematic errors to a minimum include:

(i) Accurately centering the polygon on the precision indexing table or rotating table.

(ii) Ensuring that the polygon rotates accurately in the plane of its lower face.

(iii) Setting the autocollimator so that it reads angles of tilt truly in the plane of rotation of the polygon.

(iv) Shielding as much as possible, particularly in relation to the optical path, to reduce draughts and thermal effects.

By observing the above precautions and exercising due care when taking readings it is possible, when using the highest quality equipment, to achieve very small random errors of the order of 0.01 second.

An assessment of systematic errors is necessary. This assessment is usually subjective, and based upon knowledge gained by past experience with the equipment being used. Experience in this Division shows that it is possible for systematic differences in readings of up to 0.1 second to occur between two high quality autocollimators of different type when viewing certain faces of a high quality polygon.

We have also found that a precision indexing table of the highest quality is capable of repeating its calibration within approximately 0.01 second, but that if any alterations are made to the set up, particularly with respect to the loading or clamping, changes to the calibration of up to 0.1 second may be introduced.

Some of the errors which may be encountered when using or calibrating a precision polygon are discussed by Bouillon, Tremblay & Delisle (1973) and Brezina (1974b).

(e) Variations on the Procedures Described

(i) When a simple calibration procedure is used for subdivision of a linear scale, using two microscopes set at increasing spacings, the number of scale reading positions must be reduced as the spacing is increased in order that the microscopes remain over the scale. With reference to Table 2, the differences observed are then only those below the diagonal in the square and it is normal practice to complete the square by entering these differences, with signs reversed, in symmetrical positions on the other side of the diagonal.

Some published accounts, e.g. Gilet & Watson (1953), Evans & Taylerson (1964), follow the same practice when the technique is used for a polygon or other full circle angle device. It is preferable that the full circle of observations be taken for each spacing of the observing instruments, thus obtaining differences for the whole square as previously described. The instructions issued by NRLM, Japan, in October 1980 for the international comparison of angle standards, under the auspices of Bureau International des Poids et Mesures (BIPM), specify this procedure.

(ii) In each of the calibration procedures described previously, the observational data is collected in a straightforward sequence which is readily understood. Sometimes, as in the case of the international comparison referred to above, it is recommended that this sequence be altered so that the order in which the sets of data are collected is varied in a random manner.

During a simple calibration, for example, the spacing between the two autocollimators is not sequentially increased by only one angular interval for successive sets of observations, but either increased or decreased by varying amounts until the full number of sets is obtained. Similarly, in the case of a cross-calibration between a polygon and an indexing table, the polygon is repositioned on the table by rotating it through varying numbers of angular intervals for successive sets of observations. Furthermore, in both cases, the angular setting of the table at the start of each set of observations is varied.

It should be noted, however, that within any set of observations, rotation should always be by one interval at a time through a full circle in one direction, and then repeated in the reverse direction.

We have found no significant advantage in using this random order, but under conditions less stable than those in our laboratory the amended procedure may well be desirable.

Lasers, Interferometry and the Measurement of Length

H.M.P. Stock
CSIRO Division of Applied Physics, Sydney, Australia 2070

Abstract

Although the metre is currently defined as a multiple of the vacuum wavelength of the radiation from the Kr-86 atom, a number of practical, secondary standards of greater precision are available in terms of the vacuum wavelength of the radiation from a He-Ne or other laser, the wavelength of which is stabilized by one of several possible methods. Through the technique of interferometry these "wavelength" standards can be readily used to calibrate the material standards such as gauge blocks, line scales and tapes or wires that are used as the next level of standards in the length hierarchy. The purpose of this paper is to review the theory and practice of the equipment, procedures and techniques required to ensure a successful transfer from these secondary standards to the material working standards. Particular attention is given to the calibration of line standard scales and tapes.

1. Introduction

It is now some twenty years since the first laboratory versions of LASERs and MASERs (Light (Microwave) Amplification by Stimulated Emission of Radiation) appeared in research and industrial laboratories, and some fifteen years since the first commercially produced lasers became available. Since then the number of atomic, ionic and molecular species known to exhibit laser action has risen from less than ten into the hundreds and the range of wavelengths and powers available in commercial systems has risen accordingly.

Although originally dubbed "a solution looking for a problem", lasers, because of the high specific intensity, monochromaticity and coherence of the emitted radiation, were soon being widely used throughout science, technology and industry (Gagliano *et al.* 1969). In dimensional and engineering metrology, lasers, particularly the lower power (0.5 to 5 mW) 633 nm helium-neon laser, found a ready acceptance as a light source, simplifying many measurement procedures and making possible many others (see, for example, Barrell 1957 and Balasubramanian & Wyant 1978). It was, however, in the specific area of interferometry that the laser, because of the properties of the emitted radiation, made the greatest impact in metrology, and it is to this use of the laser that this paper is addressed.

In the following discussion, reference is sometimes made to particular properties of lasers, the detailed discussion of which has been left to the Appendix.

2. Lasers as Tools in Dimensional Metrology

Although non-coherent light sources were already widely used in dimensional metrology, including interferometry, lasers, particularly the low cost low power gas lasers and the small and rugged solid state lasers, have had a great impact on the type, range and accuracy of light-based measurements. This is particularly so in the area of non-contact measurement of dimension in hostile or hazardous environments such as chemical plants or steel works.

Among the many metrology tasks that have been simplified or made possible by the use of lasers are:

- Length measurement from calibration of line thickness in micro-electronic circuits to measurements of earth strain over distances of hundreds of metres.
- Determination of the position of an edge or surface using back-scatter techniques, and, as an extension, non-contact thickness or profile determination of moving objects.
- Diameter or thickness measurement of solids in the range 0.1 mm to 500 mm using two-beam or single-beam (oscillating or diffraction) systems.
- Alignment with accuracies of better than 1 μm/m over lengths up to several hundred metres.
- Non contact measurement of surface finish parameters.
- Non contact measurement of movement and velocity of solids, liquids and gases including anemometry and in vivo measurement of blood flow.
- Measurement of angle and angular change.

In all these tasks the laser and radiation detector(s) can be linked to suitable electronics (including microprocessors) to provide real time display and/or control of the features or dimensions being measured.

A number of reviews and conference proceedings have been published over the last decade or so, many by the Society of Photo Optical Instrumentation Engineers (SPIE). Some of these are Gagliano *et al.* (1969), Balasubramanian & Wyant (1978), Cullen (1978) and Axelrod (1980).

An example of the type of non-contacting measuring instrument now being developed is the Profile Recorder recently described by Thwaite & Bendeli (1980a & b). In this instrument the laser light back-scattered from a surface and focussed on to a photodiode array is used to determine the position of a moving surface, and thus a profile, with an uncertainty of better than ±0.1 mm in 10 mm and with a data update rate of up to 2000 sec^{-1}.

Another area of dimensional metrology to which lasers have made substantial changes is that of the length standard itself. With the development of stabilized lasers in both the visible and infrared it became possible to develop a laser chain to compare the frequency of the ^{133}Cs standard (9192.63177 MHz) with the frequencies of the CH_4 and, more recently, the I_2 stabilized He-Ne lasers, and thus, through the ^{86}Kr lamp, to measure c, the velocity of light in a vacuum. Laser chains were built up at a number of national laboratories (the first being at NBS, Evanson *et al.* 1972, 1973 and NPL, Blaney 1975, 1976) and as a result of this work the 1973 meeting of the CCDM agreed to define two stabilized laser frequencies and to adopt a value for c. The agreed values were:

for the P(7), $\nu(3)$ band in CH_4, ν = 88.376 181 627 THz

for the i component in $^{127}I_2$, ν = 473.612 214 THz

and for c, c = 299 792 458 ms^{-1}.

(See also Terrien 1976 and Hall 1978 for a fuller description of this work.)

Further, as a result of this work, consideration was given to redefining the metre in terms of either a laser transition or the velocity of light, and at the 1982 meeting of the CCDM it was recommended that a new definition of the following form should be adopted:

"The metre is the length of the path travelled by light in vacuum during the fraction 1/299 792 458 of a second."

It is expected that this recommendation will be adopted by the 1983 meeting of the CGPM.

While those laboratories able to afford to construct a laser chain will be able to realize the proposed new definition of the metre and to achieve a precision of about 1 in 10^{10}, smaller laboratories will be able to rely on agreed secondary standards in the form of stabilized lasers and achieve a metre with a precision of better than several parts in 10^{10}.

Irrespective of how the metre is realized or maintained, the problem of the dissemination of the standard into commerce and industry remains. The solution to this problem is often achieved through the technique of interferometry.

3. Interferometry and the Measurement of Length

When two or more electromagnetic waves interact at a point in space the resultant wave at that point will have an intensity and phase determined by the vector combination of the amplitudes and phases of the components in the interaction. This phenomenon is known as *interference*. If the interacting waves are temporally coherent (i.e. maintain their phase relation with time) the resultant will be stable in time. If, in addition, the waves are spatially correlated and periodic a pattern of *interference fringes* will be built up, the variation in intensity across the pattern being dependent on the changes in the combined phase differences of the component waves, as well as on the amplitudes of the component waves. Where the interacting waves emanate from a single source, or from two or more sources in fixed phase relationship to one another, the interference pattern can be interpreted in terms of path differences between the component waves. As a result measurements of absolute length or changes in length can be inferred from the interference patterns or the changes in the interference patterns. This process is known as *interferometry* and the instruments used are called *interferometers*.

(a) Interferometry based on Wave Amplitude Division

The classical form of interferometry is based on interferometers of the Michelson type or on some variant such as the Twyman-Green, or Fizeau type (Steel 1967). In all these designs the amplitude of the radiation emitted from the source is divided by a beam splitter with the two component waves travelling different paths before recombination. The resultant fringe pattern, or changes in the resultant fringe pattern, as the frequency of the radiation or the length of one or both of the interferometer arms is changed, is used to infer a linear dimension. Systems of this type have been used to measure the lengths of end standards (Barrell 1957) using the method of "exact fractions" (Puttock 1982). With the introduction of the ^{86}Kr lamp similar methods were adopted for the measurement of line scales (Ciddor & Bruce 1967) replacing the mechanical comparator methods relied on until then (Thwaite & Leslie 1963, Thwaite 1965).

In all these applications the maximum length that could be measured was limited by the short coherence length of the radiation used, and the low brightness of the light sources which resulted in faint fringes of low contrast even at relatively short path differences. This necessitated skilled calibration staff and limited the extent to which electronics could be used in the measurement process.

Many of these problems were overcome with the introduction of lasers into interferometry; their much greater effective coherence length allowed measurement of much longer lengths, and while the limited range of frequencies available inhibited the use of "fractions" methods, the high brightness of the laser simplified the optical and electronic difficulties of fringe counting. Finally when stabilized lasers were introduced, the uncertainty in the laser wavelength ceased to be a significant contribution to the overall uncertainty of the measurement.

In most cases the interferometer systems developed to utilize the new technology did not differ markedly from the earlier instruments, the major changes to the underlying Michelson or Twyman Green form being the use of cube corner retroreflectors (Peck 1948, Murty 1960a, Stroke 1965) to alleviate the otherwise critical mirror alignments, and the introduction of comparatively sophisticated optoelectronic fringe counting systems.

Laser based length measuring systems were implemented by most national and many research laboratories, the length capacities ranging from millimetres (Sydenham 1972, Thwaite & King 1982) to metres (Giacomo *et al.*1972, Gillilland *et al.* 1966, Rowley & Wilson 1968) and to tens of metres (Bennett 1974, Turner & Pfitzer 1970), while industrial applications soon followed (Tanimura & Yamamoto 1972, Loewen & Burns 1971).

One problem encountered by many of these instruments, however, was the difficulty in determining the direction of motion of the movable arm, as this information is not derivable from a single point observation of the time varying fringe pattern. A number of methods have been devised to obtain this information; one method used in this laboratory (Thwaite & King 1982) is shown in Fig.1.

This device, which was constructed for the calibration of stage micrometers and similar devices up to 10 mm long, as well as micrometer microscopes, with an uncertainty of measurement of down to ±0.1 μm, is basically a folded Twyman-Green interferometer (Murty 1960b) with a composite prism beam splitter and reference arm. A $\lambda/8$ step evaporated on to the main mirror in the reference arm results in a phase difference of $\pi/2$ between the fringes from different sides of the step; this phase difference contains the information required for bidirectional counting of the fringes. The folded path arrangement results in one fringe change for each $\lambda/4$ displacement of the retroreflector, while subdivision of a fringe (estimated from the Lissajous figure display of the signals

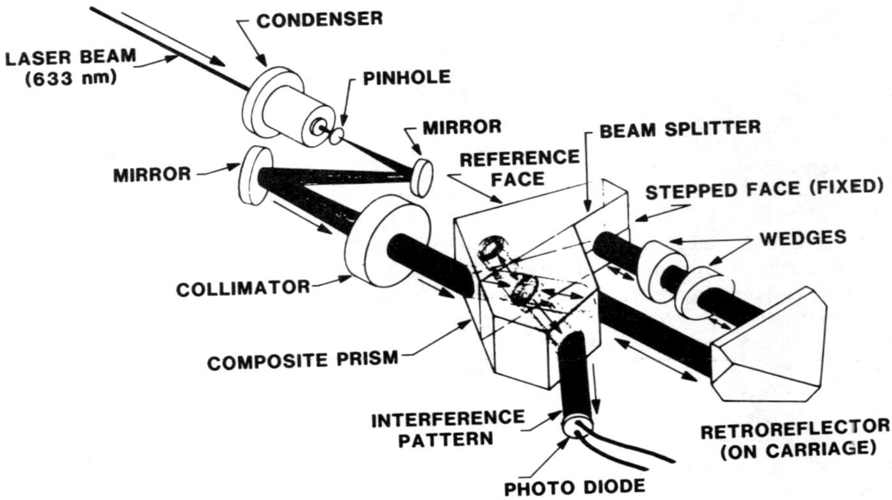

Fig.1. Compensated, bi-directional fringe-counting laser interferometer (from Thwaite & King 1982).

obtained from the two sides of the step) allows a resolution of 0.01 μm in the measurement of displacement.

Although fringe counting laser interferometers have been used successfully for measurements of displacements of up to many metres, they are subject to a number of problems at very long path differences. In particular diffraction effects (Tango & Twiss 1974) and turbulence (Sirohi *et al.* 1971) can result in further degradation of the fringe contrast. One result of reduced contrast can be that the opto-electronic system may fail to recognize fringes and errors in measurement may occur. This problem is less likely to occur in interferometers using a two-frequency laser.

(b) Interferometry with Two-Frequency Lasers

In 1962 Culshaw & Kannelaud observed that when a He-Ne laser was operated in a weak longitudinal magnetic field, the atomic transition in Neon was Zeeman split into a gain doublet corresponding to the two allowed transitions $\Delta m = \pm 1$, as shown in Fig.2. As a result, the output of an otherwise single TEM_{00} mode laser is split into two frequencies with *left and right hand circular polarizations*. (For further information on the Zeeman effect see, for example, Herzberg 1944.)

Although the separation of the Zeeman split components may be up to 1500 MHz, depending on the magnetic field, the two laser frequencies being from the same axial mode are separated by only a few MHz. This frequency difference results from the different indices of refraction (different values of c') for the two polarizations in the active medium (see, for example, Baer *et al.* 1980).

Stabilization of this type of laser, the "Zeeman split laser", is obtained in the same manner as for any other polarization stabilized laser. The error signal needed for the feedback control of the cavity length is derived from the difference in the output powers in the two orthogonal linear polarizations which are obtained by passing the two opposite circular polarizations through, say, a quarter wave plate.

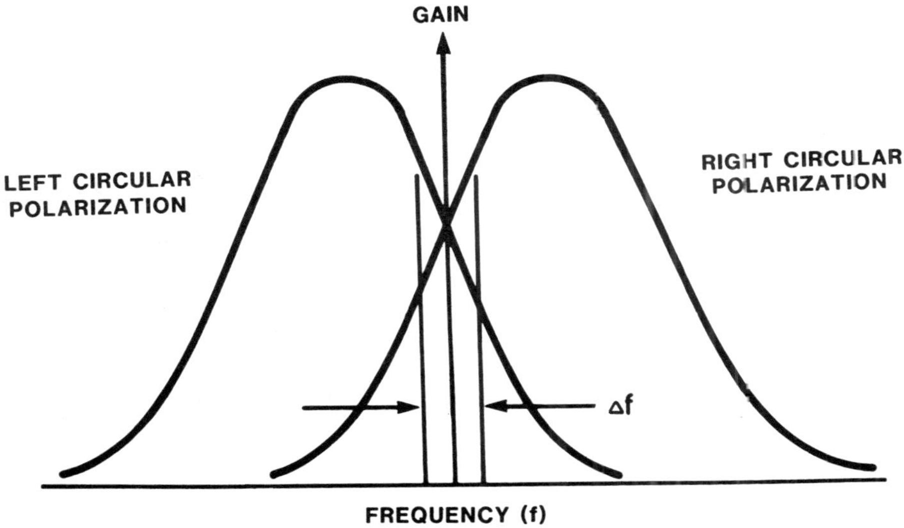

Fig.2. Zeeman split gain doublet showing the two components of the single TEM_{00} mode due to the different refractive indices of the two opposite circular polarizations. (After Liu & Klinger 1979).

A Zeeman split laser of this type lies at the heart of the two-frequency displacement measuring interferometer. As shown in Fig.3 the two circular polarizations emitted by the laser are linearized, separated by a polarizing beam splitter, and directed to the two arms of the interferometer. When both retroreflectors are stationary the beat signal between the two optical frequencies, f_1 and f_2, in the recombined beam at the photo detector is

$$S' = S'_o \sin[2\pi(f_1 - f_2)t]$$

which corresponds to the beat signal obtained at a photo detector for the unperturbed laser output,

$$S = S_o \sin[2\pi(f_1 - f_2)t].$$

If, however, the retroreflectors are moving with velocity v_1 and v_2 respectively the frequency of return beam in each arm is $f_1 + \Delta f_1$ and $f_2 + \Delta f_2$ respectively where Δf_i is the Doppler shift frequency due to the motion of the retroreflector in arm i.

The beat signal in the returned beam will then be

$$S'' = S''_o \sin[2\pi(f_1 - f_2 \pm \Delta f)t]$$

where

$$\Delta f = (\Delta f_1 - \Delta f_2).$$

If, then, the two beat frequencies S and S" are fed to a reversible counter so that one beat frequency produces up counts while the other beat frequency produces down counts the resultant net cumulative count corresponds to the distance travelled in wavelengths of light, λ, and

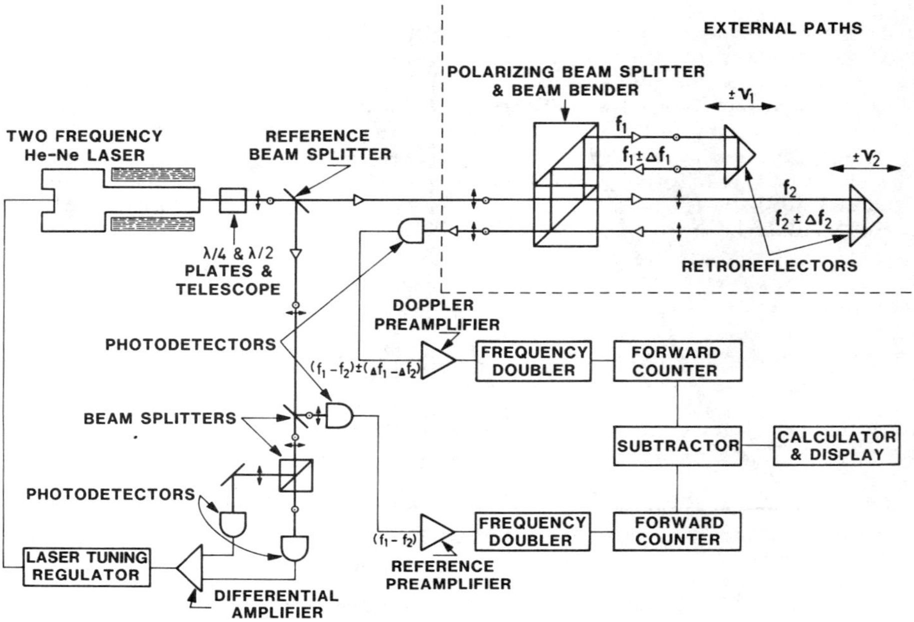

Fig.3. Two frequencies of opposite polarization originating in the laser are separated into different external paths. The beams are Doppler shifted by moving retroreflectors. The return beams are optically mixed, the difference frequency detected and after amplification, counted against a reference, and converted to net change in path difference. (After Liu & Klinger 1979).

by simple conversion into length in standard units

$$\Delta f = 2(v_1 - v_2)\frac{f}{c} = \frac{\Delta d}{\Delta t}\cdot\frac{2}{\lambda}$$

so that

$$\Sigma\Delta d = \frac{\lambda}{2}\,\Sigma\Delta f\Delta t \qquad (1)$$

where equation (1) is derived from the expression for the Doppler shift frequency for a beam reflected from a moving reflector.

A two-frequency laser measuring system of this type is manufactured and marketed by Hewlett-Packard with the identification Model 5526A Laser Measurement System. (See also Dukes & Gordon 1970, Burgwald & Kruger 1970 and Baldwin *et al.* 1971).

Like wave amplitude dividing laser interferometers, the additional degradation of the fringe contrast due to diffraction and turbulence can, at very long path differences, result in the opto-electronics of the two-frequency laser system failing to recognize fringes. The resultant loss of the beat frequency count is, however, indicative of an error condition so that errors in measurement due to loss of fringe contrast are much more likely to be recognized. (Hewlett-Packard laser systems initiate an error mode if the beat frequency is lost for any cause.)

The main limitation of the two-frequency laser is the low maximum displacement speed for the retroreflectors. As the Doppler shift frequency for a moving retroreflector is

$$\frac{\Delta f}{f} = \frac{2v}{c}$$

and as *aliasing* will occur if Δf exceeds $(f_1 - f_2)$ the maximum displacement speed is given by

$$v_{max} < \frac{c(f_1 - f_2)}{2f} .$$

Now $(f_1 - f_2)$ is determined by the different refractive indices for left and right hand circular polarizations in the plasma of the He-Ne laser discharge, and, for most systems, is of the order of 2 MHz, so that $v_{max} < 500$ mm/s. (Hewlett-Packard recommend $v_{max} \lesssim 300$ mm/s and have included an overspeed system that initiates an error mode when $\Delta f \lesssim (f_1 - f_2)$.

In this Laboratory we have implemented a commercial two-frequency laser interferometer system for the calibration of machine tools, measuring machines, line standards of length (Fig.4) and geodetic tapes (Fig.5). Quantities derived from length, such as angle (Fig.6), flatness and straightness, can also be measured with this system.

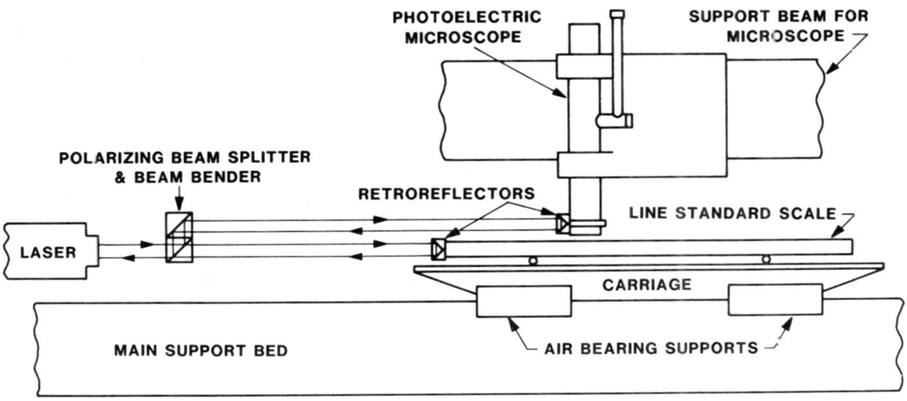

Fig.4. Schematic diagram of the system used for the calibration of line standards of length.

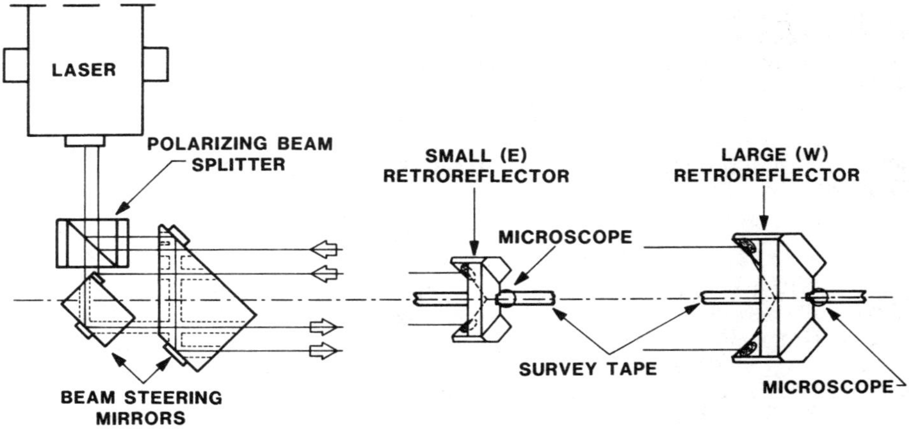

Fig.5. Schematic Diagram (plan) of the system used for the calibration of geodetic tapes.

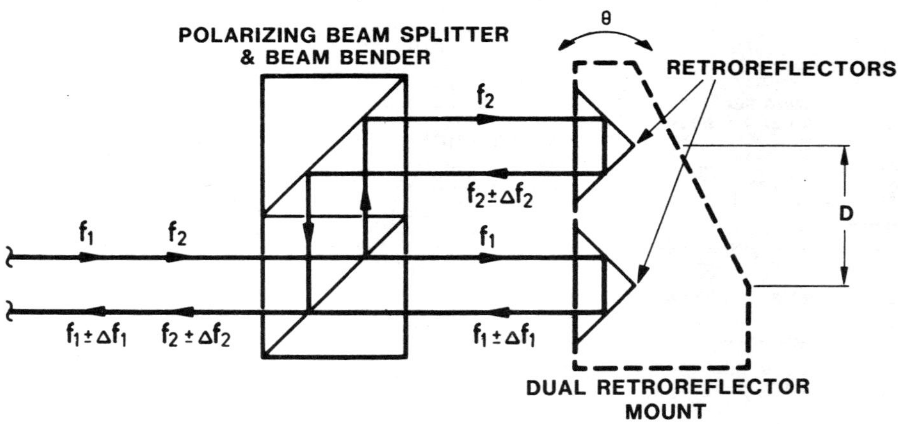

Fig.6. Schematic diagram of the system used to measure angle with a two-frequency laser interferometer.

4. Considerations relating to Length Measurement with Laser Interferometers

Although laser interferometer systems can be used to determine displacement with a resolution of 10 nm or better, the uncertainty to be associated with the final measured length is unlikely to be better than 5 in 10^6 unless a number of precautions are taken and a number of corrections applied. Some of these are discussed below.

(a) Measurement Procedures

It is desirable that, wherever possible, the measurement of an unknown length should be undertaken in a time symmetric manner so that the data can be interpreted on the basis of the *"Reduction to Mean Instant"*. This procedure, which is akin to the "principle of alternating" (standard, unknown, standard) of general metrology (Puttock 1969) helps compensate for any changes in the physical conditions of measurement.

On this basis then, the measurement of, say, an interval A/B of a line standard using a laser interferometer should take the form "position of line A relative to datum (or zero of system), refractive index (see below), temperature (see below), position of line B relative to datum, (mean instant), position of line B, temperature, refractive index, position of line A".

The analysis of this information will then yield the length of A/B at the temperature of the turnaround point.

Although this procedure, the reduction to mean instant, assumes a linearity of change in the measuring conditions with respect to time, and such an assumption is unlikely to be strictly true, it is a sufficiently close approximation for nearly all practical purposes.

It is also desirable to undertake more than one determination of any length using the same measurement protocol so as to provide a constant base for the assessment of any errors of measurement and of the uncertainty to be associated with the final result.

(b) Thermometry

As all material objects change size with temperature, it is necessary that, if the sizes of various objects are to be related or compared, measurements of size should be made at or referred to the standard temperature of 20°C. To achieve this the temperature of measurement of any material object must be determined at the time of measurement and the measured size corrected for the departure from 20°C through knowledge of the coefficients of linear thermal expansion. This is particularly necessary for measurements made with laser interferometers where the principle of comparing "like with like" is not complied with (i.e. steel with steel) so as to minimise the differential coefficient of linear thermal expansion.

The relation often used to relate length and temperature is

$$L_t = L_{20}[1 + \alpha(t - 20)]$$

or for very precisely known standards

$$L_t = L_{20}[1 + \alpha(t - 20) + \beta(t - 20)^2]$$

where L_t and L_{20} are the lengths at t and 20°C respectively and α and β are the coefficients of linear thermal expansion.

In most cases either liquid-(mercury)-in-glass thermometers (Beavis 1982) or platinum-resistance thermometers (Fallon 1971, Kemp 1977) will be used for temperature measurements in the determination of length using laser interferometers, the choice of thermometer type and precision of calibration and reading depending on the uncertainty to which the measured length is required to be known. For example, measurements to 1 in 10^6 can be satisfactorily made with mercury-in-glass thermometers (±0.1°C) if the items being measured are made of invar (36% Ni steel; $\alpha \sim 1 \times 10^{-6}$/°C), while for 1 in 10^6 for items made of steel ($\alpha \sim 11 \times 10^{-6}$/°C) or brass ($\alpha \sim 18 \times 10^{-6}$/°C), high quality thermistors or simple resistance elements (±0.02°C) would be needed. For uncertainties to 1 in 10^7 on a steel line standard, temperatures must be known to ±0.002 - 0.005°C which require high quality calibrated platinum resistance sensors with regular checks of the ice points.

Further, it is necessary, in any high precision measurement, to measure the temperature at more than one position so as to detect and allow for any temperature gradients in the object being measured. Experience at this laboratory suggests that for measurements requiring an overall uncertainty of 5 in 10^6 or better on a length of up to 1 m the temperature should be measured at at least two positions, while for geodetic lengths the temperature should be measured at no more than 4 m intervals. As a general, empirical, rule the assumption of a linear temperature gradient is not believed to introduce any unacceptable additional uncertainty in measurements of length up to 1 m, except when the product of the temperature differential and the coefficient of linear thermal expansion becomes as large as the overall uncertainty sought.

Although the use of time symmetric procedures will satisfactorily compensate for slowly varying temperatures, unacceptable errors may be introduced if the temperature change over the period of the measurement is too rapid or if the magnitude of the change is too large (see the empirical rule mentioned above). It is necessary, therefore, to allow sufficient time for the object to "soak" after any temperature change (i.e. following handling), that is, to allow the item sufficient time to come into thermal equilibrium with its surrounds. Our experience in this laboratory is that given the thermal capacity of our measuring systems "soak" times of up to 6 hours may be necessary for both scales and tapes after prolonged handling. The technique described by Miller (1970) can be used to provide a more quantitative determination of this time.

Finally, we have found it necessary both to increase the thermal inertia of scales being measured by partially enclosing the scale in a copper box, and to provide full thermal shielding so as to prevent radiated body heat (~100 W) from causing unacceptable temperature drifts. For measurements where loading or shielding is not a viable option, such as measurements of geodetic tapes, we have found it advisable to limit the duration of a measurement.

(c) Refractive Index of Air

It is usual practice, when quoting the wavelength of a laser radiation, to quote the vacuum wavelength, λ_v. However, as most laser interferometers measure displacement in terms of wavelengths, it is necessary to apply a correction for the difference between λ_v and the wavelength of the radiation in air, λ_a, under the conditions applying at the time of the measurement. As the refractive index n of the air is related to λ_v and λ_a by

$$n = \frac{\lambda_v}{\lambda_a}$$

the correction to be applied can be derived from a measurement of the refractive index of the air in the radiation path.

In the absence of a direct measurement of the refractive index using a refractometer (Terrien 1965) it is usual to measure the air pressure, air temperature and partial pressure of water vapour (humidity) and to relate these to the refractive index using the formula of Barrel & Sears (1939), and the dispersion relation of Edlen (1966)

$$(n-1)\times 10^8 = [8\ 342.13 + 2\ 406\ 030(130-\sigma^2)^{-1} + 15\ 997\ (38.9-\sigma^2)^{-1}]$$

$$\times \left[\frac{p_t}{720.775} \cdot \frac{1 + 10^{-6}\ p_t\ (0.817 - 0.0133t)}{(1 + 0.0036\ 610\ t)}\right]$$

$$- \quad f_t\ [5.7224 - 0.0457\ \sigma^2] \qquad (2)$$

where σ is the vacuum wave number in μm^{-1}, p_t is the atmospheric pressure in mmHg, t is the atmospheric temperature in °C and f_t is the partial pressure of atmospheric water vapour in mmHg, (1 mmHg = 133.25 Pa).

For the Zeeman-split two-frequency laser such as used in the Hewlett Packard interferometer system $\lambda_v \simeq 0.632\ 99$ µm so that $\sigma \simeq 1.5798$ and

$$(n-1)\times 10^6 = 0.3836\ 391\ p_t \left[\frac{1 + p_t(8.17\times 10^{-7} - 1.33\times 10^{-8}t)}{(1 + 3.661\times 10^{-3}t)}\right]$$

$$- 5.608\ 34 \times 10^{-2}\ f_t \qquad (3).$$

(Note: In their "Compensation factor handbook", Hewlett Packard give the last term in this equation, the water vapour component, as

$$3.033 \times 10^{-3} R \exp(0.057\ 627\ t)$$

where R is the relative humidity in %. This formulation is based on the approximation

$$f_{sat} = 5.408 \exp(0.057\ 627\ t)$$

for the saturation vapour pressure of water, f_{sat}, as a function of t. The error introduced by the use of this approximation should not exceed 5 in 10^8 on the final measured value of length.)

The alternative refractivity expressions of Owens (1967) and Popela (1972) are not used in this laboratory.

The distance shown on the 5505A laser display of a Hewlett Packard 5526A laser interferometer system is the product of half the nominal wavelength of the radiation, $\lambda_d/2$, and the net count since last reset, $n_\lambda = \Sigma\Delta f\Delta t$ (equation 1),

i.e.
$$L_d = n_\lambda \cdot \frac{\lambda_d}{2}$$

where
$$\lambda_d = (999\ 000 + C_s) \times 10^{-6} \cdot \lambda_v$$

and C_s is the preset compensation factor.

In practice the distance moved is the product of the net count since last reset and half the laser wavelength in air under the conditions of p, t and f applying.

$$L = n_\lambda \cdot \frac{\lambda_a}{2} = n_\lambda \frac{\lambda_a}{\lambda_v} \cdot \frac{\lambda_v}{2}$$

and $\lambda_a/\lambda_v = 1/n = (999\ 000 + C)\times 10^{-6}$ where C is the actual compensation factor that should have been set.

So

$$C = \frac{10^{12}}{(n-1)\times 10^6 + 10^6} - 999\ 000 = \frac{10^{12}}{N+10^6} - 999\ 000 \qquad (4)$$

where N is as referred to in the H.P. "Compensation factor handbook" and = $(n-1)\times 10^6$.

The actual correction to be applied to the laser display is given by

$$L = L_d + \Delta L_d = L_d(1 + \delta_d)$$

so that

$$\delta_d = \frac{L}{L_d} - 1 = \frac{999\ 000 + C}{999\ 000 + C_s} - 1$$

and since $C_s << 999\ 000$

$$\delta_d \simeq (C - C_s)\times 10^{-6}.$$

(d) Sine and Cosine Errors

There are a number of errors and additional sources of uncertainty which can creep into a length measurement with a poorly designed or aligned interferometer. The most important of these errors is the sine error (Schulze 1967) which can occur if the interferometer is insufficiently rigid and deflects under load when the scale being measured, or alternatively, the observing microscope, is displaced during the measurement.

If, as depicted in Fig.7, the microscope rotates about some horizontal axis with an angular component, θ_A, in the longitudinal direction, and with the pivot a distance h above the divided surface of the scale, during the displacement of the scale from scale markA to scale mark B, the measured length will be in error by an amount $\Delta L = L\sin\theta_A$.

For uncompensated systems with $h \sim 100$ mm and $\theta_A \sim 10$ seconds (say), $\Delta L \sim 5$ μm or 5 in 10^6 on 1 m.

This error can be minimised by restricting the lever arm, which can be achieved by folding the reference arm of the interferometer to monitor the position of the microscope, or by increasing the rigidity of the interferometer structure. These techniques have been used in the line standards comparator used in this laboratory (Fig.4).

Errors can also occur if there is a misalignment between the mechanical axis of motion and the axis of the object being measured, or between the mechanical axis of motion and the axis of measurement (laser beam). In both cases the error is a cosine error, and for small angles is proportional to $\theta^2/2$ (see Fig.8).

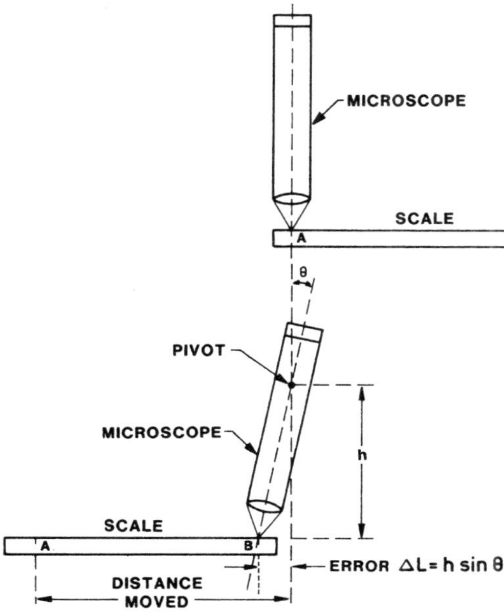

Fig.7. Occurrence of sine error due to microscope movement during measurement.

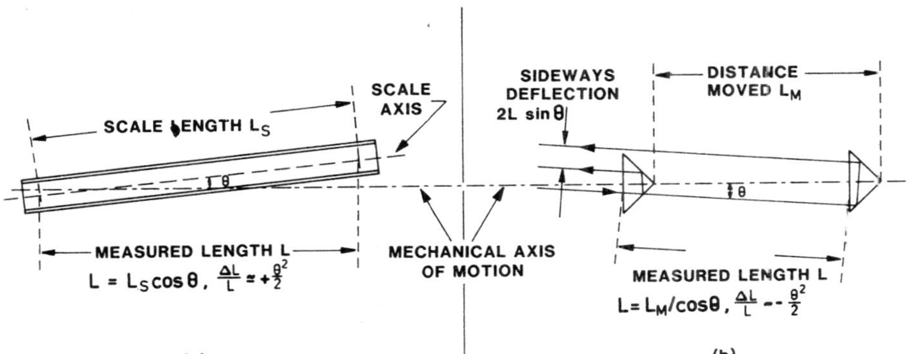

Fig.8. Cosine error in laser interferometer systems a) Misalignment between mechanical axis of motion and scale axis, b) Misalignment between mechanical axis of motion and measuring axis (laser axis).

It should be noted that, as the reflected beam from a cube corner retroreflector is always parallel to the incident beam, the measured length will *exceed* the mechanical displacement when there is a misalignment between the axes of measurement and motion (Fig.8b). The resultant error is, therefore, in the opposite direction to that normally associated with cosine errors (Fig.8a). (See also Peck 1948).

(e) Laser Frequency

Although the operational principle and the stabilization method used restrict the frequency range over which a Zeeman-split, two-frequency laser can operate, the actual frequency of oscillation can differ from the nominal value by some 1 - 2 in 10^7. Table 1 shows the measured vacuum wavelengths for four Zeeman-split lasers tested at this laboratory together with the difference between the measured and the nominal wavelengths expressed as parts per million, as well as the measured frequency separation of the two polarizations.

Table 1. Measured Vacuum Wavelengths, Departures from Nominal Wavelength and Polarization Frequency Separation for four Zeeman-split Lasers

Laser	Vacuum Wavelength (nm)	Departure from Nominal (p.p.m.)	Frequency Separation (MHz)
1	632.991 375	-0.166	2.2
2	632.991 383	-0.153	1.5
3	632.991 410	+0.018	1.5
4	632.991 426	+0.041	1.7

It would seem necessary, therefore, that, if measurements with a precision greater than 3-5 in 10^7 are to be made with a Hewlett Packard laser system, the actual vacuum wavelength of the laser radiation must be determined and that a correction for the difference between actual and nominal wavelength be applied to any measured length.

5. Assessment of the Uncertainty to be associated with a Laser Interferometric Measurement of Length

Although laser interferometers can be used to determine displacement with a resolution of 0.01 μm, this resolution is rarely maintained when interferometry is used to measure the length of a material object, the uncertainty associated with the final measured length usually being more than one order of magnitude greater. In the main, however, the causes of the reduction in precision lie not in the laser interferometer but in the manner in which the interval for a material standard is defined, and in the limitations of the subsidiary measuring equipment.

(a) Observational Uncertainties

Most working standards of length at the hierarchical level immediately below wavelength standards are either end standards such as gauge blocks and end bars, or line standards such as scales or tapes.

Calibration of high quality end standards, in which the interval of interest is defined as the separation of the mean planes of the end faces, is only rarely done with displacement measuring interferometers. High quality line standards, however, in which the interval is defined as the separation of two lines engraved on the standard, are normally calibrated using displacement type interferometers.

The majority of systems used for the calibration of line standards by laser interferometry take one of two forms. In the first, most frequently used for rigid standards, the scale being measured and the retroreflector defining one arm of the interferometer are supported on a carriage underneath an observing microscope (Fig.4). The measurement of length then consists of translation of the carriage by the distance between the two scale marks defining the interval of interest, the scale marks being set, in turn, at the focus of the microscope. In the second, most frequently used for flexible standards, the standard is fixed in space, and the observing microscope(s) together with the retroreflector(s) defining the arm(s) of the interferometer are translated from scale mark or reference mark to scale mark, with the focus of the microscopes being set in turn at the centre of the scale marks (Fig.5). In both cases, however, the achievable observational uncertainty is substantially determined by the ability of the fiducial system to recognize and repeat on the centre (or on some other predetermined and repeatable feature) of the scale mark or marks.

In addition to the division into rigid and flexible, line standards can also be divided into two main groups based on the width and quality of the scale marks. In one group, which includes geodetic tapes and draughtsman's scales, the scale marks are typically between 0.1 and 0.2 mm wide. Using optical microscopes with bifilar graticules and magnifications between X10 and X25, a skilled observer can recognize and set on the determining feature of the scale marks with a repeatability of about 1-5% of line width, i.e. between 1 and 10 μm, depending on the quality (uniformity and contrast) of the mark. In the second group, which includes H-section line standard scales and some precise stage micrometers, the scale marks are typically between 5 and 10 μm wide. Using optical microscopes with magnifications of between X50 and X200, settings with a repeatability of between 5 and 10% of line width, i.e. between 0.2 and 1 μm, can be achieved, again depending on the quality of the scale marks. Using a photoelectric microscope (Goldberg & Ritter 1964, Ciddor 1982) this can be improved to 0.1 to 0.2% of line width, i.e. 5 to 20 nm.

In some applications, however, the above method of estimating observational uncertainties is inadequate and resort must be made to more rigorous statistical methods of estimation. Following Campion *et al.* (1973) the mean and standard deviation of a measurement sequence can be assessed as follows:

If in a calibration sequence, the lengths of m intervals, each with its own mean value, $\bar{x}_i$ $(i = 1,m)$, were each measured n_i times, the estimate of the observational variance, S_o^2, of the sequence is given by

$$s_o^2 = s_{x_{ij}}^2 = \frac{\sum_{i=1}^{m}\sum_{j=1}^{n_i}(x_{ij} - \bar{x}_i)^2}{\sum_{i=1}^{m}\nu_i}$$

(where x_{ij} is the jth measurement of the length of the ith interval and $\nu_i = n_i - 1$ is the number of degrees of freedom for the ith population), provided that the measurements on the intervals were all made using the same observational procedure and under similar conditions.

Further where more than one measurement sequence has been made (say measurements on two or more scales) an estimate of the pooled observational variance can be made such that

$$s_o^2 = s_{x_{ijk}}^2 = \frac{\sum_{k=1}^{N}\sum_{i=1}^{m_k}\sum_{j=1}^{n_{ik}}(x_{ijk} - \bar{x}_{ik})^2}{\sum_{k=1}^{N}\sum_{i=1}^{m_k}\nu_{ik}}$$

where N is the number of observational sequences being pooled.

The estimated standard deviation for the mean of any population then is

$$s_{\bar{x}_{ik}} = s_o . n_{ik}^{-1/2}$$

and the uncertainty to be associated with the mean of any population is

$$s_{\bar{x}_{ik}} = t_{\nu_t, p} . s_{\bar{x}_{ik}}$$

where $t_{\nu_{t,p}}$ is the value of Student's t such that, for the number of degrees of freedom

$$\nu_t = \sum_{k=1}^{N}\sum_{i=1}^{m_k}\nu_{ik},$$

the probability that the true value of the mean lies outside the range $\pm\delta_{\bar{x}_{ik}}$ about $\bar{x}_{ik}$ is less than p.

All statements of uncertainty associated with the results of dimensional metrology carried out in this laboratory are based on $p = 0.01$; that is, there is less than one chance in one hundred that the measured value differs from the true value by more than the stated uncertainty; this convention has been adopted throughout this paper.

(b) Other Sources of Uncertainty

As noted above, the determination of the length of an interval of a line standard using a laser interferometer requires a number of procedures and measurements additional to the primary measurement of displacement. Each of these will have an inherent uncertainty which must, in turn, be included in the final estimate of the uncertainty in the answer. The following list includes some of the sources of uncertainty but is not exhaustive. To obtain a complete list for any given measurement a careful analysis of the measurement system and procedures must be undertaken.

(i) Sine error is due to angular movement of the observing microscope with change in load distribution as the scale or microscope is traversed. Given a sufficiently rigid system the movement of the microscope can be limited to less than 1 μrad (0.2 seconds) while the reference arm retroreflector mounted on the microscope can be within one retroreflector aperture diameter of the microscope focus, say, within 50 mm. The error on a metre movement will then be less than 5 in 10^8.

(ii) Alignment Errors. As can be seen in Fig.8 any misalignment between the axis of motion of the moving component and the scale axis and/or the measuring axis will lead to a cosine error. With careful equipment design each of these misalignments can be restricted to less than 0.1 mrad so that the total cosine error on a metre movement will then be less than 1 in 10^8. An additional alignment error can occur in a system in which the microscope axis is at an angle, β, to the normal to the axis of motion if the alignment between the scale axis and the axis of motion changes as the scale is translated. If the scale alignment changes by an angle α the resultant "scale tilt" error will be proportional to $\alpha\beta$ so that for $\alpha \simeq 1$ μrad and $\beta \simeq 0.01$ μrad the resultant error on a metre movement will be less than 1 in 10^8. (Additional details about scale tilt errors are to be found in Ciddor & Bruce 1967).

(iii) Thermometry. There are a number of sources of error in the determination of the temperature of the item being measured and in any subsequent referral of the measured length to 20°C. The most important of these are: the uncertainty in the calibration of the thermometers themselves; the uncertainty in reading of the thermometer; the uncertainty in the interpolation of the temperature interval using an assumed linear temperature gradient; the uncertainty in the interpolation to the temperature at the "mean instant" through the assumption of time symmetry and linear temperature drift; and the uncertainty in the applied length correction to 20°C due to any uncertainty in the value for the coefficient of linear thermal expansion. Each of these must be evaluated for the measurement in question as indicated in Table 3.

(iv) Refractive Index of Air. In the absence of a refractometer, refractive index corrections are usually obtained from measurements of air pressure, temperature and water vapour content and one of the standard formulae relating these to refractivity. As can be seen from equations (3) and (4), minor changes in pressure, temperature and humidity in the vicinity of standard conditions (760 mmHg, 20°C, 10 mmHg H_2O) lead to changes of about 0.36×10^{-6}/mmHg, 0.93×10^{-6}/°C and 0.056×10^{-6}/mmHg H_2O in the observed length as measured by laser interferometry at 633 nm. The uncertainties to be included in the overall uncertainty in the measured length, thus depends on the uncertainties in the calibration and use of the instruments used to determine air pressure, temperature and humidity. Typically, air pressure can be determined with an uncertainty of ±0.04 mmHg using calibrated aneroid or Fortin barometers, air temperature to ±0.01°C using calibrated thermistor or resistor elements and humidity to ±0.4 mmHg H_2O using a psychrometer of the Assmann type (Wylie & Lalas 1981). The resultant contributions to the uncertainty in the final result are then ±1.5, ±1.0 and ±2.2 × 10^{-8} respectively.

Although Edlen's expression does not explicitly include an allowance for the effects of atmospheric CO_2, some provision is made for this gas in that the equation (7a) is based on spectroscopic measurements made in 'standard' air, i.e. air containing 0.03% CO_2. If, however, the concentration of CO_2 varies from the assumed 0.03% an additional uncertainty of about ±5 × 10^{-8}/0.03% CO_2 may need to be included.

A commercial instrument for the direct optical measurement of refractivity is expected to become available shortly (Grace 1980). Whether this instrument is a viable alternative to the indirect method is not clear at this stage.

(v) Laser Frequency. From equation (5) it can be seen that the differential change in frequency with change in cavity length that can be expected in a 300 mm long TEM_{q00} single mode He-Ne laser is about 1.5 MHz/nm. It is not unreasonable, therefore, to expect that, except for the most highly stabilized laser systems, minor variations in the frequency of the laser radiation will occur. We have measured the frequencies of our two-frequency lasers and found that, while the frequency stability of these lasers is high, there is a variation of frequency around the mean of the type shown in Table 2.

It can be seen, therefore, that it is necessary to include a component of ±4 in 10^8 due to laser frequency variation in any final assessment of the uncertainty to be associated with a length measurement made using these lasers.

Table 2. Measured frequency stability of a Zeeman-split two-frequency laser

Interval	Root Allan Variance	Variation (p-p)	Relative Variation (p-p)
1 s	15-50 kHz	-	-
10 s	50-80 kHz	-	-
100 s	150-250 kHz	1 MHz	2 in 10^9
10 min	-	10 MHz	20 in 10^9
1 hour	-	20 MHz	40 in 10^9
Long term	-	40 MHz	80 in 10^9
Cold switch-on repeatability	-	20 MHz	40 in 10^9

(c) Combination of Uncertainties

There is, as yet, no internationally agreed or accepted rule for the combination of uncertainties where the individual sources of error may have differing probability distributions. The practice often adopted in this laboratory is to treat *all* uncertainties irrespective of origin as if their variances were derived from normally distributed populations and to combine them as a square root of the sum of squares, so that

$$\delta_t^2 = \sum_j \delta_j^2.$$

Table 3 shows, as an example, the results of such an evaluation and combination of errors for the calibration of a 1 m line standard. It must, however, be noted that this procedure may, for any given probability level for the individual uncertainties, lead to a value which is a slight underestimate of the true probability level for the combined uncertainty (Wagner 1969). Whether or not it would be profitable to use the more elaborate methods of combination outlined by Wagner (1969) or Ku (1966) must be evaluated with respect to the requirements of the individual experiment under review.

Recently the Working Group on the Statement of Uncertainties in their recommendation to the CIPM (Recommendation INC-1 1980, Giacomo 1981) suggested that the old classifications of "random" and "systematic" uncertainties should be dispensed with and replaced by "statistically" and "non-statistically" evaluated uncertainties which would be combined according to their variances and number of degrees of freedom. Whether this proposal is appropriate to the situation where probabilistic uncertainty estimates are required has yet to be evaluated. Further, the methods of implementation of the recommendation have yet to be specified.

Table 3. Estimates of the uncertainties to be combined to give the overall uncertainty in a measurement

Source	Magnitude	Proportional uncertainty in length
Observational Error (1 m interval)	±0.065 μm	6.5×10^{-8}
Sine error	1 μrad, 50 mm	5.0×10^{-8}
Alignment error	0.1 mrad/axis	1.0×10^{-8}
Scale tilt error	0.01 rad & 1 μrad	1.0×10^{-8}
Refractivity, pressure	0.04 mmHg	1.5×10^{-8}
temperature	0.010 °C	1.0×10^{-8}
humidity	0.4 mmHg	2.2×10^{-8}
Calibration of resistance thermometers	0.003 °C, 11×10^{-6}/°C	3.3×10^{-8}
Reading of thermometers	0.002 °C, 11×10^{-6}/°C	2.2×10^{-8}
Departure from linear thermal gradient	0.002 °C, 11×10^{-6}/°C	2.2×10^{-8}
Departure from time symmetry & linear drift	0.002 °C, 11×10^{-6}/°C	2.2×10^{-8}
Uncertainty in coefficient of thermal expansion	0.07×10^{-6}/°C, ±0.5 °C	3.5×10^{-8}
Compression of scale by air pressure	740-770 mmHg	0.7×10^{-8}
Distortion of scale by retro-reflector weight	2×90 g	0.5×10^{-8}
Uncertainty in laser frequency	40 MHz (p-p)	4.0×10^{-8}
Sum of squares "non-observational" errors		89.49×10^{-16}
Uncertainty due to "non-observational" errors		$\pm9.5\times10^{-8}$
Combined total uncertainty		$\pm11.5\times10^{-8}$

An analysis of this type would normally be carried out for any calibration system the results of which were to be reported internationally, or to which legal meaning was to be given; see for example Birch (1980) or Tanimura *et al.* (1980).

6. Case Studies

Lasers and laser interferometers are used widely throughout the Division of Applied Physics for the measurement of linear dimension, and some of these uses have been mentioned above. In particular, Hewlett Packard type 5526A laser interferometers are used for the calibration of precise line standards of length, both rigid scale and flexible tape types, and the systems used to perform these measurements will now be described.

(a) Case Study A: Rigid Line Standard Scales

The Division's line standards subdivisional comparator for the calibration of rigid line standard scales of between 0.1 and 1 m in length is shown schematically in Fig.4.

In this system the scale to be measured is supported on rollers at the designated points such as the Airy or Bessel points, within and in thermal contact with a copper box such that, although lightly constrained the scale is free to expand or contract with temperature changes. A mounting containing a retroreflector is clamped at one end while a counterweight is clamped to the other end.

The copper box, which is surrounded by an aluminium radiation shield, rests on two wooden blocks, 590 mm apart, which act as pivot and wedge for adjusting the scale surface with respect to the object plane of the observing microscope. In turn, the wooden blocks rest on rollers normal to the longitudinal axis of the scale and are constrained by adjustable spring loaded clamps, allowing both box and scale to be translated in the horizontal plane and to be pivoted about a vertical axis. These horizontal adjustment clamps are attached to the platform of a sliding air table which rides on a 3 m long cast iron bed. The air table and bed have been described by Sim *et al.* (1971, 1972) though the original cable and pulley system used to control the coarse positioning of the air table has since been replaced by a lead screw and nut. The air table also supports the inner of the two aluminium clad polyurethane-foam thermal shields, the outer shield being supported from the floor and surrounding the whole instrument. The laser, remote polarizing beam splitter and beam bender are supported from the 3 m cast iron bed, the arms of the interferometer being parallel to one another.

The microscope used in this comparator is an improved version of that described by Goldberg & Ritter (1964) and by Ciddor (1971). The microscope is supported on a rigid beam which in turn is kinematically supported on brackets attached to the main 3 m support bed. The reference beam cube corner retroreflector is clamped to the column containing the microscope objective with its pole about 50 mm above the object plane and about 40 mm from the optical axis of the microscope. As the differential movement of the microscope with respect to the carriage and bed due to change in load distribution when the carriage is traversed is less than 0.5 μrad, the sine error introduced by this positioning of the retroreflector is estimated as being less than

2.5 in 10^8 on a metre. Further by visual setting of the ruled surface of the scale both vertically and horizontally to the field of view of the microscope and its graticule, alignments of better than 50 μrad can be achieved. Similarly autoreflection and minimum sideways deflection techniques (see Fig.8b) are used to achieve an alignment of interferometer to mechanical axis to better than 100 μrad (see also Hewlett Packard 1971), the overall alignment error of the system contributing an estimated 1 in 10^8 on the length.

Scale temperatures are inferred from measurements of the resistances of two platinum thermometers in thermal contact with the copper box and in known position relative to the scale. The resistance of each thermometer is measured with an a.c. resistance bridge (Thompson & Small 1971). Given the uncertainty of calibration of the thermometers, the resolution of the thermometers, the degree of thermal contact between the thermometers and scale, the thermal inertia of the system, the observed temperature gradients and the observed rate of temperature drift, it is believed that the temperature of any interval of the scale is known at the time of measurement to within 0.005°C.

The refractive index of the air in the optical paths is inferred from measurements of air pressure, temperature and humidity using a calibrated aneroid barometer, a calibrated thermistor and an Assmann psychrometer respectively together with Edlen's formulae. It is believed that the uncertainty in the final measured length due to refractive index is less than 5 in 10^8.

Using this experimental system and a time symmetric observational sequence such as that outlined previously, and including reversal of the scale on the carriage to minimise any effect of asymmetry in the scale marks defining the interval, and rotation of the microscope through 180° to minimise any effect due to asymmetry of the microscope optical or mechanical systems, measurements of the metre interval of an H-section line standard have been made with an overall uncertainty of ±0.2 μm (2 in 10^7) based on four measurements in each of the four measurement sets. Similarly, measurements of the 1 m interval of a brass bevelled edge scale such as is used as a working standard by many of the weights and measures offices of the Australian States have been made with a final uncertainty of ±2 μm (±2 in 10^6) based on eight measurements only and without scale reversal or microscope rotation.

(b) Case Study B: Flexible Line Standards

The Division's geodetic base for the calibration of flexible line standards such as surveyors' tapes and wires of between 4 m and 50 m in length, the optical system of which is shown schematically in Fig.5, has already been referred to by Puttock (1982).

The optical and mechanical components of this geodetic base have been constructed on and are supported by a single free-standing reinforced-concrete beam approximately 2 m high, 0.6 m wide and 70 m long. Running along the entire length of the upper surface of this beam is a two-rail track made from 25 mm diameter chromium plated steel, and with a gauge of about 300 mm. The rail sections, approximately 3 m long, are supported on sleepers at about 0.5 m intervals and the system has been aligned optically such that the generator of the front,master rail lies within an (imaginary) cylinder of diameter 0.5 mm.

At one end of the concrete beam there is a fixed platform which supports a Hewlett Packard 5526A laser. The output light beam from this laser is divided into its two frequencies by a polarizing beam splitter and the two oppositely polarized beams are directed by beam steering mirrors into light paths parallel to the generator of the front rail. The reflecting components defining the measuring arms of the interferometer are two cube corner retroreflectors. These retroreflectors have been constructed from front surface aluminium-on-glass mirrors optically contacted together so that, in the course of reflections, the laser beams are folded up and out of the optical plane, thus straddling the survey tape; the two reflected beams are returned to the beam steering optics, are recombined at the beam splitter and the beat frequency detected by the photodiode in the laser head.

It should be noted that the two retroreflectors are of different sizes so that the outer laser beam passes around the smaller retroreflector to be reflected at the larger, while the inner beam is completely reflected at the smaller and does not reach the larger. As a result, the two arms of the interferometer are independent and movement of either retroreflector causes a Doppler shift in the frequency of the laser beam incident on that reflector with a consequent measurement of the relative displacements of the reflectors.

Associated with each retroreflector is a microscope mounted with its focus coincident with the virtual pole of the reflector. As a result, a tape scale mark brought into focus at the microscope is also at the pole of the reflector, thus minimizing the sine error that would otherwise be caused by any rotation of the observing system.

Each retroreflector-microscope pair is mounted on a carriage which is free to move along the track, the system being configured such that both pole/microscope focus pairs move along a common line parallel to the generator of the track. The tape under calibration is supported in the specified configuration (single, double or multiple catenary) under the specified tension by "friction free" rollers which are mounted on a double aluminium support track fixed to the front face of the concrete beam. The tape is also positioned so that the scale marks defining the interval of interest are in focus at the microscopes and are in a predetermined relationship (tape edge or centre) with the common axis of the retroreflector movements. Then by use of an appropriate sequence of carriage movements and microscope settings, first on the same scale mark and then on both scale marks, the separation of the scale marks defining the unknown interval can be determined.

During this sequence of movements the tape temperature and refractive index of the air path are also determined, the former from measurements of the air temperatures in close vicinity to, and at a number of points (say every 4 m) along, the tape and the latter from the air temperature measurements together with measurements of air pressure and humidity. Temperatures are measured with calibrated mercury-in-glass thermometers, while pressure and humidity are determined with a calibrated aneroid barometer and an Assmann psychrometer respectively. Based on the same criteria as for the measurements associated with line standard scales it is believed that the temperature of a tape interval at the time of measurement is known to within 0.1°C and that the uncertainty in the final measured length due to refractive index is less than 1 in 10^7.

Using this experimental system and a time symmetric measurement sequence of the form

"determine initial separation of microscopes,
{move microscopes} {temperatures, pressure, humidity}
{push
determine final separation of microscopes
<mean instant>
{pull
redetermine final separation of microscopes
{move microscopes} {temperatures, pressure, humidity}
redetermine initial separation of microscopes",

measurements of the "as positioned" length of the interval of a high quality invar tape ($\alpha \sim 1 \times 10^{-6}/°C$) can be made to within 5 in 10^7 on 50 m and on a high quality steel tape ($\alpha \sim 10 \times 10^{-6}/°C$) can be made to within 3 in 10^6 on 50 m; these values are based on four measurements of the interval. (The procedure known as "Push-Pull", in which the tape is moved slightly towards one or other end thus breaking the friction between the tape and the support rollers, is used to minimise the effect of friction on the applied tension.)

It should be noted that the results of the calibration of the length of an interval of a standard tape are normally reported assuming the tape was supported in a specified and exactly reproducible manner, under a particular and exact tension, at 20°C and at standard gravity, $g = 9.806\ 65\ ms^{-2}$. In practice, these conditions cannot normally be duplicated exactly and corrections must be applied to reduce the "as positioned" measurement of length to the length at standard conditions. Using the theory of tape suspension developed by Hotine (1939) it is usually possible to determine the required corrections with an associated uncertainty of about 10% of the actual correction. It is believed, therefore, that after the inclusion of these corrections and their uncertainties, it is possible to determine the length of a tape interval of an invar tape to within ±1 in 10^6 and of a steel tape to within ±5 in 10^6 on a 50 m interval.

7. Conclusion

Lasers and laser interferometry are now powerful techniques in dimensional and engineering metrology with uses right throughout these disciplines. Their impact, however, has been greatest on those applications requiring the highest precision, the standard of length and the calibration of the material standards which form the next level of the hierarchy of length, line standard scales and tapes. This paper has attempted to review the field at an introductory level with sufficient references to enable the reader to follow up any point. Some extra amplification, however, has been given to the pitfalls that exist and the uncertainties of measurement that can be introduced in any calibration of a line standard scale or tape.

Acknowledgements

The assistance and encouragement of Messrs. P.E. Ciddor, M.E. Purss, M.J. Puttock and P.J. Sim of this Division are gratefully acknowledged.

References

Allan, D.W. (1966). Proc. IEEE, 54, 221-230.

Axelrod, N.N. (1980). *Prod. Eng. (USA)* 27, 58-62.

Baer, T., Kowalski, F.V. & Hall, J.L. (1980). *Appl. Opt.* 19, 3173-3177.

Balasubramanian, N. & Wyant, J.C. (Eds.) (1978). 'Advances in Optical Metrology', Proc. S.P.I.E. 153, pp.166.

Baldwin, R.B., Gordon, G.B. & Rude, A.F. (1971). *Hewlett-Packard J.* 23(4), 14-20.

Balhorn, R., Kunzmann, H. & Lebowsky, F. (1972). *Appl. Opt.* 11, 742-744.

Barnes, J.A., Chi, A.R., Cutler, L.S. *et al.* (1971). IEEE Trans.: Instrumentation & Measurements, IM20, 105-120.

Barrell, H. (1957). 'The Bases of Measurement', 1957 Sir Alfred Herbert Paper, Institution of Production Engineers Journal, February 1958, pp. 36.

Barrell, H. & Sears, J.E. (1939). *Phil.Trans.Roy.Soc.*, A258, 1-64.

Beavis, M. (1982). 'Liquid in Glass Thermometry: Care and Use in Measurement', CSIRO Division of Applied Physics Technical Paper No.6, Sydney.

Bennett, S.J. (1974). *Survey Review* 22(172), 270-275.

Birch, K.G. (1980). NPL Report MOM 48, pp.25.

Blaney, T.G., Bradley, C.C., Edwards, G.J., Jolliffe, B.W., Knight, D.J.E. & Woods, P.T. (1975). *Nature* 254, 584.

Blaney, T.G., Edwards, G.J., Jolliffe, B.W., Knight, D.J.E., & Woods, P.T. (1976). *J. Phys. D.: Appl.Phys.* 9, 1323-1330.

Brown, N. & Ciddor, P.E. (1980). *Aust. Physicist* 17, 42-44.

Burgwald, G.M. & Kruger, W.P. (1970). *Hewlett-Packard J.* 21(12), 14-16.

Campion, P.J., Burns, J.E. & Williams, A. (1973). 'A Code of Practice for the Detailed Statement of Accuracy', H.M.S.O., London.

CCDM (1973). Com.Int.Poids Mes., Proc.Verb.Com.Consult.Defin. du Metre, 5th Session (Paris:Offilib.), M14-16.

Ciddor, P.E. (1971). *Metrologia* 7, 135-142.

Ciddor, P.E. (1982). 'Photoelectric Setting Microscopes' in *Advances in Optical and Electron Microscopy* 9 (in press).

Ciddor, P.E. & Bruce, C.F. (1967). *Metrologia* 3, 109-118.

Cullen, D.L. (1978). SME Technical Papers (Inspection & Quality Control), IQ78-346.

Culshaw, W. & Kannelaud, J. (1964). *Phys. Rev. A.* 136, 1209-1221.

Desai, J.N., Chandrasekhar, T. & Madhavan, R. (1979). *J. Phys. E.: Sci. Instrum.* 12, 1040-1042.

Dukes, J.N. & Gordon, G.B. (1970). *Hewlett-Packard J.* 21(12), 2-8.

Edlen, B. (1966). *Metrologia* 2, 71-80.

Einstein, A. (1917). *Phys.Zeit.* 18, 121-128.

Evanson, K.M., Day, G.W., Wells, J.S. & Mullen, L.O. (1972). *Appl.Phys.Lett.* 20, 133.

Evanson, K.M., Wells, J.S., Petersen, F.R., Danielson, B.L. & Day, G.W. (1973). *Appl.Phys.Lett.* 22, 192-195.

Fallon, W. (1971). *Instrum. Prac.*, July, 400-405.

Fox, A.G. & Li, T., (1961). *Bell Syst. Tech. J.* 40, 453-488.

Gagliano, F.P., Lumley, R.M. & Watkins, L.S. (1969). Proc. IEEE 57 (2), 114-147.

Giacomo, P. (1981). *Metrologia* 17(2), 73-74.

Giacomo, P., Hamon, J., Hostache, J. & Carré, P. (1972). *Metrologia* 8, 72-82.

Gillilland, K.E., Cook, H.D., Mielenz, K.D. & Stephens, R.B. (1966). *Metrologia* 2, 95-98.

Goldberg, J.L. & Ritter, H.J. (1964). *Microtecnic* 18, 117-124.

Grace, W.H. (1980). 'Advances in Laser Engineering and Applications', Proc. S.P.I.E. 247, 192-199.

Hall, J.L. (1978). *Science* 202, 147-156.

Heavens, O.S. (1964). 'Optical Masers', Methuen & Co. Ltd, London.

Herzberg, G. (1944). 'Atomic Spectra and Atomic Structure', Dover, New York.

Hewlett-Packard (1971). 'Operator Handbook, Laser Systems Manuals', Santa Clara, p.11.

Hotine, M. (1939). *Empire Survey Review* 5, 2-36.

Javan, A., Bennett, W.R. & Herriott, D.R. (1961). *Phys.Rev.Lett.* 6, 106-110.

Kemp, W.R.G. (1977). 'International Practical Temperature Scale of 1968 (Amended Edition of 1975)', National Measurement Laboratory, CSIRO, Sydney.

Kogelnik, H. & Li, T. (1966). *Appl. Opt.* 5, 1550-1567.

Ku, H.H. (1966). *J.Res.NBS* 70C(4), 263-273.

Ladenburg, R. & Levy, S. (1934). *Z. Physik* 88, 461-468.

Liu, L.S. & Klinger, J.H. (1979). 'Interferometry' (G.W. Hopkins Ed.), Proc. S.P.I.E. 192, 17-26.

Loewen, E.G. & Burns, R.H. (1971). Annals of the C.I.R.P. 19, 597-602.

Miller, J.R. III (1970). *Instrum.Tech.*, Dec., 56-60.

Murty, M.V.R.K. (1960a). *J.Opt.Soc.Am.* 50, 7-10.

Murty, M.V.R.K. (1960b). *J.Opt.Soc.Am.* 50, 83-84.

Owens, J.C. (1967). *Appl. Opt.* 6, 51-59.

Peck, E.R. (1948). *J. Opt. Soc. Am.* 38, 1015-1024.

Popela, B. (1972). *Optica Acta* 19, 605-612.

Puttock, M.J. (1969). 'Notes on Engineering Metrology', William Brooks & Co., Sydney.

Puttock, M.J. (1982). These proceedings.

Rowley, W.R.C. & Wilson, D.C. (1968). Proc.Instn.Mech.Engrs. 183(3D), 29-33.

Schawlow, A.L. & Townes, C.H. (1958). *Phys.Rev.* 112, 1940-1949.

Schulze, R. (1967). 'Optical Instruments', Handbuch der Physik XXIX, 804-807, Springer-Verlag, Berlin.

Siegman, A.E. (1971). 'An Introduction to Lasers and Masers', McGraw Hill Book Co., New York.

Sim, P.J., Ross, H.A. & Taylor, R.J. (1971). 'Harold Armstrong Conference on Production Science in Industry', Melbourne, 57-68.

Sim, P.J., Ross, H.A. & Taylor, R.J. (1972). Mech. & Chem. Eng. Trans., Inst. Eng. Australia MC8(1), 79-83.

Sirohi, R.S., Nayyar, V.P. & Virdi, S.P.S. (1971). *J.Phys.D.:Appl.Phys.* 4, 485-488.

Steel, W.H. (1967). 'Interferometry', Cambridge University Press.

Stroke, G.W. (1965). *J. Opt. Soc. Am.* 55, 330-331.

Sydenham, P.H. (1972). *J.Phys.E.: Sci. Instrum.* 5, 721-733.

Tango, W.J. & Twiss, R.Q. (1974). *Appl. Opt.* 13, 1814-1819.

Tanimura, Y. & Yamamoto, A. (1972). *Bull. Japan Soc. of Prec. Engg.* 6, 45-50.

Tanimura, Y., Fujinuma, N. & Sakurai, Y. (1980). *Bull. N.R.L.M.*, Series 42, July, 1-8.

Terrien, J. (1965). *Metrologia* 1, 80-83.

Terrien, J. (1976). *Rep. Prog. Phys.* 39, 1067-1108.

Thompson, A.M. & Small, G.W. (1971). Proc. IEE 118 , 1662-1666.
Thoneman, F.F. (1968). 'Fundamental Laser Physics' in 'Lasers and Plasma Physics', 3rd AIP Summer School, Adelaide.
Thwaite, E.G. (1965). *Aust. J. Phys.* 18, 401-408.
Thwaite, E.G. & Bendeli, A. (1980a). International Conf. on Mfg.Eng., Melbourne, 25-27 August 1980, 393-396.
Thwaite, E.G. & Bendeli, A. (1980b). *Australian Electrical World*, Oct., 43-44.
Thwaite, E.G. & King, H. (1982). *Prec. Eng.* 4, 15-18.
Thwaite, E.G. & Leslie, R.T. (1963). *Aust. J. Phys.* 16, 82-107.
Tolman, R.C. (1924). *Phys. Rev.* 23, 693-709.
Troup, G. (1963). 'Masers and Lasers', 2nd Ed., Methuen & Co. Ltd, London.
Troup, G. (1967). 'Optical Coherence Theory - Recent Developments', Methuen & Co. Ltd, London.
Turner, R. & Pfitzer, E.K. (1970). *Metrologia* 6, 94-97.
Wallard, A.J. (1973). *J. Phys. E.: Sci. Instrum.* 6, 793-807.
Wagner, S. (1968). *PTB Mitteilungen* 79(5), 343-347.
White, A.D. & Rigden, J.D. (1962). Proc. I.R.E. 50, 1697.
Wylie, R.G. & Lalas, T. (1981). 'The WMO Psychrometer', CSIRO Division of Applied Physics Technical Paper No.3, Sydney.

APPENDIX

A. Introduction to the Theory of Lasers

As noted previously, it is the properties of high brightness, monochromaticity and coherence of the radiation from a laser which lead to the usefulness of lasers in length interferometry. To see how these properties arise it is necessary to consider the theory of laser action. This discussion is, of necessity, very simplified; there are a number of more detailed discussions available in the literature (see, for example, Heavens 1964, Troup 1963, Siegman 1971 or Thoneman 1968).

The first demonstration of a working laser (in about 1960) lagged by some forty years the description by Einstein (1917) of the working principles, and, indeed, had Ladenburg & Levy (1934) enclosed the active medium in which they demonstrated the existence of inverted populations and stimulated emission (Tolman 1924) in a neon gas discharge in an optical resonator, as did Schawlow and Townes in 1958, a continuous wave gas laser may have been working in 1935.

(a) The Einstein Equations

The process of stimulated emission, the radiative de-excitation of an atom (ion, molecule) in the presence of radiation, is generally discussed by reference to an atom having two electronic levels of the same degeneracy of energy E_m and E_n ($E_n > E_m$) between which radiative transitions of energy $h\nu$ (= $E_n - E_m$) are allowed. For an electron in the upper state the probability of spontaneous de-excitation $n \to m$ in a time interval dt is $dP_{nm} = A_{nm}dt$ where A_{nm} is constant.

Similarly, the probabilities of stimulated excitation $m \rightarrow n$ and de-excitation $n \rightarrow m$ in the presence of a radiation field of energy density per unit frequency $\rho(\nu)d\nu$ depend on $\rho(\nu)$ and are

$$dP_{nm} = \rho(\nu)\ B_{nm}dt \qquad \text{transition } n \rightarrow m$$

and

$$dP_{mn} = \rho(\nu)\ B_{mn}dt \qquad \text{transition } m \rightarrow n,$$

where B_{nm} and B_{mn} are constants.
Under conditions of equilibrium the number of transitions $m \rightarrow n$ equals the number of transitions $n \rightarrow m$, so that

$$\rho(\nu)B_{mn}\ n_m dt = A_{nm}\ n_n dt + \rho(\nu)B_{nm}\ n_n dt, \tag{A1}$$

the relation between the probability of transition and the number of transitions being dependent on the number of electrons in the initial state of the transition.

For a system in thermal equilibrium with a radiation field the distribution of electrons in any state follows a Boltzman Law, $n = n_o \exp(-E/kT)$, while the energy density of the radiation field is given by Planck's Law

$$\rho(\nu)d\nu = \frac{8\pi\nu^3}{c^3} \cdot \frac{1}{\exp(h\nu/kT)-1}\, d\nu. \tag{A2}$$

Thus, (A1) becomes

$$\rho(\nu)B_{mn}\exp(\frac{-E_m}{kT}) = (A_{nm} + \rho(\nu)B_{nm})\exp(\frac{-E_n}{kT})$$

or

$$\rho(\nu) = \frac{A_{nm}}{B_{mn}\exp(h\nu/kT) - B_{nm}} \cdot \tag{A3}$$

Comparing (A2) and (A3) shows that $B_{mn} = B_{nm}$ and that $A_{nm} = (8\pi h\nu^3/c^3)B_{nm}$. A_{nm} and B_{nm} are known as the Einstein coefficients of spontaneous and induced (stimulated) emission respectively.

For a system in thermal equilibrium with a radiation field the ratio of spontaneous to stimulated emission is given by

$$R = \frac{A_{nm}}{\rho(\nu)B_{nm}} = \exp(\frac{h\nu}{kT}) - 1.$$

For $\nu = 4.74 \times 10^{14}$ Hz ($\equiv$ 633 nm) and T = 300 K, $R \sim \exp(76) \sim 10^{33}$. Thus, for conditions of thermal equilibrium the proportion of stimulated emissions is negligible for transitions in the visible region.

Equality of the rates of stimulated and spontaneous emission for thermal equilibrium occurs at a frequency of $\nu = kT\ln 2/h = 4.33 \times 10^{12}$ Hz ($\sim$ 70 μm), or for $\nu = 4.74 \times 10^{14}$ Hz, at $T = h\nu/k\ln 2 \sim 33\ 000$ K.

Clearly, therefore, if a device is to utilize the phenomenon of stimulated emission in the optical region, some means must be found to enhance greatly the number of stimulated emissions relative to the spontaneous ones. It can be seen in equation (A1) that the instantaneous power emitted by an assembly of excited emitters is

$$\frac{dW}{dt} = h\nu_{nm}\rho(\nu_{nm})B_{nm}(n_n - n_m) + h\nu_{nm}\ n_n\ A_{nm} \tag{A4}$$

and thus that net stimulated power out can only be achieved when $n_n > n_m$. Since under equilibrium conditions the population of a lower state always exceeds that of a higher one, the condition $n_n > n_m$ is described as a *population inversion*. Such an inversion is also known as a *negative temperature* since if the Boltzman relation is considered $n_n > n_m$ implies $T < 0$. It must be noted that a population inversion or negative temperature implies a *non-equilibrium* condition.

(b) Attainment of Inverted Populations

As noted above, the rate of stimulated emission will exceed that of spontaneous emission and that of absorption only when $n_n > n_m$. In practice a number of methods have been devised for achieving such a population inversion, the processes being known generally as *pumping*. For the purposes of this discussion we will consider the pumping method used for the He-Ne laser. From the energy level or *Grotrian* diagrams of He and Ne (Fig.A1) it can be seen that if the metastable 2^1S_o and 2^3S_1 levels of He can be excited, *superelastic* collisions between the excited He^m atoms and ground state neon atoms should lead to the excitation of the $3s_2$ and $2s_2$ levels of neon respectively. (Such collisions are sometimes called 'collisions of the second kind').

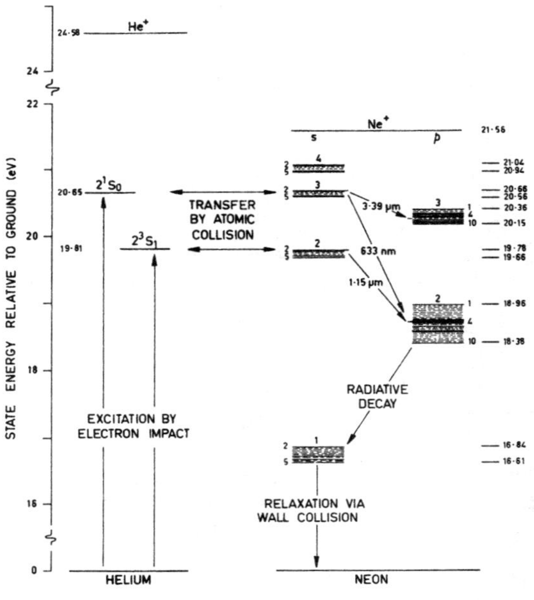

Fig.A1. Energy level diagram for helium and neon showing the important energy levels for the He-Ne laser, and the observed laser transitions (after Javan *et al.* 1961).

These s levels in neon can then decay or be stimulated to decay to a lower state of excitation with this lower level of the transition generally decaying radiatively to the 1s state with a short time constant (for example, the 3p states decay to the 1s state by radiating at about 360 nm). The 1s state in turn decays to the ground state, thereby completing the chain. Therefore by creating a high population of He metastables in a He-Ne gas mixture, pumping of the upper states of the lasing transition in neon can be achieved with a population inversion.

(This pumping procedure was first implemented by Javan *et al.* (1961) for laser action for the 1.15 μm transition and by White & Rigden for the 633 nm transition. Since then laser action has been achieved for the 3.39 μm and other transitions as well.)

In practice then, laser activity in a He-Ne system is obtained by initiating and maintaining an electric discharge (either dc or rf) in a He-Ne mixture (typically 90% He and 10% Ne). In such a discharge excitation of the He metastable states occurs via electron impact (as does excitation of many other excited states including the continuum, ionization, state) while the gas pressure (1-5 mmHg) is such that collisional de-excitation of the He^m (including the superelastic excitation of the upper states of the Ne lasing transitions) predominates over de-excitation via wall collisions. The discharge is usually run inside a capillary tube and at relatively low currents (10-50 mA) so as to optimise the de-excitation rate of the Ne 1s states; these de-excitations occur via wall collisions. For further details of the practical systems used, including a discussion of the rate equations, see Siegman (1971).

(c) Line Widths, Doppler Broadening and Gain Curves

If the energy levels n and m were perfectly sharp, the emitted radiation from a transition would have a frequency exactly ν_{nm}. These levels, however, are never perfectly sharp, being broadened by many causes. For example, from the Heisenberg uncertainty principle relating the spread of energy in an excited state, $\Delta E = h\Delta\nu_{nm}$ and the average lifetime spent by an atom in the upper state, Δt

$$\Delta E.\Delta t \simeq h/2\pi$$

we obtain

$$\Delta\nu_{nm} \simeq 1/\Delta t.2\pi .$$

Yet, as pointed out above, the average time spent by an electron in an excited state is directly related to the spontaneous transition probability, $dP_{nm} = A_{nm}dt$, so that the mean lifetime of an atom in an excited state is related to the $A_{nm}(=1/dt)$ so that

$$\Delta\nu_{nm} \simeq A_{nm}/2$$

the line shape having a Lorentzian form,

$$\frac{A_{nm}}{2\pi^2} \cdot \frac{1}{(\nu-\nu_{nm})^2 + (A_{nm}^2/4\pi^2)} .$$

As has been noted previously, laser action can occur when an inverted population interacts with a radiation field, and such an inverted population can be obtained by pumping as in a He-Ne discharge. In such a discharge, however, the emitting atoms are not stationary, but are in rapid motion so that the apparent resonance frequency of the atom is *Doppler* shifted; if the speed of the atoms in the direction of observation is, say, $v_x(\ll c)$, then the observed frequency will be

$$\nu_{obs} = \nu_{nm} + \delta\nu_{nm} \qquad \text{where} \qquad \delta\nu_{nm} = \frac{v_x}{c}\,\nu_{nm},$$

(c is the velocity of light).

In a gas discharge based laser the distribution of atomic (and molecular) velocities is governed by the Maxwell-Boltzmann equation, such that the number of atoms dN with an x component of velocity between v_x and $v_x + dv_x$ is given by

$$dN(v_x) = N_o \left(\frac{M}{2\pi kT}\right)^{\frac{1}{2}} \exp(-Mv_x^2/kT)\cdot dv_x$$

where N_o is the total number of atoms in the assembly, M is the mass of the emitting or absorbing atom, k is Boltzmann's Constant and T is the temperature associated with the kinetic motion of the gas atoms.

It can be shown (Siegman 1971) that the Doppler broadened line has the form

$$(\pi \ln 2)^{\frac{1}{2}} \frac{A_{nm}}{2\pi\,\Delta\nu_d} \exp\left[-2\ln 2\left(\frac{\nu-\nu_{nm}}{\Delta\nu_d}\right)^2\right]$$

where $\Delta\nu_d = (8kT\ln 2/Mc^2)^{\frac{1}{2}}\,\nu_{nm}$ is the full line width between the half power points.

For the He-Ne laser oscillating on the 633 nm transition with a gas temperature (elevated by the energy losses in the electric discharge) of about 400 K the Doppler width is about 1500 MHz, $(\Delta\nu_d/\nu_{nm} \sim 3\times10^{-6})$, which is about 10 times larger than the lifetime broadening discussed above, and about 5 times larger than the other main form of broadening, collisional broadening (Siegman 1971).

Therefore, laser oscillation can occur at any frequency within the broadened line provided that there are enough excited atoms with the correct velocity and a sufficiently high radiation density at that frequency. For this reason, the broadened line is also known as the *gain curve*.

(d) Amplification in Optical Resonators

As noted previously (Equation A4) both stimulated and spontaneous emission contribute to the net instantaneous power emitted by an assembly of excited atoms, the output power due to stimulated emission being proportional to the Einstein coefficient, B_{nm}, the radiation density, $\rho(\nu_{nm})$, and the differential number density of excited states, $(n_n - n_m)$, while the ratio between stimulated and spontaneous emission is given by $R = \rho(\nu_{nm})B_{nm}/A_{nm}$. Therefore, for any given laser, the output power due to stimulated emission can be increased independently of that for spontaneous emission only by increasing the radiation density. In practice, this increase in the radiation density is often achieved by enclosing the active medium in a resonant optical cavity. These cavity resonators usually consist of two flat or curved mirrors set up facing one another so that the optical wave bounces backwards and forwards between the two reflectors. Resonators of the Fabry-Perot (standing wave) or Mach-Zehnder (travelling wave) form have low loss paraxial modes which can be exactly calculated. The low-loss mode of a Fabry-Perot resonator defines a distribution of light energy which preserves its shape as it bounces between the two end mirrors. Such low loss modes exist for various combinations of mirror curvatures, R, and cavity lengths, L, (Fox & Li 1966).

For a confocal resonator, (R=L) the frequency spectrum of the resonator has the form

$$f_{qnm} = [q + (n + m + 1)/2] \frac{c'}{2L} \tag{A5}$$

where q is the number of half wavelengths between the mirrors, n is the azimuthal mode number (half the number of angular zeros), and m is the radial mode number (half the number of radial zeros), and m and n are not to be confused with the upper and lower states of the lasing transition. c' is the velocity of light in the active medium at the frequency of the mode.

The frequency spectrum of a confocal resonator will, therefore, contain a number of *axial* modes as shown in Fig.A2(a) with spacing c'/2L with *transverse* modes spaced c'/4L apart.

For a near-planar resonator with two identical and nearly flat end mirrors the resonant spectrum is given by

$$f_{qnm} = [q + (n + m + 1)\frac{2L}{\pi R}]\frac{c'}{2L} \tag{A6}$$

and the frequency spectrum of the resonator will then contain *transverse* modes separated by c'/πR while the axial modes are still separated by c'/2L as shown in Fig.A2(b).

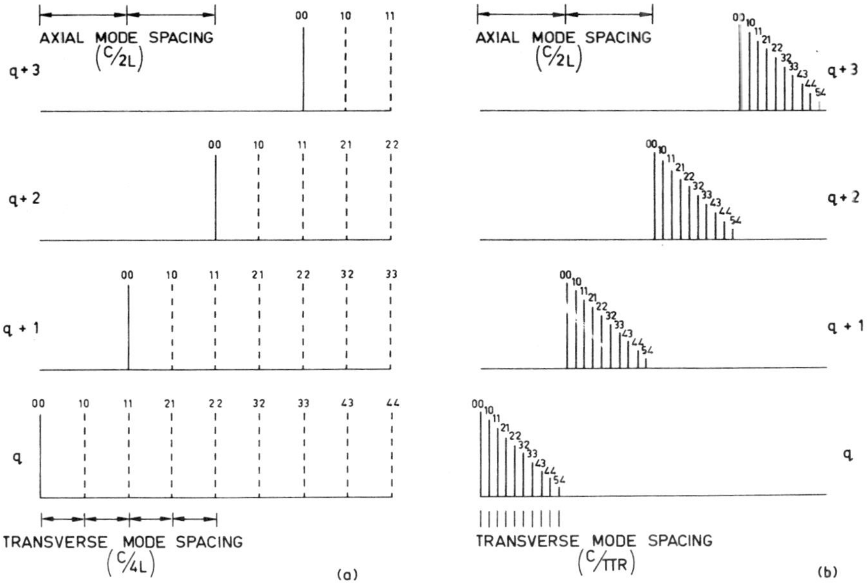

Fig.A2. Axial and transverse modes in an optical resonator, a) Confocal resonator, b) near planar resonator. (After Siegman 1971).

For further information on optical resonators the reader is referred to Siegman (1971). Details of the computer techniques used to calculate resonator modes can be found in the paper by Fox & Li (1961).

The output spectrum of a laser oscillating in an optical resonator can then be obtained from the convolution of the frequency spectrum of the resonator with the Doppler broadened gain curve of the active medium. A typical spectrum for a long laser is shown in Fig.A3.

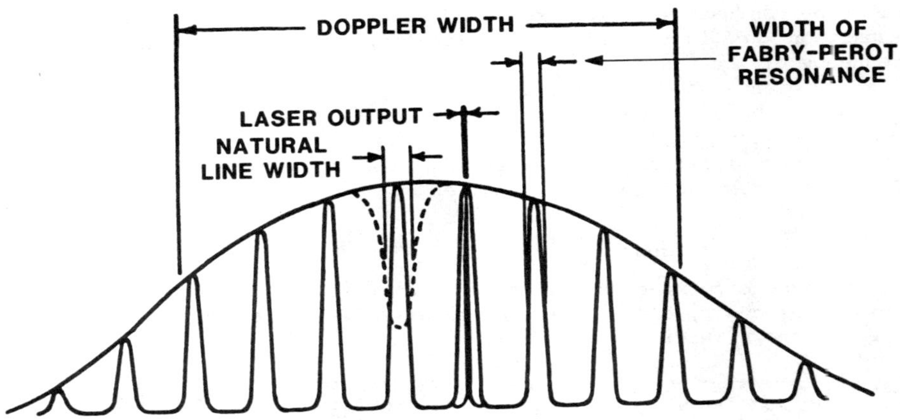

Fig.A3. Cavity and Doppler line widths in a long He-Ne laser system. (After Heavens 1964).

As can be seen in Fig.A3, the result of placing the active medium in an optical resonator is that the output of the laser is confined to a number of axial and transverse modes. The power density $\rho(\nu)$ is, therefore, greatly magnified at the frequencies of the modes so that stimulated emission predominates over spontaneous emission and the system can lase at the frequencies of the modes. In addition, it can be seen that by appropriate selection of L and R single or multimode output can be chosen and the output frequency can be tuned across the gain curve.

One other effect that arises from enclosing the active medium in an optical resonator is that in many cases the output beam from the laser has a *Gaussian* power distribution, and as a result the properties of Gaussian beams must be taken into account when designing any optical system for the laser. For further details see Kogelnik & Li (1966).

(e) Coherence

Coherence is, regretably, much more difficult to describe than the other aspects of lasers, and the reader is referred to Troup (1967) for a more detailed discussion.

However, a very simplistic view can be obtained by considering the interaction of a photon, represented as a wave function, with an excited atom. After the interaction the wave mechanical description must contain, if stimulated emission has occurred, a different excited state or an unexcited state and two photons. Analysis of the interaction process shows that the two photons must be identical in frequency and phase. We have, therefore, a situation that at any point in the active medium the photons of a given frequency all have the same phase, and that this *coherence* is preserved in the output laser beam.

In the above, the essential features of laser action have been sketched semi-quantitatively, a more rigorous treatment of any of the topics touched on being inappropriate in this discussion. The equations, themselves, are valid as a first approximation and lend themselves to refinement.

B. Stabilization of Lasers

The output frequency (or frequencies) of a laser can be tuned through the Doppler broadened gain curve by appropriate choice of the optical resonator parameters. The reverse, however, is also true: the frequency of the laser will change if there are changes in the parameters of the optical resonator. Because of this, and because of other limits to frequency reproducibility such as pressure broadening, frequency (mode) pulling and isotope effects (Ne^{20} or Ne^{22}), it is necessary to stabilize the laser output frequency of a laser used in metrology if uncertainties better than 1 in 10^6 are required.

Reduced to its fundamentals, frequency stabilization of a laser is obtained by servo control of the separation of the mirrors defining the optical resonator, the refinements being associated with the manner in which the feedback or control signal is obtained. Many methods have been devised for obtaining the stabilizing control signal, of which Lamb dip stabilization, polarization stabilization and saturated absorption stabilization are among the more important.

The Allan variance (Allan 1966, Barnes *et al.* 1971)

$$\langle\sigma^2(2,\tau)\rangle = \frac{1}{2N}\sum_i^N(\nu_i - \nu_{i-1})^2, \qquad \text{(A7)}$$

or its square root, is often quoted to indicate the stability of a laser. In this equation (A7), τ is the time over which each sample of frequency is averaged, N is the number of pairs of observations, and ν_i and ν_{i-1} are the measured frequencies of the ith sample pairs. Typically, for $\tau \leqslant 10$ s, 100 measurement pairs are required to achieve a satisfactory confidence level, while Allan variances with $\tau > 10$ s are normally computed using fewer sample pairs.

(a) Lamb Dip Stabilization

In a short laser oscillating in a single TEM_{00} mode of frequency $\nu = qc'/2L$ the resonant standing wave comprising the mode in the cavity can be regarded as being made up of two travelling waves propagating in opposite directions along the cavity. Each of these components will interact with that group of excited atoms which has a longitudinal velocity component such that the natural transition frequency is Doppler shifted to the mode frequency. The resultant stimulated emission will augment the field of the travelling wave and deplete the population of excited atoms in the group. Because of the two field components with their opposite directions of propagation two population groups will be depleted, these groups being symmetrically placed about the centre of the gain curve. This phenomenon is known as *'hole burning'*.

When the emitted frequency coincides with the natural frequency (centre of the gain curve) the population being depleted is that group with negligible longitudinal velocity and, as both field components are contributing to the depletion of the same group, the number of emitting atoms is reduced more strongly and a sharp minimum is superimposed on the gain curve. This minimum is known as the *'Lamb Dip'*.

If a small ac signal is applied to the servo system controlling the cavity length, the output of the laser will change frequency in response to the input signal. The amplitude and phase of the output frequency change depends on where on the gain curve the laser is oscillating. At the centre of the Lamb dip, however, the modulation of the laser output at the input frequency vanishes and the error signal to control the cavity length can be derived by standard Phase Sensitive Detection techniques. For further details on Lamb dip frequency control of lasers see the review by Wallard (1973).

(b) Polarization Stabilization

It is, of course, not feasible to use Lamb dip stabilization for lasers for which the Lamb dip is blurred out by, say, isotope effects or for lasers which are so long that laser action can occur on more than one mode. It is in the latter case that polarization stabilization has sometimes been used.

In a polarization stabilized laser the resonant cavity is configured so that laser oscillation occurs for two axial modes; the modes separated by $\Delta\nu = c'/2L$. In such a situation each axial mode is usually orthogonally polarized to the adjacent mode and has a fixed direction of polarization. Therefore by separating the two modes with a polarization divider such as a Wollaston prism (Balhorn *et al.* 1972) or with a rotating polarizer (Desai *et al.* 1979) an error signal can be derived from the difference in the output powers in the two modes, and this error signal can be used to control the length of the cavity.

(c) Saturated Absorption Stabilization

The stability of He-Ne lasers stabilized by Lamb dip methods has been variously reported as being between 1 in 10^8 and 2 in 10^7 (Terrien 1976) and similar results have been ascribed to polarization based systems. While stabilities of this order are clearly adequate for most metrological purposes, far greater stability is required before the laser radiation could be used for length or frequency standards. To achieve the stabilities required for these applications an alternative stabilization technique is required and that most frequently used is saturated absorption.

In a laser system stabilized by saturated absorption, a cell containing an absorbing gas or vapour (such as methane or iodine) is included inside the optical cavity but separate from the tube containing the lasing medium. The cavity length is then adjusted until the frequency of the radiation emitted from the laser coincides with an absorption line of the gas. The power of the radiation emitted from the laser, which depends on the relative population of emitting atoms as described by the gain curve, is then decreased by absorption in the absorbing gas. As in the case of the Lamb dip system the effect of this absorption is to deplete the two populations of absorbers with which the two travelling waves comprising the overall standing wave in the cavity are interacting, these groups being symmetrically placed about the center of the Doppler broadened absorbing line. If, however, the laser frequency coincides with the absorbing frequency of those molecules having a zero component of velocity in the direction of the laser beam, both travelling wave components deplete the same population group with the result that the proportion of the power absorbed from the laser beam shows a sharp dip and the emitted power shows a sharp peak at the frequency of the unbroadened absorbing transition superimposed on the overall depressed gain curve. This feature is sometimes called the *inverse Lamb dip*.

If then, in a manner analogous to that described above under Lamb dip stabilization, a small ac perturbing signal is applied to the servo system controlling the cavity length, an error control signal can be derived from the output of the PSD; in this case, the null occurs at the maximum in the inverse Lamb dip. An alternative method of obtaining the error control signal relies on the zero crossing of the third derivative of the laser response to the applied perturbation (Wallard, 1973) and this method is frequently preferred as it minimises any effect of the slope of the background absorption.

Saturated absorption has been used to stabilize the He-Ne lasers at 3.39 μm against methane and at 633 nm against iodine, as well as a number of other laser types, such as the 514 nm argon ion laser and the 564 nm mercury laser against iodine, and the He-Xe laser at 3.5 μm against formaldehyde (Hall 1978). For He-Ne lasers frequency stabilities ranging from 1 in 10^{11} to 3 in 10^{12} have been reported while reproducibilities of between 1 in 10^{10} and 1 in 10^{11} have been achieved (Brown & Ciddor 1980). Absolute determinations of wavelength, however, have been limited to 4 in 10^9 by the uncertainty in the ^{86}Kr lamp implementation of the current length standard.

The Measurement of Surface Roughness

E.G. Thwaite
CSIRO Division of Applied Physics, Sydney, Australia 2070

Abstract

It is necessary to accurately measure the roughness of surfaces over a wide range of values for many reasons. Specialised stylus profilometers, which must have a number of specified characteristics if consistent results are to be obtained, are principally used for making these measurements and they require calibration in terms of the unit of length. The characteristics of stylus instruments and their calibration are discussed with extensive reference to the Australian Standard for Surface Texture, AS 2536-1982. This Australian standard or an equivalent standard will need to be read in conjunction with this paper.

1. Introduction

We are all aware that the state of roughness of a surface influences a wide variety of its physical properties. Often quoted examples are: friction and wear, electrical and thermal contact resistance, corrosion resistance, sterile properties, and radiation absorption properties.

In many industrial processes it is necessary to produce surfaces with roughnesses that fall within specified limits and which are of a certain type. In constructing explanations for the behaviour of surfaces and testing the conclusions it is necessary to be able to measure their roughness in a consistent way.

There are then two broad reasons for wanting to measure surface roughness:

i) to provide quantitative data to explain functional behaviour;
ii) as a means of process control.

The types of surfaces that require measurement cover a wide spectrum, from the smooth highly polished surfaces used in optics (laser mirrors are a good example) to surfaces produced by conventional machining techniques such as planing.

The range of magnitudes of structures on surfaces is a continuum in their heights and their distribution over the surface, from heights and spacings measured in millimetres to tenths of nanometres.

When we talk about the 'topography of a surface' we mean the distribution of heights across it. Microgeometry and roughness are also used as substitute terms for topography.

A 'surface' is the transition region between the bulk properties of one material and those of another. The location of a surface depends on the means used to detect it; there is an operational definition

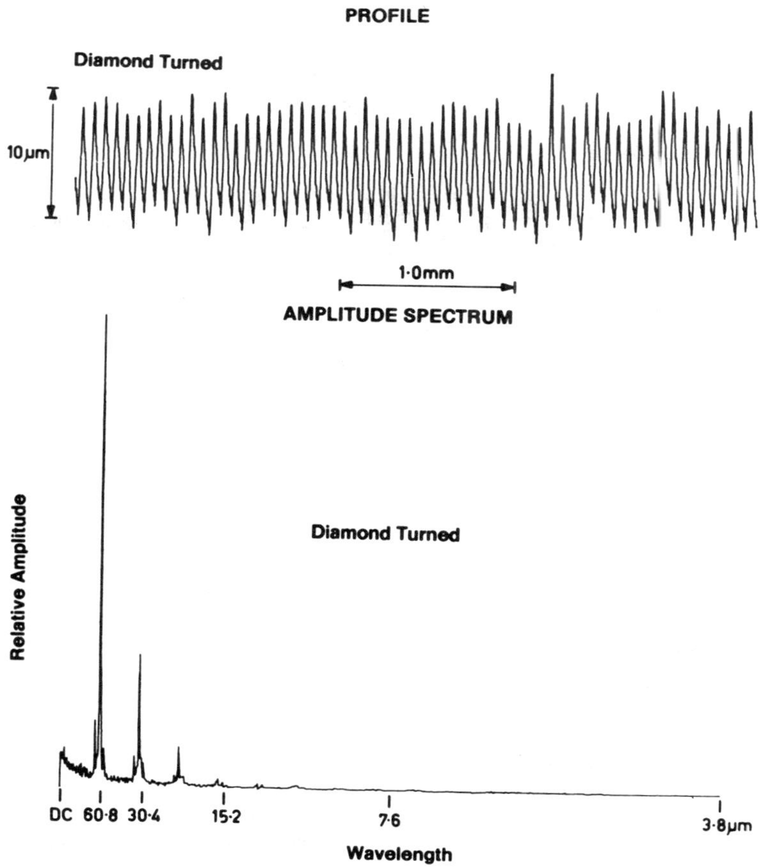

Fig.1. Surface profile and amplitude spectrum of a turned surface.

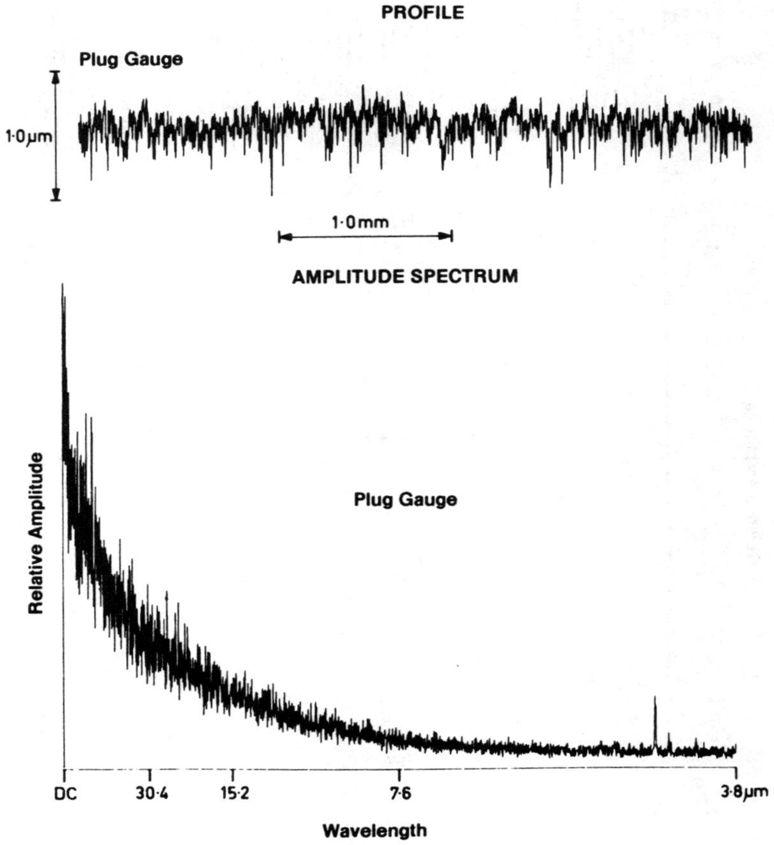

Fig.2. Surface profile and amplitude spectrum of a ground surface.

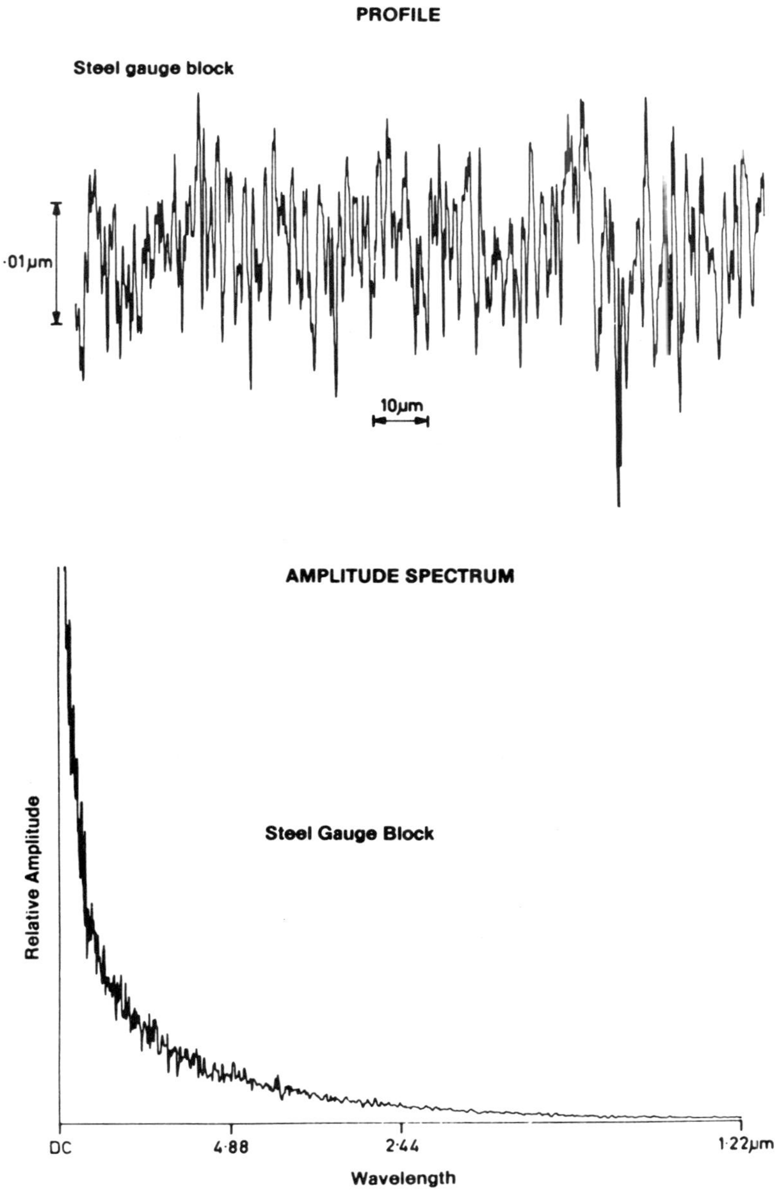

Fig.3. Surface profile and amplitude spectrum of a lapped surface.

involved and there can be major or minor differences in interpretation of the surface location. This is just another way of saying that the location ascribed to a surface is dependent upon the nature of the instrument used to detect it. For most technical surfaces the difference resulting from different measuring techniques will be unimportant but there are occasions when this is not so. For example, if it were necessary to know very precisely the maximum depth of scratching on a surface, different measuring techniques can give very different values.

Figures 1, 2 and 3 illustrate distinctive types of surfaces. One is a periodic surface which was produced by a turning process and the other two random surfaces produced by grinding and lapping respectively. Other surfaces will be made up of composites of these two types and will have both random and periodic features. These random and periodic characteristics are present throughout the roughness range of surfaces.

Another obvious property of surfaces is the homogeneity of the distribution of surface structures, what is called their 'lay'. A turned surface has a periodic profile in one direction and an essentially random one in a direction at right angles. Most machining operations give rise to characteristic lays.

Many surfaces have a Gaussian distribution of heights. It is the height distribution of the highest asperities that is of functional significance in many instances and this distribution is often claimed to be Gaussian also. The height distribution is sometimes skew, particularly when it results from sequential finishing operations.

As an example of the types of surface that require their roughness determined, Table 1 lists the approximate range of parameters for surface asperities on tribological surfaces, i.e. surfaces which are paired in bearings or which rub against each other.

Table 1 Tribological Surface Asperity Parameters

Density	$10^2 - 10^6/mm^2$
Heights	2.5 - 7.5 μm
Spacing	1 - 75 μm
Slopes of sides	5 - 10°, some less than 2°, others more than 35°
Radii	10 - 20 μm

From the values in Table 1 we can conclude that an asperity is a rare event on a surface and the 'aspect' of the structures is low. This low aspect characteristic is apparent when the profile of a surface is drawn with equal magnifications in the vertical and horizontal directions. It has many implications both for the functional behaviour of surfaces and in the measurement of their topography. The introduction of this sparce "asperity model" of surfaces, in the 1930s, was one of the great innovations in the study of surface contact and its consequences are still being worked out in detail.

In the study of surface topography, it is often difficult to reach clearly stated conclusions,but despite this, understanding of the influence of topography or surface roughness has been at a level to influence practical technology for many years.

The remainder of this paper will deal mainly with the stylus profiling method of measuring surface roughness. In doing this extensive reference will be made to the recently published Australian Standard for Surface Texture, AS 2536, which contains a wide range of reference material as well as the basic standard specifications. It is intended that the Australian standard should be read concurrently with this work.

There are a number of introductory and reference works on the topic of roughness measurement which should receive special mention here. These are:

a bibliography by Thomas and King (1977),
a general introduction to the topic by Dagnall (1980),
a background to high resolution techniques by Young (1971), and
a work on numerical interpretation, Thwaite (1978).

An extensive bibliography covering stylus and other methods will be found in Appendix G of AS 2536.

2. Standard Specifications

Principally for the purpose of process control but also to provide a consistent basis for technical roughness measurement all industrial nations, through their industrial standards organizations, have developed standards documentation which specifies conditions governing the measurement of surface roughness. This approach is necessary because, as we have already discussed, the assessment of topography is dependent upon the means used. As also mentioned, the Standards Association of Australia (SAA) has very recently published a comprehensive standard on the subject.

There is a large measure of agreement between national standards related to surface texture which has been brought about by the activities of the International Standards Organization (ISO). There are a number of ISO standards which define nomenclature and measuring methods and which have been accepted as the basis for national standards.

A list of ISO standards and the national standards of most industrial countries is given in Appendix G of AS 2536.

National standards pay most (almost exclusive) attention to the stylus technique of measurement. This is because although optical microscopy, light scattering, electro-optics, capacitive, pneumatic, and other methods can give information on a variety of surface properties, stylus profiling techniques are still currently superior for producing comprehensive quantitative descriptions of surface topography.

3. Surface Roughness

When we speak of a rough surface we need to define the scale of roughness as otherwise we could equally be talking about structures at the atomic level or gross features millimetres in size. There is a convention in nomenclature that is fairly universally accepted that describes surface topography as follows:

Total topography { Form
Surface Texture { Flaws
Waviness
Surface Roughness

Form refers to the very long wavelength components of a surface, *waviness* to the topography that can be constructed from intermediate wavelength components, and *surface roughness* is the fine structure. For completeness flaws are also mentioned to make sure that their contribution to structure is assessed in a consistent way.

To be able to measure a surface in quantitative terms the boundary between roughness and waviness must be defined with sufficient precision to achieve the uncertainty required in the measurements. This is done by specifying the characteristics of a high-pass filtering process which excludes the long spatial wavelength components in a controlled manner. This filtering process is the most important of the essentially arbitrary conditions which have to be imposed to make surface roughness measurable. It will be discussed in more detail later.

There is no generally accepted definition of the boundary between waviness and form though the subject is being actively pursued at the present time.

4. Surface Roughness Parameters

There is an extensive range of possible parameters that may be used to typify surface roughness. Many of the quantities which occur in the literature of mathematical statistics or the theory of random processes are possible candidates. Only a small number of these are often used, however.

There are strong reasons for restricting the number of parameters used in process control. The necessity to avoid overcomplicating shop-floor communication and the need to restrict proliferation of instrumentation are probably the most important. On the other hand, there is a continuing search for parameters that will correlate with the functional properties of surfaces.

The choice of parameters for use in process control has come about for historical reasons to some extent. It was not nearly as easy 30 years ago to process squares electronically as it is today and this is the major reason for the use of an arithmetic average rather than a standard deviation as the principal parameter in current practice.

(a) Parameters

The parameters that are most frequently used to characterize rough surfaces and which have internationally accepted definitions are:

Arithmetic mean deviation, R_a (2.2.2)*;
Root mean-square deviation, R_q (2.2.3);
Maximum peak-to-valley roughness height, R_{ma} or R_y (2.2.4);
Ten-point height, R_z (2.2.5);
Bearing length ratio, t_p (2.2.8).

It should be especially noted that these parameters are defined in terms of profiles not surfaces and either use a *reference line* or are specifically related to a *cut-off length*.

The way in which the reference line and the cut-off are defined is of paramount importance to the realisation of the measurements.

*These numbers refer to the appropriate clause in the Australian Standard 2536.

(b) Reference Lines

Formally the reference line is defined as:

Reference line - the curve or straight line defined by convention in relation to the nominal profile or other datum, which serves as a reference for the definition of parameters.

In practice there are three reference lines in use:

Electrical reference line (1.2.11) - a line established by the filtering action of an instrument on the electrical signal which represents the measured surface profile.

Centreline (1.2.12) - a reference line representing the form of the traced profile and parallel to its general direction throughout the cut-off length, such that the sums of the areas contained between it and those parts of the profile which lie on each side of it, are equal.

Least-squares centre line (1.2.13) - a reference line having the form of the nominal profile within the limits of the cut-off, and placed so that within the cut-off the sum of the squares of the deviations of the profile from the reference line is a minimum.

The electrical reference line is usually the appropriate reference since it is built into *most* electronic stylus instruments. The centreline is defined because it can be used simply on graphical profile records. The least-squares reference line is readily implemented digitally.

(c) Cut-off, λ_B (1.2.23)

In usage cut-off has two meanings:

(i) the value of the wavelength (λ), defined by convention as the longest wavelength transmitted by an instrument, i.e. in the signal used to assess the roughness;

(ii) the length of the segments into which a profile record is divided for the purpose of making a geometric determination of surface roughness.

This usage differs slightly from current ISO practice in that a second term *sampling length* is sometimes used in place of the term cut-off, particularly when referring to geometric determination. Sampling length is a confusing term and is very often taken to mean sampling in the statistical sense - it has not been used in the Australian standard for this reason.

(d) Other terminology

A range of basic terminology definitions will be found in the relevant standards but reference should be made here to a group of parameters and statistical functions which are widely used and which will become more and more common in the literature. These are:

peak count, profile peak density, average spacing of peaks, average spacing of irregularities, amplitude density function, skewness, kurtosis, auto-correlation function, wavelength spectrum.

Further details will be found in Appendix D of AS 2536 or a standard text on mathematical statistics.

5. Instrumentation and Methods of Measurement

The study of the fine structure of surface topography became possible when instruments with sufficient magnification to resolve the vertical structures were developed. The introduction of precision stylus profiling instruments in the late 1920's made it possible to establish the nature of topography comprehensively and on a quantitative basis.

A wide variety of instrumentation has been developed to meet various needs but only a few types are in general use for studying topography quantitatively. These are:

Visual microscopy. There are many microscopic techniques for the examination of surfaces - bright field illumination, dark field illumination, the Normaski and other shearing systems which give contrast related to local shape etc., and taper sectioning. The visual methods of greatest interest are the Schmaltz profile microscope and interference microscopy. These methods are described in greater detail in Appendix C of AS 2536 than is possible here.

Schmaltz profile microscope - This instrument is used for examining fairly rough surfaces. It has a peak-to-valley range of approximately 0.5 to 50 μm. A profile is produced by imaging an illuminated slit or a dark wire on to a surface from a projecting device with axis 45° to the surface normal. The projected image is viewed with a second microscope again at 45° to the surface normal but on the other side of the normal to the projecting optics. The instrument is available from a number of sources.

Interference microscopy - Both two-beam and multiple-beam systems are available. Instruments with magnifications of X750 can give height differences by visual interpolation to better than 0.05 μm. Other systems are Normaski and shearing (Zeiss Epival Interphako).

General references to these optical techniques will be found in Appendix G of AS 2536.

Electron microscopy - High resolutions. It is natural to think of high resolution imaging devices such as transmission and scanning electron microscopes. These instruments, however, do not readily provide detailed quantitative data on topography. They are more adapted to the examination of features. Transmission electron microscopes can give resolutions of better than 3 nm and scanning instruments better than 10 nm with improvement of these values with the right specimen.

Profilometers - Contact profiling instruments have a fine stylus attached to a displacement transducer. As the stylus is drawn across and in contact with a surface an electrical analog of the surface is generated. This analog is usually recorded graphically and processed electronically or digitally to give the surface parameters. These instruments are described in much more detail below.

6. Contact Profiling Instrument

Contact profiling devices using a sharp stylus attached to a transducer to trace profiles of a surface are by far the most common instruments. Magnifications in commercial instruments go as high as

LONG WAVELENGTH TRANSMISSION CHARACTERISTICS—NOMINAL SINUSOIDAL FREQUENCY RESPONSE

Wavelength mm	Amplitude transmission, percent				
	Cut-off λ_H, mm				
	0.08	0.25	0.8	2.5	8.0
0.008	99.6	—	—	—	—
0.01	99.4	—	—	—	—
0.025	96.8	99.6	—	—	—
0.05	88.4	98.7	—	—	—
0.08	75.0	96.7	99.6	—	—
0.1	65.7	94.9	99.4	—	—
0.25	23.5	75.0	96.8	99.6	—
0.5	7.1	42.8	88.4	98.7	—
0.8	2.2	22.6	75.0	96.7	99.6
1.0	—	15.7	65.7	94.9	99.4
2.5	—	2.9	23.5	75.0	96.8
5.0	—	—	7.1	42.8	88.4
8.0	—	—	2.9	22.6	75.0
10.0	—	—	—	15.7	65.7
25.0	—	—	—	2.9	23.5
50.0	—	—	—	—	7.1
80.0	—	—	—	—	2.9

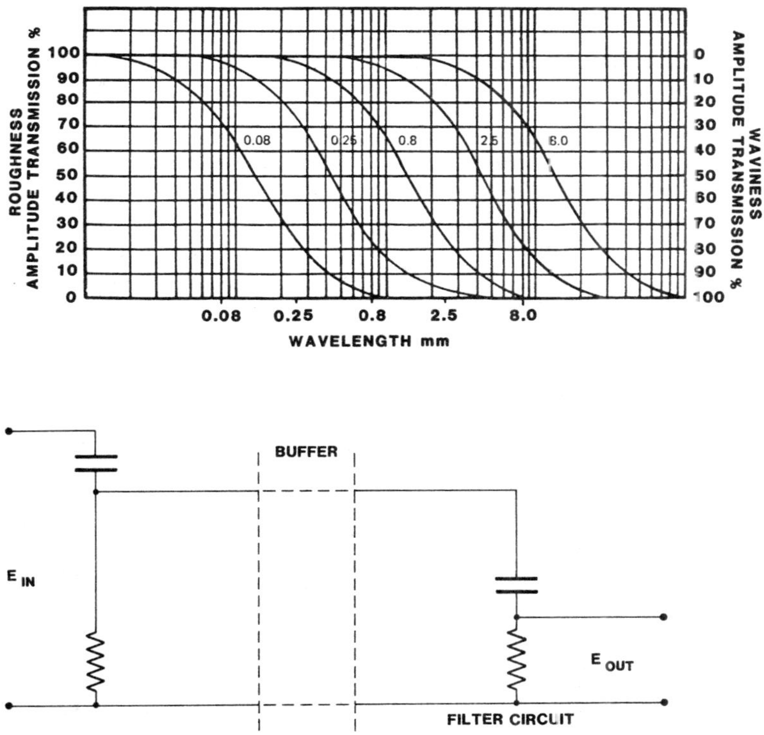

Fig.4. Standard instrument long wavelength filter.

X200 000, and there are also low magnification instruments. In their simplest form, stylus instruments give a measure of roughness in terms of the arithmetic mean deviation (R_a) of the profile from a reference line. They may also have provision for producing a graphical record of the surface.

There is a class of stylus instruments which can be used to obtain profiles with very high vertical resolutions. With these, magnifications of $X10^6$ are obtainable giving resolutions of better than 2 nm. They have specialised uses such as the measurement of optical finish and can also be used for measuring the thickness of thin films. Apart from saying that their displacement transducers need to be calibrated, they will not be mentioned further here.

To ensure reproducibility of results and consistency, it is necessary to standardize a number of features of instruments in an arbitrary way. As has already been mentioned, all major industrial nations have standard specifications which fix features and characteristics such as stylus radius and force, the minimum traversing length and the all-important bandwidth or cut-off.

The essential instrument characteristics which are specified by international agreement are referred to below. Other features, such as the range of magnification, are sometimes specified as a matter of convenience rather than as essential for technical reasons.

Measuring traverse length (4.2) - This needs to be specified to ensure that the signal is present for sufficient time for the filtering circuits to operate without significant switching disturbance. An additional requirement is that the longer the signal the more representative it will be. The longest wavelength that can be analysed in the signal is also related to this length. The measuring traverse length of an instrument can be readily measured.

Transmission characteristics (4.3) - The instrument bandpass or cut-off is of critical importance and is closely specified in the standards.

The standard instrument long wavelength filter has characteristics corresponding to the attenuation properties of two independent RC networks of equal time constant. The circuit of this type of filter is illustrated in Fig.4.

The long wavelength roughness cut-off (λ_B) is assessed at an amplitude transmission of 75% of the maximum transmission and is chosen from the following range of values:

0.08, 0.25, 0.8, 2.5, and 8.0 mm.

An instrument is considered to have a satisfactory cut-off if the specified value lies within 70 to 80% of the maximum transmission. A full description of the other filtering requirements for instruments will be found in section (4.3) of AS 2536.

While it is imperative that the filtering meets the requirements of the specifications and it is not an easy matter to fully test them in assembled commercial instruments from basic principles, producing the filters within specification is not normally considered a problem.

Stylus shape and force (4.4) - It is important that the stylus shape is within the tolerances laid down and since the stylus is subject to wear, it needs to be checked from time to time.

The checking is usually done using a shadow projector or a visual microscope. In both cases high resolving powers are required. With projectors, only the best quality instruments at high magnification are just adequate, while with microscopes a magnification of at least X100 is necessary and a high numerical aperture objective is required.

An alternative method of checking the stylus is to use a specially prepared fine grating, a 'stylus check', which gives a specified value of a roughness parameter only when the stylus is not too badly worn. When the stylus is worn it does not penetrate the structure of the grating fully and a low reading results. This method has been advocated for many years but as yet is not widely used because suitable gratings are not readily obtainable.

When available, a scanning electron microscope can be used to examine the stylus. If the stylus arm cannot be fitted into the microscope a replica in gold has been used.

If the force on the stylus is too great it will scratch the surface of quite hard materials, if too low it may bounce. Since the tip radius is small very high pressures can be obtained at small loads. The specified forces range from 0.7 mN for a radius of 2 μm to 16 mN for a radius of 10 μm with additional restrictions on the rate of change of the measuring force. A beam balance or torsional balance or other sensitive force measuring device is required to make a check of these forces. Because of the type of construction used it is unlikely that the values specified will be exceeded.

Skid dimensions (4.5) - Very often stylus instruments are used with a skid or shoe fitted to the stylus arm. The skid acts as a support of the stylus arm and greatly reduces the sensitivity of the instrument to vibration. Provided the radius on the skid surface is not too small, the use of a skid will not drastically affect the transmission characteristics of an instrument. For this reason the skid radius is required to be at least 50 times greater than the cut-off being used.

7. Instrument Calibration

Profilometers are essentially displacement measuring devices and require calibration at regular intervals to maintain their accuracy. The two functions, profile recording and parameter measurement, are often calibrated separately even though both derive their signal from the stylus transducer.

The means used to perform this calibration depends upon the magnification and the accuracy required. In most cases, calibration is performed on one or two ranges and the instrument attenuators are relied upon for the other ranges.

In most cases the calibrations are done using calibrated grooves and R_a roughness standards which are often supplied with the instrument. In some cases the instrument manufacturers advocate calibration of the transducer only and rely on the stability of the processing electronics to give correct parameter values.

The grooves and R_a standards in turn need calibration and this is done using a calibrated instrument or alternatively in the case of grooves by optical interference using an interference microscope.

The instrument transducer may also be calibrated by using gauge blocks at the low magnifications or block gauges and a lever arm at higher magnifications. These two methods are described in Appendix E of AS 2536. R_a standards are calibrated by taking the direct readings from a calibrated master instrument, evaluating graphical records of profiles by planimetry, or digitally recording profiles and computing

Transducer Calibration - Static

Gauge block,
Lever arm,
or
Calibration
groove.

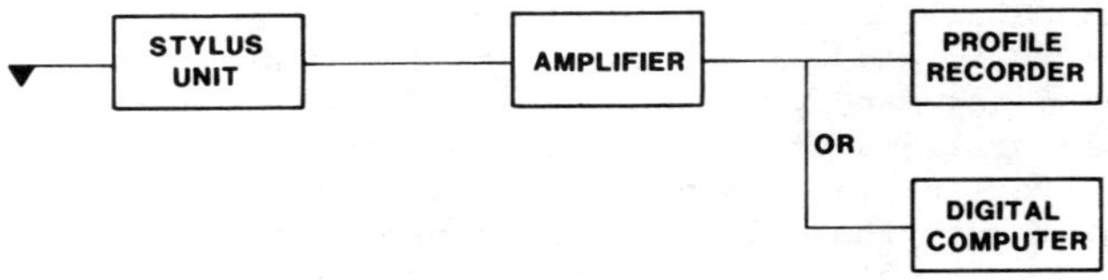

Parameter Unit Calibration

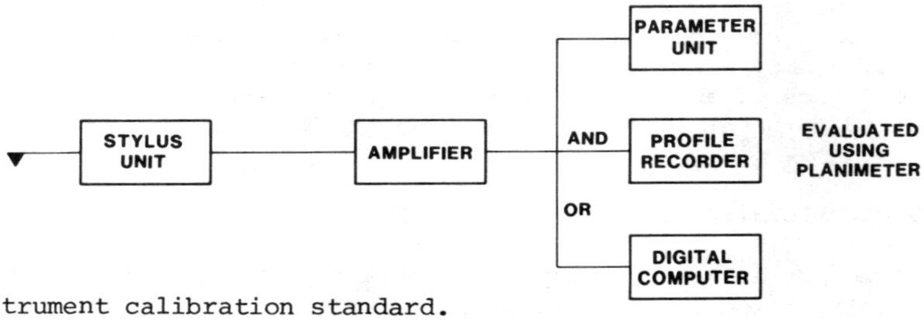

R_a instrument calibration standard.

The parameter unit is calibrated by using a known standard or by determining the R_a value of an unknown signal by means of a calibrated recorder or digital computer.

Groove Calibration

Calibrated stylus instrument

or

Interference microscope.

Fig.5. Alternative calibration methods.

the appropriate parameter value. These various alternative methods of calibration are illustrated in Fig.5.

A specialised method of instrument calibration based on a feedback controlled vibration table has been developed at the Division of Applied Physics. As well as giving a static calibration of the stylus transducer and parameter calibrations, it can be used to investigate filtering characteristics in detail, Bendeli, *et al.* (1974). A precision wedge has also been developed for static stylus transducer calibration, Thwaite (1974). Neither of these instruments is available commercially.

Further details relating to calibration techniques will be found in Appendix E of AS 2536.

The accuracy achievable in the calibration of precision grooves and instrument calibration standards (roughness standards) depends upon the technique used, their inherent geometrical trueness, and their nominal value.

Grooves having approximate depths of 2.5 μm and 0.5 μm are available commercially and values of 0.05 μm are also obtainable. The most precise calibration techniques used on a regular basis give uncertainties of approximately ±3% for groove depths in the range 2.5 to 0.5 μm.

While instrument calibration standards with R_a values in the range 5 to 0.05 μm are obtainable commercially, no completely satisfactory standards are generally available as yet. The larger and smaller values are particularly difficult to manufacture. The main problem is uniformity of value across the surface of the standard. For a good quality R_a standard having a nominal value of 1 μm R_a an uncertainty for the mean measured value of ±0.04 μm is achievable based on a sample of 25 values. The individual measured values would be distributed with a standard deviation of typically 0.01 μm R_a for this class of standard. These values can be judged against an instrument uncertainty of ±7% that is laid down in some national standards. Calibration results for a pair of grooves and an R_a instrument calibration standard are appended.

The accuracy achievable in measuring a specimen is highly dependent upon its nature, particularly in relation to its uniformity. As a guide, at 1 μm R_a, repetitions in mean values within 2% are achievable. At smaller values much larger amounts will have to be accepted.

8. Conclusion

Stylus profiling instruments must conform to a number of specified characteristics if consistent measurements of surface roughness are to be obtained. These characteristics have been reviewed and the basic means of calibrating both the instrumentation and calibration standards in the form of grooves and R_a instrument calibration standards described. Extensive references have been made throughout to the recently published Australian Standard for Surface Texture, AS 2536 (1982).

References

'Australian Standard for Surface Texture', 2536 (1982). Standards Association of Australia, Sydney.

Dagnall, M.A. 'Exploring Surface Texture', Rank Taylor Hobson, Leicester, U.K.

Bendeli, A., Duruz, J. and Thwaite, E.G. (1974). *Metrologia* 10, 137-143.

Thomas, T.R. and King, M. (1977). 'Surface Topography in Engineering', Vol.3, (British Hydromechanics Research Association, Fluid Engineering), Cotswold Press Ltd, U.K.

Thwaite, E.G. (1974). 'Proc.Int.Symp. on Metrology'(Insymet'74), Bratislava, Czechoslovakia, 5-8 November.

Thwaite, E.G. (1978). *Wear* 51, 253-267.

Young, R.D. (1971). *Physics Today*, 42-49.

Appendix 1

Calibration results for a pair of grooves and an R_a instrument calibration standard

This roughness standard has been examined at the National Measurement Laboratory and the following results obtained.

Visual Examination:

The roughness standard and the two sets of grooves were examined using a microscope with a magnification of X160. This examination revealed one small defect at the centre of the standard groove marked 2.46 μm.

R_a Standard:

The arithmetical mean deviation (R_a value) at 0.8 mm cut-off has been determined, by reference to a master standard, at 25 locations distributed evenly over the outlined area using a calibrated 'Talysurf 5' profile instrument with a nominal stylus radius of 2.5 μm. At 10 additional locations, 4096 ordinates were recorded by a numerical process at a spacing of 1.0 μm and the R_a value calculated after numerically applying a filter with a standard instrument cut-off of 0.8 mm. The analysed profiles were taken in the direction of the arrows on the outlined area.

From the recorded data, the following results were obtained:

Marked R_a value	0.78 μm
Mean measured R_a value	0.788 μm
Standard deviation	0.003 μm
Range of measured R_a values	0.782 μm - 0.794 μm.

The value quoted above for the mean measured R_a value has an estimated uncertainty of ±0.03 μm.

Groove Depths:

Using the 'Talysurf' instrument described above, determinations were made of the 'total height' of the central of each of the two sets of three grooves. Profiles of these grooves were recorded at 5 locations and the depths calculated from the recorded profile ordinates. From these values the following results were obtained:

(a) Marked groove depth 2.46 μm
Mean measured groove depth 2.45 μm
Range of measured groove depth 2.40 μm - 2.48 μm

(b) Marked groove depth 0.33 μm
Mean measured groove depth 0.340 μm
Range of measured groove depth 0.333 μm - 0.343 μm.

The values quoted above for the mean depth of grooves have an estimated uncertainty of ±3%.

Profile graphs of the R_a standard were taken normal to the lay at a horizontal magnification of X100 and a vertical magnification of X5000. A sample length of graph is given in Figure 6. The profile graphs showed no significant irregularities.

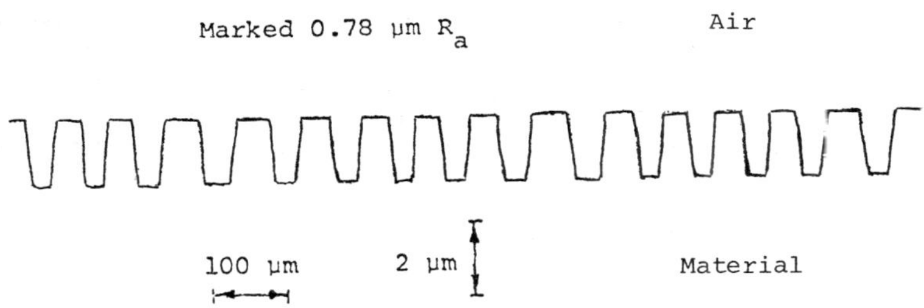

Fig.6.

Should Calibration Facilities be Based on Primary or Secondary Temperature Standards?

T.P. Jones
CSIRO Division of Applied Physics, Sydney, Australia 2070

Abstract

The methods of establishing temperature calibration facilities based on primary and secondary standards are described briefly and the uncertainties attainable using these calibration facilities are discussed. A comparison of national temperature needs with these uncertainties will determine whether calibration facilities should be based on primary or secondary standards. It is suggested that all national laboratories involved in an initial establishment of temperature standards should base calibration facilities on secondary standards. In general, industrially developing countries should not attempt to establish primary standards of temperature.

1. Introduction

A national measurement system is based on a hierarchy of calibrated standards of physical measurements. Primary standards of measurement are generally realized in terms of fundamental laws of physics. These primary standards are used to calibrate secondary standards which are transportable and can be used to calibrate measuring instruments for practical use. The establishment of primary physical standards requires extremely sophisticated instrumentation, specialized knowledge and many years of experimentation. Because of the cost and difficulty of establishing primary standards the question must be asked whether this is necessary, or whether the use of secondary standards can provide a satisfactory national measurement system.

This question is considered for standards of temperature. The methods of establishing calibration facilities based on primary and secondary standards are described briefly and the uncertainties attainable using these calibration facilities are compared.

2. The International Practical Temperature Scale of 1968

The International Practical Temperature Scale (IPTS) is recognized as the scale on which all national standardizing laboratories base their temperature measurements. It is under constant review and is modified as necessary, generally at intervals of about 10 years; the name is also changed to distinguish between scales. The current scale is referred to as the International Practical Temperature Scale of 1968 (Amended Edition of 1975), which is generally referred to as IPTS-68.

Details of IPTS-68 are given by Kemp (1977). A section of the Introduction of IPTS-68, which is written to give a correct description of the units and terms used is quoted below:

> "The unit of the fundamental physical quantity known as thermodynamic temperature, symbol T, is the kelvin, symbol K, defined as the fraction 1/273.16 of the thermodynamic temperature of the triple point of water.
>
> For historical reasons, connected with the way temperature scales were originally defined, it is common to express a temperature in terms of its difference from that of a thermal state 0.01 kelvins lower than the triple point of water. A thermodynamic temperature, T, expressed in this way is known as a celsius temperature, symbol t, defined by
>
> $$t = T - 273.15 \text{ K} \quad (1)$$
>
> The unit of Celsius temperature is the degree Celsius, symbol °C, which is, by definition, equal in magnitude to the kelvin. A difference of temperature may be expressed in kelvins or degrees Celsius.
>
> The International Practical Temperature Scale of 1968 (IPTS-68) has been constructed in such a way that any temperature measured on it is a close approximation to the numerically corresponding thermodynamic temperature. Moreover, such measurements are easily made and are highly reproducible; in constrast, direct measurements of thermodynamic temperatures are both difficult to make and imprecise.
>
> The IPTS-68 uses both International Practical Kelvin Temperatures, with the symbol T_{68} and the International Practical Celsius Temperatures with the symbol t_{68}; the relation between T_{68} and t_{68} is the same as that between T and t, ie,
>
> $$t_{68} = T_{68} - 273.15 \text{ K} \quad (2)$$
>
> The units of T_{68} and t_{68} are the kelvin, symbol K, and the degree Celsius, symbol °C; that is, the names of the units are the same as those used for the thermodynamic temperature T and the Celsius temperature t".

In general, kelvin temperatures, T_{68}, are used below 0°C and Celsius temperatures, t_{68}, are used above 0°C.

The IPTS-68 is based on the assigned values of the temperatures of a number of reproducible equilibrium states (defining fixed points) and on standard instruments calibrated at these temperatures. Between the fixed point temperatures, interpolation formulae relate indications of the standard instruments to values of IPTS-68. The standard instruments specified are

(1) the platinum resistance thermometer, from 13.81 K to 630.74°C
(2) the platinum 10% rhodium-platinum thermocouple, from 630.74 to 1064.43°C and
(3) instruments using Planck's Radiation Law above 1064.43°C.

The interpolation formulae are complicated, because of the nature of the practical instruments. The definition of the IPTS-68 is summarized in Table 1.

Table 1. International Practical Temperature Scale of 1968 (Amended Edition of 1975)

<table>
<tr><td colspan="8">INTERNATIONAL PRACTICAL TEMPERATURE SCALE OF 1968
(AMENDED 1975)</td></tr>
<tr><td>TEMPERATURE RANGE</td><td>13.81 K</td><td>20.28 K</td><td>54.361 K</td><td>90.188 K</td><td>273.15 K
0.0°C</td><td>630.74°C</td><td>1064.43°C</td></tr>
<tr><td>FIXED POINT</td><td>T.P
H_2</td><td>B.P
H_2</td><td>T.P
O_2</td><td>B.P
O_2</td><td>T.P
H_2O
(0.0100°C)</td><td></td><td>F.P
Au</td></tr>
<tr><td>STANDARD INSTRUMENT</td><td colspan="5">PLATINUM RESISTANCE THERMOMETER</td><td>PLATINUM, 10% RHODIUM / PLATINUM THERMOCOUPLE</td><td>INSTRUMENT BASED ON PLANCK'S LAW</td></tr>
<tr><td rowspan="2">INTERPOLATION FORMULA</td><td colspan="4">$W(T_{68}) = W_{CCC-68}(T_{68})^* + \Delta W_i(T_{68})^*$</td><td rowspan="2">$t_{68} = t'^* + 0.045 \frac{t'}{100°C} \times \left[\frac{t'}{100°C} - 1\right] \times \left[\frac{t'}{419.58} - 1\right] \times \left[\frac{t'}{630.74} - 1\right] °C$</td><td rowspan="2">$E(t_{68}) = a + b\,t_{68} + c\,t_{68}^2$</td><td rowspan="2">$\frac{L\lambda(T_{68})}{L\lambda(T_{68}Au)} = \frac{\exp\left[\frac{c_2}{\lambda T_{68}(Au)} - 1\right]}{\exp\left[\frac{c_2}{\lambda T_{68}} - 1\right]}$</td></tr>
<tr><td>$\Delta W_1(T_{68})^*$</td><td>$\Delta W_2(T_{68})^*$</td><td>$\Delta W_3(T_{68})^*$</td><td>$\Delta W_4(T_{68})^*$</td></tr>
</table>

* Kemp (1977)

The realization of the IPTS-68 requires very careful experimental techniques. Supplementary information on apparatus, methods and procedures to aid in successful realization of the IPTS-68 is given in the text of the IPTS-68. However this is not sufficient to enable the realization of the IPTS-68, and reference must be made to numerous scientific papers relating to important features in the realization of this temperature scale. It is not the intention of this paper to detail satisfactory procedures to establish the IPTS-68, rather, what is intended is to stress the very large commitment necessary to complete the project successfully. To highlight the expertise and the time involved in the establishment of IPTS-68, two examples of the necessary equipment will be reviewed briefly.

The first, a typical furnace and crucible assembly used to realize a fixed point is shown in Fig. 1. The details of construction and methods of use are described fully by Jones (1967). It should be noted that suitable equipment is not obtainable commercially and must be constructed in the laboratory. The aim is to melt or freeze very pure metals in a uniform temperature zone. To obtain satisfactory fixed points the metal must remain 99.999% pure, which means that the metal must not be contaminated with use. The metal sample should be uniform in temperature during the change of state and, consequently, end heaters are necessary in the furnace to compensate for end losses of heat. The uniformity of the furnace temperature must be carefully adjusted by varying the power of individual heaters. The period necessary for the initial establishment of a fixed point is conservatively estimated as two years of a scientist's time.

The second, a photoelectric pyrometer used to extrapolate the IPTS-68 above 1064°C at the CSIRO Division of Applied Physics, is shown in

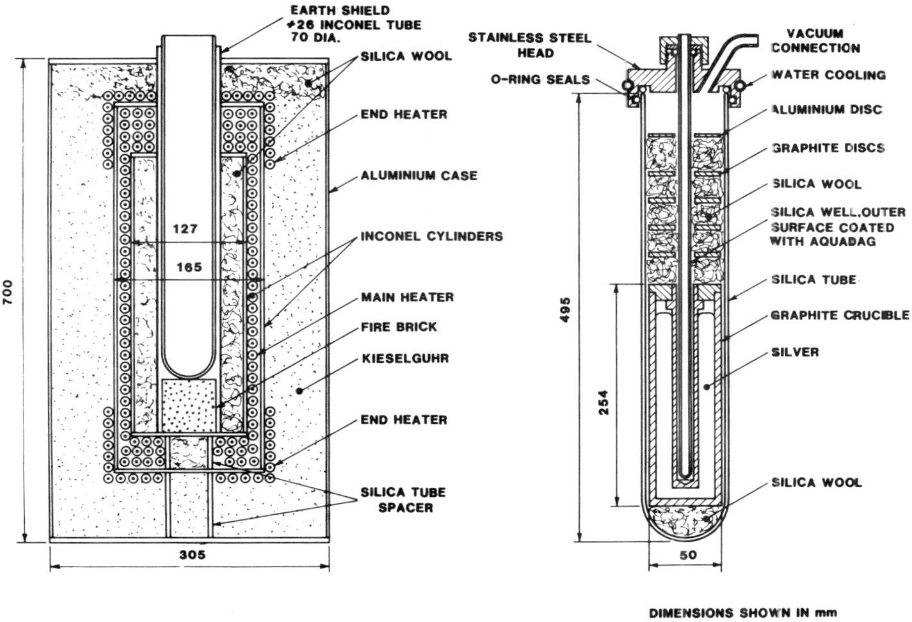

Fig. 1. Freezing-point calibration furnace and crucible assembly

Fig.2. The details of construction, method of calibration and use to establish the IPTS-68 are described fully by Jones and Tapping (1981). Pyrometers such as this are not availab_e commercially and must be designed and constructed by the user. It is designed to

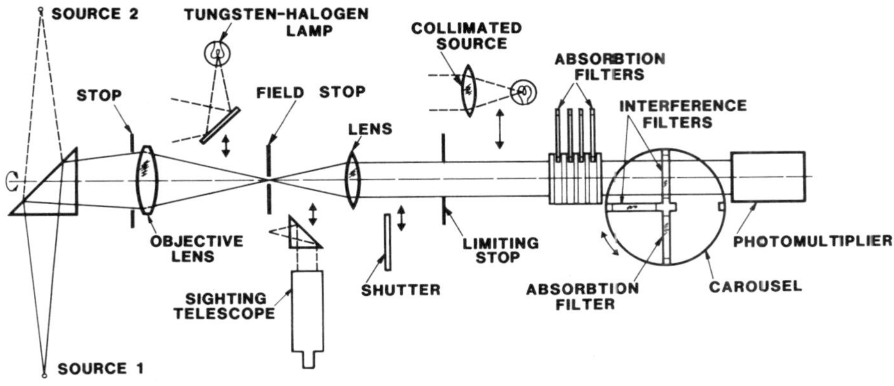

Fig. 2. Photoelectric pyrometer optical arrangement

compare the spectral radiances of two sources at an effective wavelength which must be known. Thus, besides the difficulties in optical and mechanical design and construction, the spectral characteristics of all optical components and detectors must be carefully measured. It is conservatively estimated that it would take a scientist 5 years to realize the IPTS-68 above 1064° C, even satisfactory in another laboratory.

The uncertainties which can be attained in the best possible realization of the IPTS-68, for different temperature ranges, are given in Figs. 3 to 6. The uncertainties are 99% confidence limits and relate to the calibration of primary standard interpolation instruments.

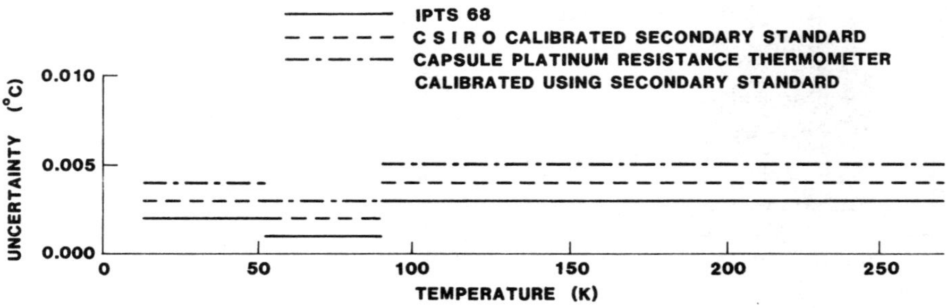

Fig. 3. Uncertainties in calibration of capsule platinum resistance thermometers

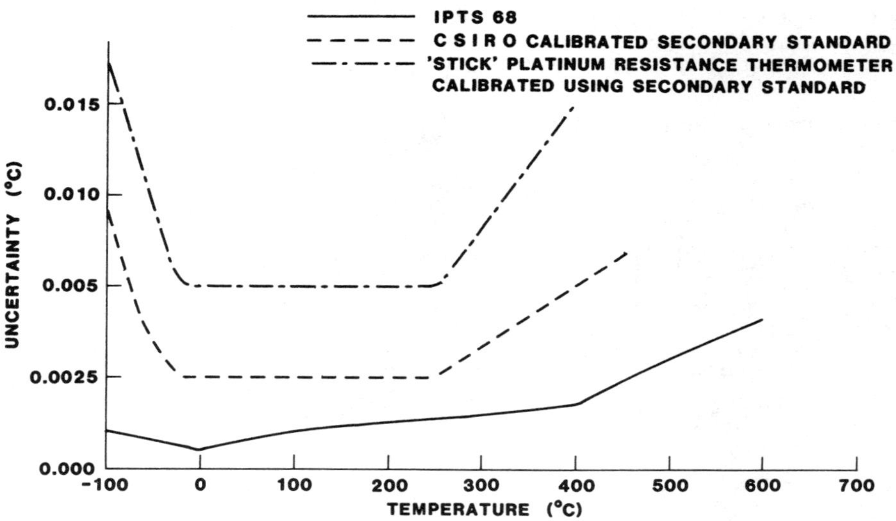

Fig. 4. Uncertainties in calibration of 'stick' platinum resistance thermometers

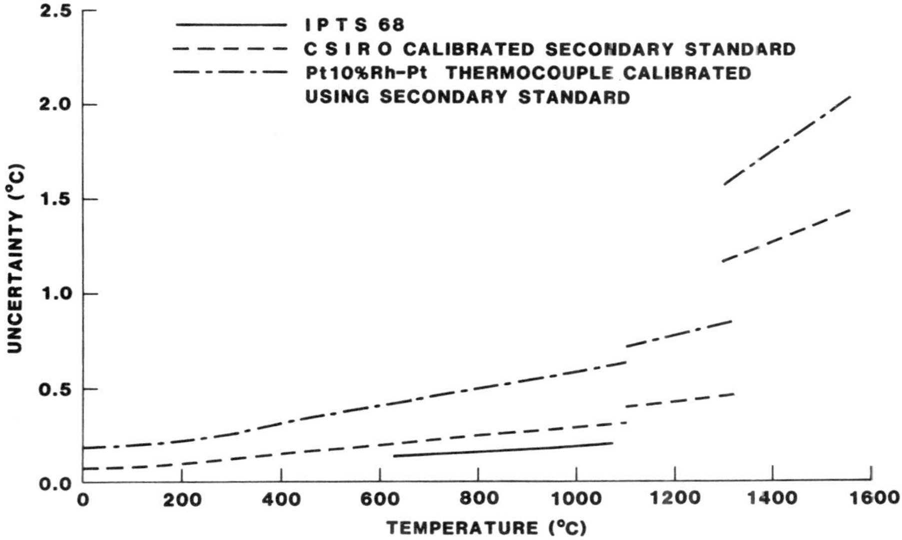

Fig. 5. Uncertainties in calibration of platinum 10% rhodium-platinum thermocouples

3. Secondary Temperature Standards

One of the main arguments in favour of basing national measurement system on secondary standards is that it eliminates the long time involved in developing primary standards. To take full advantage of the time savings in establishing secondary standards it is desirable to use commercially available equipment. Whereas, in some cases, it is possible to construct equipment, eg furnaces and baths, which perform better than commercially available equipment, in most cases this is not necessary, as their use does not significantly reduce calibration uncertainties. All the following discussions and evaluations are based on the maximum use of commercial equipment.

In all fields except optical pyrometry, the calibration of temperature-measuring instruments using secondary temperature standards is performed by an 'intercomparison method'. In this method both the standard and the instrument being calibrated are placed in an enclosure, and the temperature of the enclosure, measured by the standard, is compared with the indication of the instrument being calibrated. The fundamental requirement is that both instruments are at the same temperature, and this must be proven before calibrations begin.

The form of the calibrating enclosure depends on the temperature being realized. In the temperature range 13.81 K to 273.15 K, the enclosure generally consists of a copper block positioned within radiation shields in a vacuum insulated cryogenic system. Baths containing liquids are used in the temperature range -30 to 250° C. Alcohol is the filling liquid from -30 to 10° C, water from 5 to 90° C

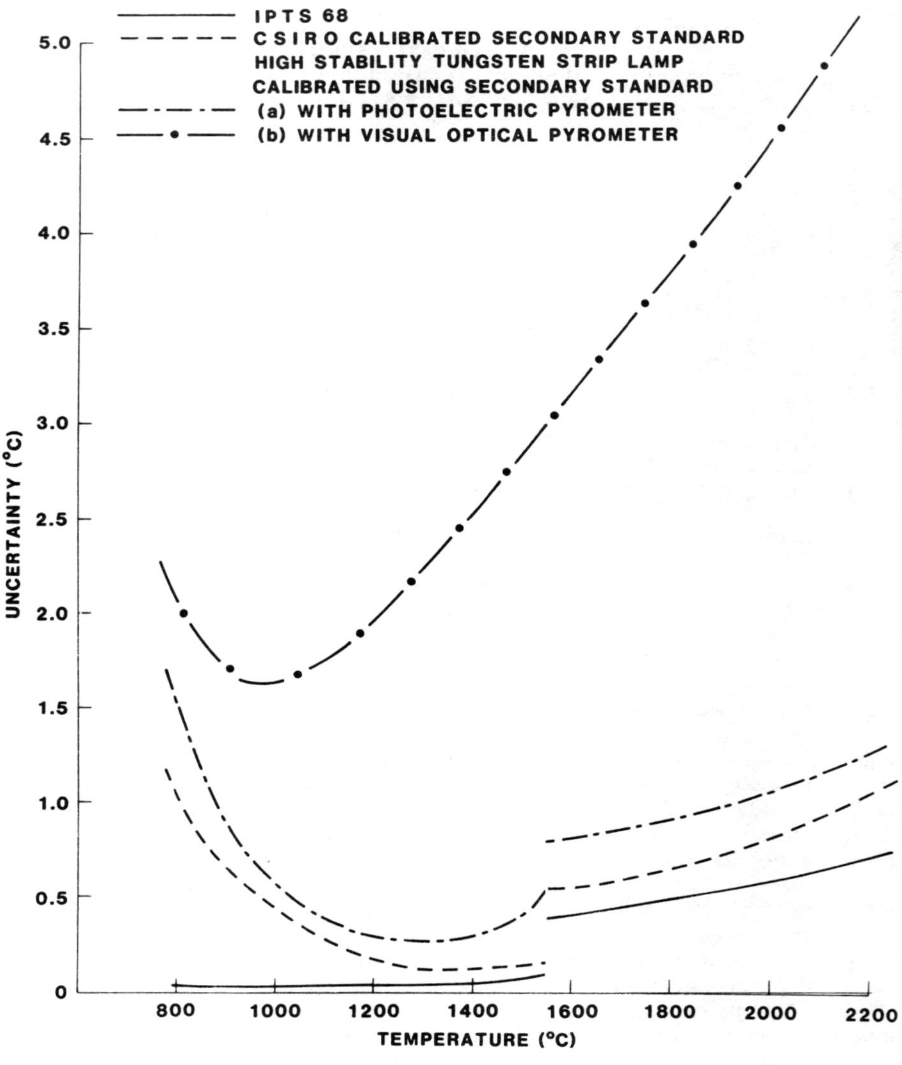

Fig.6. Uncertainties in calibration of high stability tungsten strip lamps

and silicone oil from 50 to 250°C. The liquids in the baths are circulated or stirred to ensure temperature uniformity. Furnaces are used from 30 to 1800°C. End-heaters, metal blocks and heat-pipes are used to improve temperature uniformity. The experimental procedures to ensure satisfactory uniformity of temperature enclosures are given in detail in other papers (Beavis 1982, Bentley 1982, Connolly 1982).

A satisfactory secondary temperature standard should be commercially available, convenient to use and capable of achieving a sufficient accuracy. In the temperature range 13.81 K to 273.15 K, a capsule platinum resistance thermometer is recommended as the temperature standard. In the range 200 K to 630°C the optimum instrument is a 'stick' platinum resistance thermometer containing an element in which the platinum wire is encapsulated in ceramic material. The length of the thermometer depends on the temperature at which it is to be used. Platinum 10% rhodium-platinum or platinum 13% rhodium-platinum thermocouples can be used as secondary standards in the range 0 to 1550°C. For the calibration of optical pyrometers, the optimum standard is a tungsten strip lamp in which the spectral radiance temperature is given in terms of the current passing through it. The details of the correct measurement techniques using secondary temperature standards are given in associated papers (Beavis 1982, Bentley 1982, Connolly 1982, Tapping 1982).

Typical uncertainties which are likely to be stated in CSIRO reports on the calibration of secondary temperature standards are given in Figs. 3 to 6. These uncertainties are 99% confidence limits and apply for specified periods of use. The uncertainties given in Figs. 3 to 6 for the calibration of secondary standards are, in some cases, dependent on specified calibration checks being performed. In the case of platinum 10% rhodium-platinum thermocouples, oxidation of rhodium is the main cause of drift, and this oxidation can be reversed by briefly heating the thermocouple at 1100°C. In the case of a platinum resistance thermometer, its resistance can be measured at the melting point of ice and corrections for drift applied. However for many reasons, eg contamination, severe physical shock or use at too high a temperature, it is possible for the calibration to become invalid. Consequently it is essential that regular checks be made of the validity of the standard. In general, it is necessary to have more than one standard so that they can be compared with each other. Three is the optimum number of standards, of which one is used in most of the calibrations performed and is compared periodically with one or both of the other secondary standards. An intercomparison of three standards should indicate if a change in calibration has occurred and identify which standard has changed. The secondary standards should be recalibrated when the expiry date for the validity of their calibration is neared, and their use should be regulated to ensure that only one is away from the laboratory at any time. Costs of individual standards may necessitate the reduction of the number of secondary standards to two, which should then be used in a similar way to that suggested for three.

The uncertainties which apply to instruments of the same type as the secondary standards, when calibrated in facilities based on secondary standards are also shown in Figs. 3 to 6. The magnitudes shown represent the best that can be attained: the uncertainties of other types of temperature measuring instruments may be larger, depending on their performance when tested. In the case of optical pyrometry, two uncertainty curves are shown for the calibration of tungsten strip lamps. One represents a visual optical pyrometer used to compare the radiance of tungsten strip lamps, whilst the other represents a comparison using a photoelectric pyrometer.

4. Discussion

Calibration facilities based on primary temperature standards require a major commitment when compared to those using secondary standards, which are relatively easy to establish. As a consequence, national standardizing laboratories should assess national temperature needs very carefully before deciding to establish primary temperature standards. The highest accuracies in temperature measurements required by research and industry should be determined for all temperatures. This determination should be realistic and should take into account the accuracies that can be achieved with commercially available temperature measuring instruments. The magnitude of the uncertainties should be compared with those shown in Figs. 3 to 6 to decide whether temperature calibration facilities should be based on primary or secondary standards. As can be seen from these figures, the uncertainties for the calibration of temperature measuring equipment using primary standards are less than a factor of two lower than those which would be obtained using secondary standards. In most cases, a reduction of this magnitude would not warrant the establishment of primary standards.

Calibration facilities based on secondary standards should always be established by national standardizing laboratories before commencing the realization of the IPTS-68 in any of its ranges. This permits the laboratory to commence calibration of temperature measuring instruments in a much shorter time and thus initiate the establishment of a national measurement system. In addition, the experience and knowledge gained in the techniques of temperature measurement is essential if the laboratory is to establish primary temperature standards. All equipment used in secondary-standards calibration facilities will remain in use even if primary standards are established.

5. Conclusion

The costs and difficulties involved in the maintenance of primary temperature standards require that national standardizing laboratories should carefully review their necessity. Calibration facilities based on secondary standards satisfy the needs for all calibrations, except those requiring the ultimate in precision, and thus satisfy the needs of many nations. If a laboratory does decide to establish primary temperature standards, it should first establish a facility using secondary standards to perform calibrations whilst the primary standards are being established.

References

Beavis, I.M. (1982). These Proceedings.
Bentley, R.E. (1982). These Proceedings.
Connolly, J.J. (1982). These Proceedings.
Jones, T.P. (1967). *Metrologia,* 4, 80.
Jones, T.P. and Tapping, J. (1981). *Metrologia,* 18, 23.
Kemp, W.R.G. (1977). International Practical Temperature Scale of 1968 (Amended Edition of 1975). Australian Version of the Official French Text. CSIRO.
Tapping, J. (1982). These Proceedings.

Technique for the Calibration of Liquid-in-Glass Thermometers

Margaret Beavis
CSIRO Division of Applied Physics, Sydney, Australia 2070

Abstract

The methods used in the Division to calibrate liquid-in-glass thermometers are described and details given of methods of preconditioning, the use of a reference temperature, reading techniques and the processing of results. Special attention is given to precautions that must be taken to minimise uncertainty.

1. Introduction

The instrument most commonly used for measuring temperature is the liquid-in-glass thermometer. It has the advantages of cheapness, portability and ease of reading, and does not require elaborate auxiliary equipment. However, to get the best results from thermometers and to provide an accurate calibration, a knowledge of the behaviour of the thermometers themselves and the application of a number of specialized techniques is required. Since these techniques are of such importance, they are given in detail. A list of the necessary equipment is set out in Appendix 1. The following points should also be appreciated:

(1) Mercury is the preferred liquid filling as it is readily visible, has a suitably large and reasonably constant coefficient of expansion, does not wet glass, has a distinct meniscus and can be used in a very fine glass capillary giving an open scale.

(2) Below the freezing point of mercury (-38°C), mercury-thallium is used (to -55°C) and below that alcohol or toluene (to -80°C) and pentane (to -200°C). Spirit thermometers cannot be used to the same accuracy as mercury or mercury-thallium thermometers.

(3) The accuracy obtainable from a calibrated thermometer depends on the fineness and regularity of the graduations, frequency of pointing (marks made by the manufacturer at fixed temperatures,* the graduations between the pointing marks being made by means of a dividing machine) and the thermal properties of the component glass. The latter vary with the particular batch of glass used and account

* In recent years some manufacturers have been marking thermometers such as clinicals with a minimum of pointing, relying on the improved uniformity of precision tubing.

for most of the sources of error and change in the calibration of a thermometer. For this reason the thermal history of a thermometer is important and, in fact, this history should be controlled by conditioning the thermometer according to a definite method before use.

2. Preliminary Examination

Before tests are commenced, each thermometer is examined for flaws, bubbles of gas trapped in the bulb, breaks in the liquid column, faults in marking, strain, uneven graduations, etc. (see Section 12). A permanent fault such as uneven graduation lines is sufficient reason for rejecting a thermometer for test.

3. Effect of Thermal History

Glass acts somewhat like a supercooled liquid in that its molecules are comparatively free to move, even at ambient temperatures. When glass is heated it expands but on subsequent cooling it does not contract completely to its original volume for some time, the length of the time being dependent on the type of glass, the temperature to which it was heated and the rate of cooling. This effect occurs each time a thermometer is heated and is called the 'temporary depression of zero' (Fig. 1).

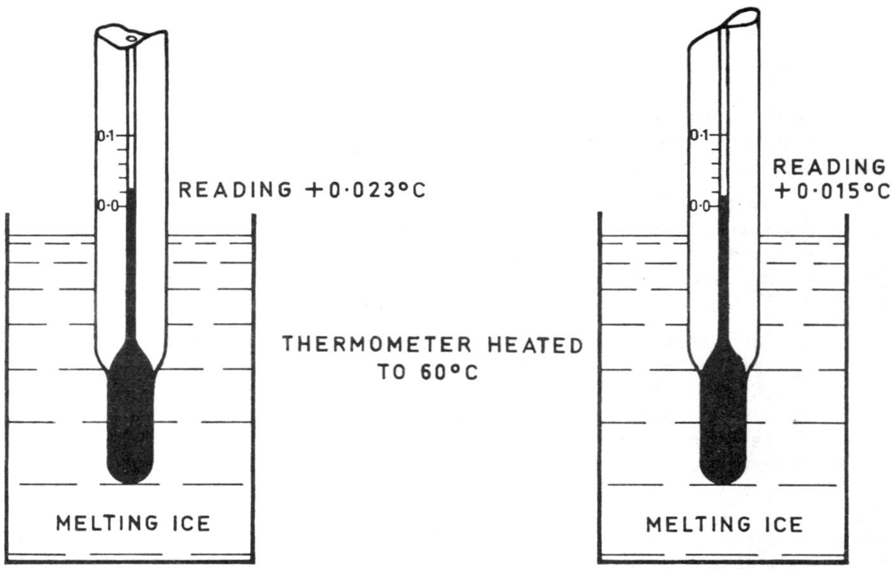

Fig. 1. Effect of heating on ice point readings.

In addition to this, the bulb is continuously recovering from strains introduced during manufacture, when the glass was heated to

500 or 600 degrees Celsius. When the thermometer is new, the rate of this recovery (contraction of the bulb) is relatively rapid and, although the rate decreases with time, the amount of contraction must be measured every year or so for the whole life of the thermometer. This continuous, irreversible change is called the 'secular change' and Figs 2a and 2b show typical curves for a new, badly annealed thermometer and a well-used, well annealed thermometer respectively.

Since the secular change is due to contraction of the bulb, its effect is to raise the reading of the thermometer. Therefore the corrections which have to be made to the thermometer readings to obtain true temperature will have to be made increasingly more negative. However, it is not necessary to repeat the whole calibration as the change will apply equally to all points on the scale. The easiest point at which to measure the change is at 0°C (ice point) and for this reason high precision thermometers are provided with an auxiliary scale in the region of 0°C when this temperature is not part of the normal working range. In this case, there is a contraction chamber between the auxiliary scale and the main scale. The change observed in the correction at the ice point is applied to the corrections at all other points on the thermometer scale as in Table 1. A record is kept of these ice point corrections, preferably in the form of a graph, so that the history of the thermometer is known and any discrepancy will be apparent.

Table 1. Incorporation of secular change

Calibration corrections 1979		Ice point correction 1981	Updated calibration corrections 1981
0.0°C	- 0.005°C	-0.007°C	-0.007°C
80.0	+ 0.023		+0.021
80.5	+ 0.017		+0.015
81.0	+ 0.019		+0.017
81.5	+ 0.014		+0.012
82.0	+ 0.022		+0.020
	etc.		etc.

In measuring the secular change, the effect of temporary depression must be minimised (even when the temperatures involved are ambient) and it is therefore necessary to precondition thermometers before calibration and before each measurement at the ice point. Suitable methods are given in Section 5, methods A or B applying to thermometers of the highest precision, method C applying to thermometers which are used above 100°C and method D applying where either the least possible uncertainty is not required or the thermometer is not of the highest precision.

If, however, the thermometer has a calibration report or certificate in which the treatment before calibration is described, then that treatment should be used.

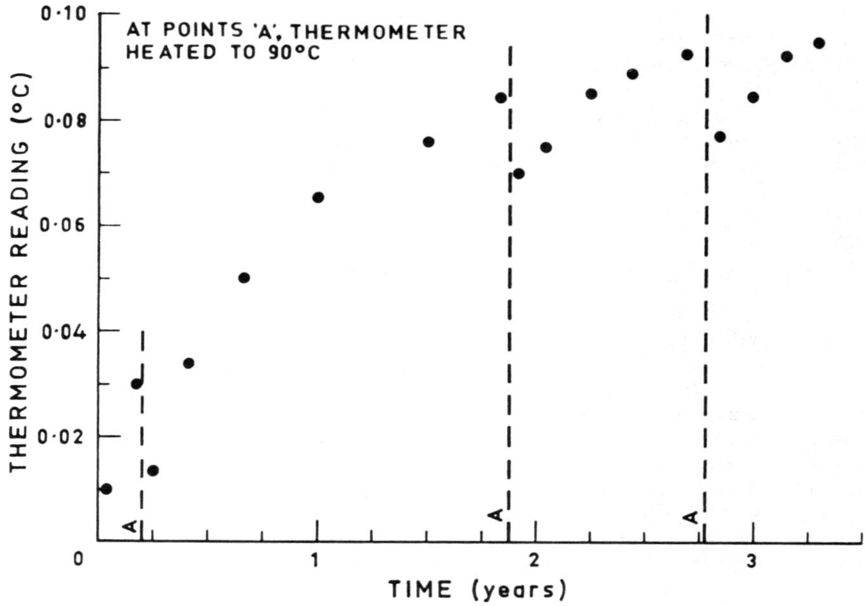

Fig. 2a. Typical secular change of a new, badly annealed thermometer.

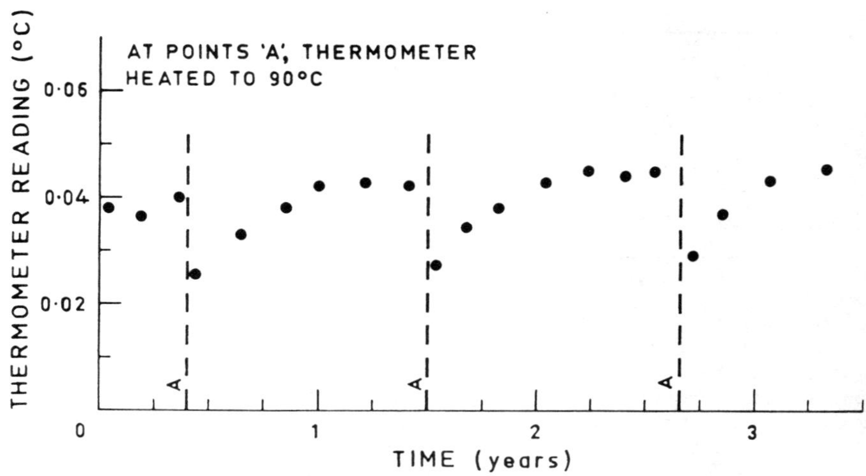

Fig. 2b. Typical secular change of a well-used, well annealed thermometer.

For a thermometer conditioned according to methods B, C or D, the secular change should be determined whenever the thermometer is used or at intervals of 1 or 2 months until the annual rate of secular change is less than half the assessed uncertainty. From this time, a determination at intervals of 6 months should suffice.

For a thermometer conditioned according to method A, it may not be practicable to measure the secular change at each 1 or 2 months as the time of recovery from temperatures above 50°C may be several months. The ice point correction is therefore measured at intervals over several months before calibration until the annual rate of secular change is less than one quarter of the least possible uncertainty to be expected from a thermometer of its type. Subsequently, the thermometer is used for 6-12 months between secular change determinations.

4. Ice Point Determination

When a thermometer is stored at ambient temperatures before its ice point correction is measured, the correction obtained is dependent on the temperature of storage. The effect can be greater than 0.01°C.

Therefore, when an uncertainty of less than 0.05°C is involved, the thermometer is washed with distilled water and stored just below or at 0°C for several hours (say overnight) before readings are taken at the ice point. When a greater uncertainty is involved, the thermometer is washed with distilled water and placed in an ice pot for about 5 min before readings are taken. An insulated porcelain or stainless steel vessel or a Dewar flask[1] fitted with an outlet tap or siphon is suitable for this purpose.* The thermometer is supported at two widely-spaced points so that it is vertical. Ice[2] made from distilled water[3] is crushed[4] finely (see Fig. 3a) and thoroughly mixed with distilled water until the ice is glassy-looking (see Fig. 3b). The ice-water mixture is then placed in the pot to fill it and the mixture is drained and re-mixed until there is no free water left but the ice is completely wet. When one set of thermometers has been read and removed, any free water is drained off, the pot is refilled with the ice-water mixture and the contents again well mixed. (A flat coil of stainless steel wire with a handle at right angles to the coil makes a useful stirrer[5].)

The thermometers are immersed to at least 5 centimetres above the top of the mercury column. Before each one is read, it is lifted until the mercury is just visible. It is then held in the fingers, raised a centimetre or so and, with the wrist supported, firmly 'dumped' down to the reading position. This is repeated two or three times until there is no change in the reading. Without being moved, the thermometer is wiped with filter paper to remove any water which could distort the reading. The reading is then taken. Figs 3c and 3d illustrate steps in the technique.

The aim of the 'dumping' is to avoid 'stiction', that is, the tendency of the mercury to stick in the capillary. This happens especially in thermometers which have a fine bore. The magnitude of the effect is unpredictable but may amount to a whole division. A record is kept of all readings taken so that the history of the bulb is known and any anomalies may be noticed immediately.

* Superscript numbers indicate the equipment listed in Appendix 1.

Fig. 3a. Ice point determination: finely crushed ice.

Fig. 3b. Ice point determination: finely crushed ice mixed with distilled water.

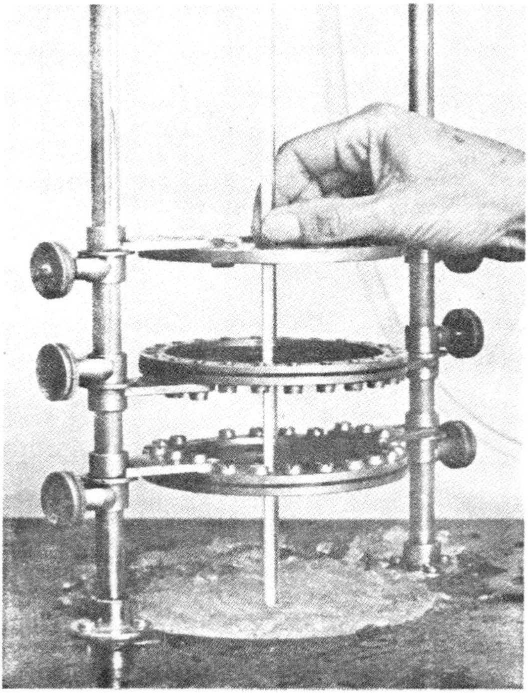

Fig. 3c. Ice point determination: thermometer raised.

Fig. 3d. Ice point determination: thermometer dumped down to reading position.

5. Preconditioning

Method A

Thermometers similar to B.S. 1900:1976 from SR2/2C to SR7/101C are kept for some weeks (months if necessary) without heating and readings are taken at the ice point as described above, say each week, until the reading is constant (Fig. 4 shows a typical recovery curve of a thermometer which had previously been used at 90°C). When the reading is constant, the thermometer is ready for calibration.

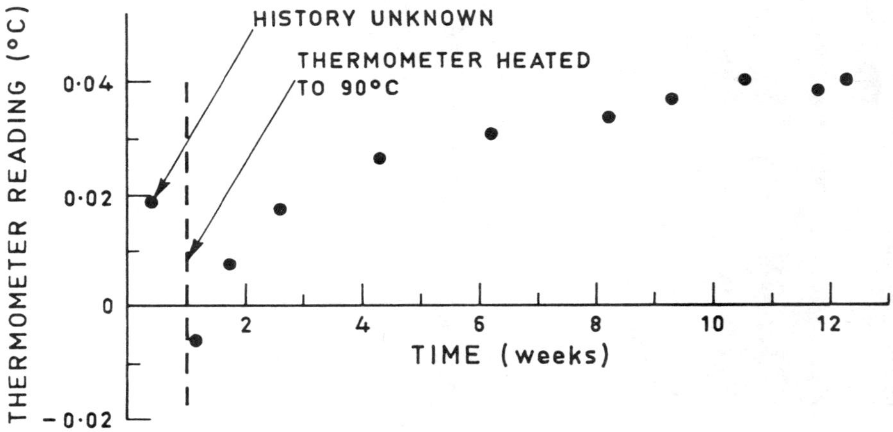

Fig. 4. Recovery from heating.

Each time the secular change is to be measured (at least once a year) the thermometer is again kept, without heating, until the reading at the ice point is constant.

Method B

This method entails greater uncertainty because the thermometer is not in a stable state when the ice point correction is measured but is suitable when it is not practicable to have thermometers out of use for long periods as is necessary with method A. The thermometer is heated to its highest test temperature for about half an hour, cooled, washed with distilled water and stored just below or at 0°C for, say, 16 hours. The ice point reading is then taken and the thermometer is calibrated.

Each time the secular change is to be measured, the thermometer is treated in the same way (including exactly the same storage time) before readings are taken at the ice point.

Method C

Thermometers similar to B.S. 1900:1976 from SR8/151C to SR12A/505C, and all other thermometers used above 100°C, are heated to the highest temperature at which they are to be used (or a few degrees above this temperature) and kept at this temperature for a few hours.

They are then allowed to cool over at least 12 hours. This can sometimes be done in the oil bath in which they are to be tested but, if the mercury in the thermometers is found to distil, they are cooled slowly in a furnace[6] designed to keep the bulb of the thermometer at the desired temperature while having the contraction chamber at or near room temperature.

An ice point determination is then made on the thermometer, leaving the thermometer in the ice-water mixture for about 5 min before readings are taken. (It is usually not necessary to store these thermometers at 0°C as the effect of changes in ambient temperature are usually less than 1/5th of the least uncertainty which would be assigned to the calibration of such a thermometer (see Appendix 2).

This procedure is repeated until the ice point reading is constant. A typical example of the results obtained from this treatment is shown in Fig. 5. The thermometer is then calibrated. After each use it is cooled in the same way as before calibration and ice point readings are taken every few months in order to determine the secular change.

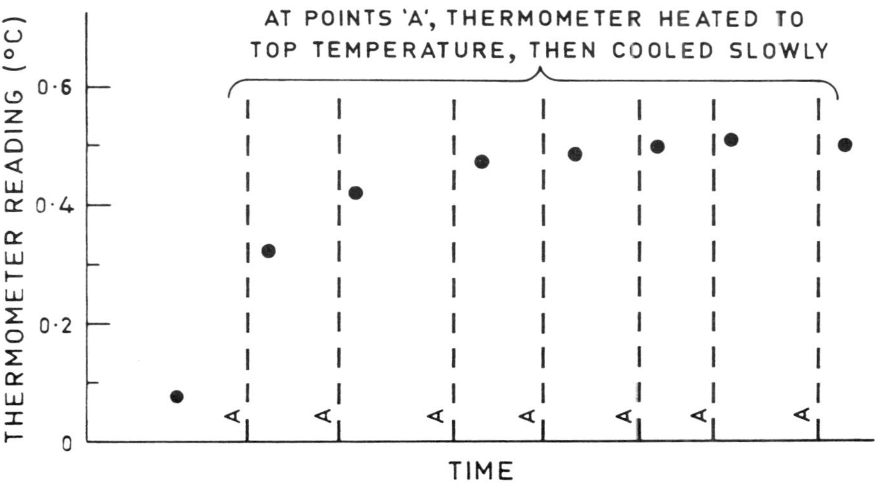

Fig. 5. Progressive changes in the ice point reading following heating to the top temperature.

Note: If the thermometer is not graduated at 0°C (eg type B.S. 1900:1976 SR11/452C), another temperature is chosen as the reference temperature.

A summary of preconditioning methods A, B, C and D is given in Appendix A.

6. Calibration Procedure

Thermometers of types similar to B.S. 1900:1976 SR1/30C to SR5/100C and SR8/151C to SR12A/505C are calibrated according to the procedure given in I below.

I. *Least Possible Uncertainty*

(a) *Ice point determination*

The ice point reading of the thermometer is taken as the first step in calibration and, when conditioning method B has been used, immediately after conditioning.

(b) *Comparison points*

At all other temperatures (with the possible exception of 100°C (see Section 7) the thermometers are calibrated by comparing their readings with those of a platinum resistance thermometer[7] which has been calibrated in terms of the International Practical Temperature Scale. The readings of the resistance thermometer are made on a resistance bridge and the comparison is carried out in baths[9,10] or in a block[11], all of which are designed and operated so as not to have temperature gradients throughout of more than one-fifth of the uncertainty required.

(c) *Resistance bridge*

For the most accurate liquid-in-glass thermometer calibrations the Division of Applied Physics uses an ac resistance bridge designed and built within the Division.* The advantage in using this bridge is that the sensitivity can be stepped down to accommodate different rates of rise in temperature of the calibration bath. There are other bridges on the market (and some pending) which may be suitable.

When the accuracy required from the bridge and resistance thermometer is not better than 0.025°C, a Leeds & Northrup 8078 Portable Precision Temperature Bridge can be used.

If a dc bridge is used instead of an ac bridge, subsections *(i)* and *(j)* below should be modified accordingly.

(d) *Arrangement of thermometers*

In order to facilitate the description of the calibration procedure, consider thermometers of type SR5/80C. The glass thermometers to be tested are placed in the appropriate well-stirred bath, with one or two mercury-filled divisions visible above the liquid level, and turned so that the mercury column appears to just touch the ends of the graduation lines, as shown in Figs 6 and 7.

(e) *Choice of test points*

In general, a thermometer is tested at intervals of 50 divisions. A test at more points than this is generally not warranted as it would not greatly decrease the uncertainty of the whole calibration (see subsection *(k)* below). However, it is important that tests be made at, and between, the manufacturer's 'pointing' marks.

(f) *Limits of test point*

When several thermometers are calibrated together and readings are taken in a sequence, it is impossible to have the thermometers all

** Instruments (type BR150/10) built to the same design may be obtained from J J Lloyd Instruments Ltd., Brook Avenue, Warsash, Southhampton SO3 6HP, England.

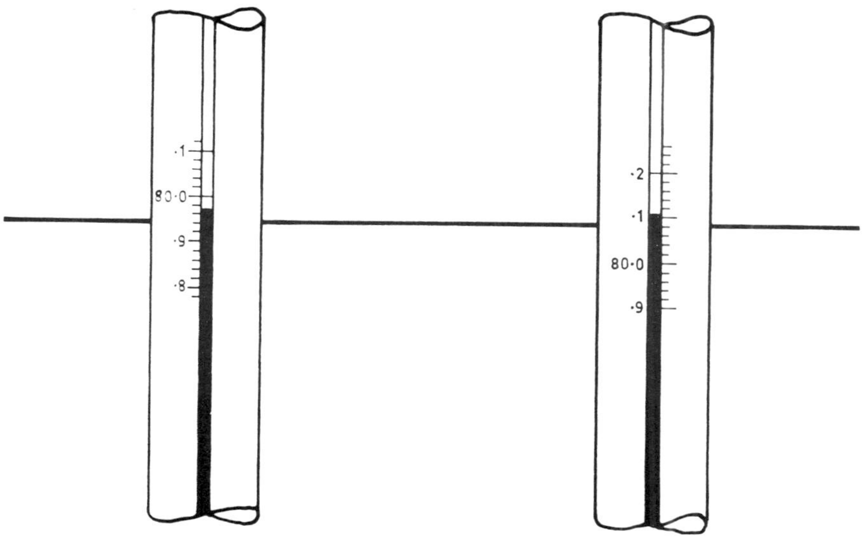

Fig. 6. Thermometers arranged in reading position.

registering exactly at the test point. As long as the readings are taken close enough to the test point, they may be considered to be at the test point. Readings taken within one-tenth of the interval from the test point to the next test point are satisfactory. With the thermometer under consideration being tested at each 1°C, the readings at 80.0°C are taken between 79.9 and 80.1°C. Unless there is some abnormality in the divisions marked on the thermometer, the differences between the corrections at two points 5 divisions apart will not usually be significant.

When a thermometer is read separately rather than in a sequence of several thermometers, it is possible to take the resistance measurement just as the mercury reaches the middle of a graduation line. In this case, the thermometer is read at graduation lines below, at, and above the test temperature.

(g) Temperature control

The bath is heated until its temperature is just below that at which readings are to be taken. It is then adjusted so that its temperature will rise at a rate small enough to avoid any lag of the resistance thermometer behind the liquid-in-glass thermometer, or *vice versa*. A rate of from 0.002°C to 0.004°C per min is appropriate when the uncertainty is to be not greater than 0.005°C. (When a greater uncertainty is acceptable, the rate may be increased appropriately.)

If the temperature were rising too rapidly, the instrument which had the slower response would lag behind the other and the difference between the readings of the two instruments would become greater for a while and then steady to a constant difference. If the temperature were rising too slowly, the mercury might rise in jumps instead of evenly. If the temperature fell or was constant, the thermometer would be subject to stiction as described in Section 4.

After readings are taken at this temperature, the bath is heated to a slightly higher temperature and the process is repeated. The taking of several different readings at each test temperature gives more knowledge of the uncertainty involved.

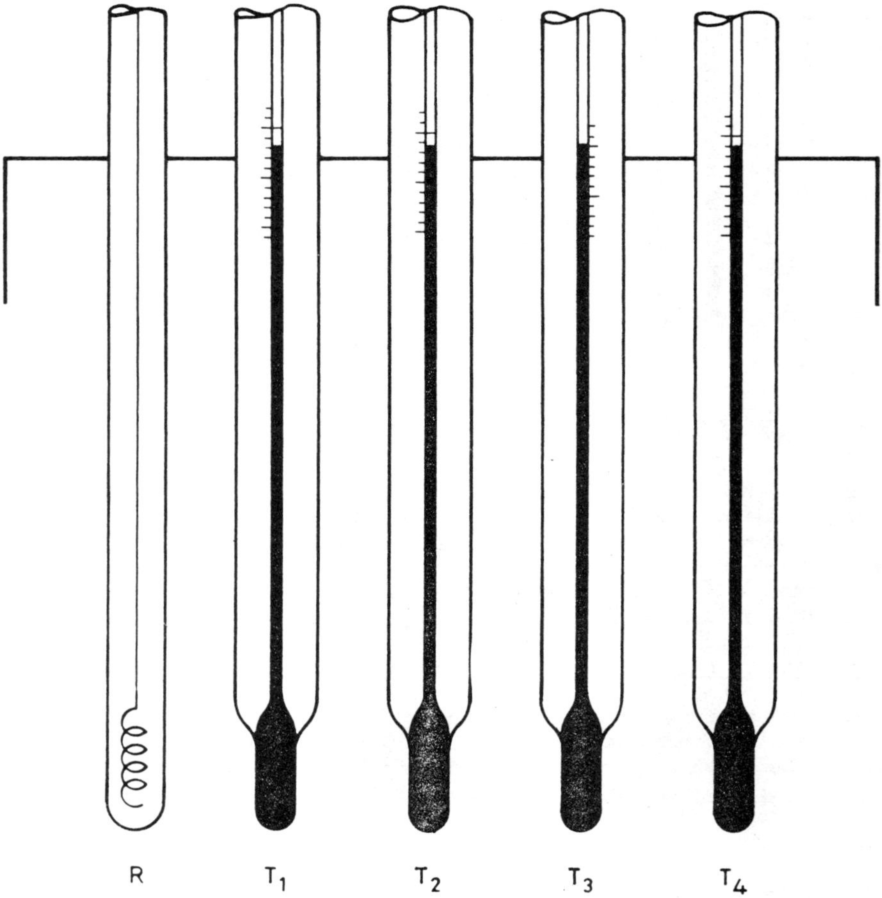

Fig. 7. Arrangement of resistance thermometer and glass thermometers.

Figure 7 shows the arrangement of the resistance thermometer and the glass thermometers. Before readings are taken the liquid-in-glass thermometer should be dumped to ensure that there is no stiction resulting from previous slight overheating above the test point.

(h) Reading

With practice, it is possible to estimate to less than one-tenth of the smallest graduations on a well-made thermometer and this figure has been used as the uncertainty of reading in subsection *(k)* below.

(i) Resistance measurement

An ac ratio bridge (see Section 6*(c)* above) is used to measure the resistance of the platinum thermometer. The resistance, R_o, of the calibrated thermometer is first measured at 0°C* and then it is placed in the bath with the glass thermometers. (The R_o value is measured whenever the thermometer is moved to a different bath or has been heated above 300°C.)

The readings are taken at a constant rate in order (1) R, (2) T1, (3) T2, (4) T3, (5) T4 and then, (6) T4', (7) T3', (8) T2', (9) T1', (10) R'.

Since the bath temperature has been set to rise at a constant rate and the times between the readings are equal, the average of each set of readings corresponds to the same time, ie.

$$\frac{R+R'}{2}, \frac{T1+T1'}{2}, \frac{T2+T2'}{2}, \frac{T3+T3'}{2}, \frac{T4+T4'}{2}$$

should all correspond to the same temperature.

(j) Calculation

Table 2. Example set of readings during calibration procedure I

Resistance reading Ω	Thermometer readings T1 °C	T2 °C	T3 °C	T4 °C
(1) 33.4907	(2) 79.951	(3) 79.982	(4) 79.954	(5) 79.978
(10) 33.4909	(9) 79.955	(8) 79.984	(7) 79.955	(6) 79.978
Average 33.4908	79.953	79.983	79.955	79.978

The temperature of the bath is calculated from the resistance reading by means of the appropriate IPTS formula.

* See Division of Applied Physics Technical Paper No. 5 "The Melting Point of Ice as a Reference Temperature for Mercury-in-glass Thermometers, Thermocouples and Platinum Resistance Thermometers".

Table 3. Example of results of calibration procedure I

Resistance reading Ω	Temperature calculated from resistance °C	Thermometer readings T1 °C	T2 °C	T3 °C	T4 °C
	(a) Complete set of readings at 80.0°C				
Observer AY			Observer BX		
33.4907	79.962	79.951	79.982	79.954	79.978
09		55	84	55	78
33.4920	79.975	66	94	66	90
22		67	96	67	90
33.4933	79.988	78	80.007	79	80.003
35		80	08	79	03
Observer BX			Observer AY		
33.4949	80.005	79.993	80.019	79.992	80.019
51		94	22	93	19
33.4963	80.019	80.008	35	80.006	31
65		09	36	07	31
33.4975	80.031	20	48	19	44
77		21	49	20	44
	(b) Averages of results				
	79.962	79.953	79.983	79.955	79.978
	79.975	79.967	79.995	79.967	79.990
	79.988	79.979	80.008	79.979	80.003
	80.005	79.994	80.021	79.993	80.019
	80.019	80.009	80.036	80.007	80.031
	80.031	80.021	80.049	80.020	80.044
	(c) Corrections				
		+0.009	-0.021	+0.007	-0.016
		+0.008	-0.020	+0.008	-0.015
		+0.009	-0.020	+0.009	-0.015
		+0.011	-0.016	+0.012	-0.014
		+0.010	-0.017	+0.012	-0.012
		+0.010	-0.018	+0.011	-0.013
	Average:	+0.009	-0.019	+0.010	-0.014
	Range:	0.003	0.005	0.005	0.004

The set of readings as in Table 2 is repeated twice by each of two observers (Table 3*(a)*) and the average of the six corrections is taken (Table 3*(b)*). The final corrections to the thermometers can then be listed (Table 3*(c)*). This complete procedure is repeated at each 1°C and the result is tabulated as below:

Correction to be applied to T1 to obtain true temperature

T1°C	Correction °C
0.0	+0.022
80.0	+0.009
81.0	+0.004
82.0	+0.008
83.0	+0.018
	etc.

When the calibration has been completed at the highest test temperature, the ice point correction is re-determined - in the case of method A after washing the thermometer with distilled water and storing it at ⩽ 0°C for approximately 16 hours; in the case of method B after washing with distilled water and storing at ⩽ 0°C for the same length of time as before calibration; in the case of method C after cooling slowly, washing with distilled water and leaving in the ice pot for about 5 minutes.

In the example, say this correction is +0.050°C (see Table 4).

(k) Uncertainty

Because the measuring process involves a physical property and a physical measuring device, results can be obtained only to a certain accuracy. For this reason, the corrections to be applied to an instrument to obtain the true temperature are quoted not as exact figures but with a tolerance which is calculated in such a way that there is 99% confidence that the measured scale corrections are within the range of figures bounded by the quoted tolerance. In this text the word 'uncertainty' is used to denote this tolerance.

In subsequent usage of the thermometer, a major source of error is the secular change in the bulb volume. This is not taken into account in assessing the quoted uncertainty of the calibration and must be determined at the recommended intervals and applied to the quoted scale corrections. The uncertainties involved in this determination, together with uncertainties in the measurement being made (due to temperature variations in the fluid whose temperature is being measured, limit of reading, etc.) must be combined (by taking the square root of the sum of the squares) with the quoted figure to obtain a total uncertainty of any measurement of temperature.

Apart from the secular change, the sources of uncertainty in a thermometer calibration are:

(1) temperature variations in the comparison bath;

(2) uncertainty of standard and measuring equipment;

(3) uncertainty of reading the thermometer;

(4) thermal instability of bulb glass during measurement;

(5) change in temperature of expansion chamber;

(6) stickiness due to capillarity;

(7) temporary depression of subsequent lower readings;

(8) irregularity of capillary between test points;

(9) difference in graduations and capillary diameter of the auxiliary scale from those of the main scale;

(10) change in ambient pressure;

(11) difference in temperature of expansion chamber when secular change is measured;

(12) uncertainty of measurement of temperature of emergent liquid column of thermometer used at partial immersion.

The uncertainty due to each of these factors could be estimated separately and combined (taking the square root of the sum of the squares) to obtain a total uncertainty.

However, when large numbers of thermometers are being calibrated, it is more practicable to assess the least uncertainty which may be expected of thermometers of given range and graduation and to then examine the results of each calibration to see if any of the factors is excessive for that particular case. The figures used by DAP are tabled in Appendix 2 and details of the separate factors given in Appendix 3.

For the thermometer in the example, the temporary depression of the reading at 80°C will be only a fraction of the temporary depression measured at 0°C. The uncertainty of the calibration is 0.02°C.

The corrections are now as shown in Table 4, the figures being rounded to approximately one-fifth of the uncertainty.

Table 4. Tabulated corrections for T1
Uncertainty 0.02°C

Reading °C	Correction °C	Temperature °C
-0.020	+0.020	0.000
80.000	+0.010	80.010
81.000	+0.005	81.005
82.000	+0.010	82.010
83.000	+0.020	83.020
etc.		
-0.050	+0.050	0.000

Because the secular change is measured easily at the ice point, it is convenient to subtract the ice point correction at the time of calibration and add the ice point correction at the time of use. This is the same as using the original correction and adding the secular change measured at the ice point (see Table 5). The easiest way of doing this is by means of a graph. The ice point correction at the time of calibration is subtracted from each of the corrections and

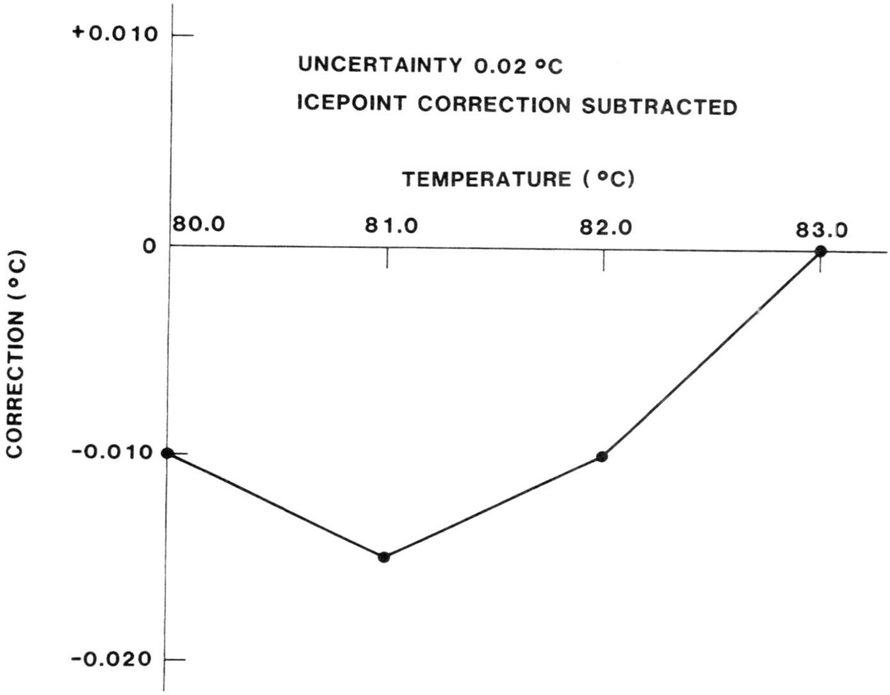

Fig. 8. Calibration graph of thermometer T1

the result drawn as in Fig. 8. Then, when the thermometer is to be used, the current ice point correction is added to the corrections as read from the graph (using linear interpolation between calibration points). So the corrections at the time of use would be as given in Table 5.

Table 5. Secular change in T1

(1) T1 reading °C	(2) Calibration Correction 1977 °C	(3) Graph °C	(4) Ice point corrections 1978 °C	(5) Ice point corrections 1981 °C	(6) Updated corrections* 1978 °C	(7) Updated corrections* 1981 °C
0.0	+0.020		+0.010	+0.005		
80.0	+0.010	-0.010			0.000	-0.005
81.0	+0.005	-0.015			-0.005	-0.010
82.0	+0.010	-0.010			0.000	-0.015
83.0	+0.020	0.000			+0.010	+0.005
etc.						

* Graph value (3) plus ice point corrections (4), (5).

II. Calibration to Greater Uncertainty

Sometimes temperatures do not have to be known to the least uncertainty which is possible with liquid-in-glass thermometers. In this case a less finely divided thermometer may be more suitable for the purpose.

Such a thermometer may be calibrated by comparison with glass thermometers whose calibrations have uncertainties of not more than half that to be expected of the thermometer being calibrated. The calibration procedure is similar to that of subsection I above, with the following additions:

(1) Thermometers to be used above 100°C are preconditioned by heating to the top test temperature and cooling slowly as in Section 5 method C above.

(2) If the uncertainty expected from the test is less than or equal to 0.02°C up to 100°C, the thermometer is preconditioned as in method A or method B before the correction at the ice point is measured.

(3) When an uncertainty greater than that in (2) is adequate or is to be expected from the type of thermometer being tested up to 100°C, the thermometer may be conditioned according to method D.

(a) Ice point determination

Unless method A or B has been used, the thermometer is left in the ice pot for about 5 min before readings are taken.

(b) Comparison points

The thermometer is tested at and between pointing marks, at intervals of approximately 50 scale divisions, eg a thermometer with range 0 to 100°C divided to 0.1°C would be tested at each 5°C.

The thermometers are arranged in the bath as follows:

S1 T1 T2 T3 T4 S2

Two liquid-in-glass standards S1 and S2 are used. If their corrected readings do not agree to within the combined uncertainties, they are examined for faults, put in order (see Section 12 below) and checked at the ice point. The bath is heated until its temperature is just below the test point and it is set to rise at such a rate that the test thermometers are seen to rise very slowly.

Readings are taken in forward then reverse order,

(1) S1	(2) T1	(3) T2	(4) T3	(5) T4	(6) S2
(12) S1'	(11) T1'	(10) T2'	(9) T3'	(8) T4'	(7) S2'

and the readings averaged,

$$\frac{S1+S1'}{2}, \quad \frac{T1+T1'}{2}, \quad \frac{T2+T2'}{2}, \quad \frac{T3+T3'}{2}, \quad \frac{T4+T4'}{2}, \quad \frac{S2+S2'}{2}.$$

The corrections to the standards are then applied and the average of the corrected readings taken as the bath temperature. The corrections which have to be applied to the test thermometers to obtain the true temperature are calculated from this.

An example set of readings is shown in Table 6. The corrected standard readings of 4.956°C and 4.954°C agree to within the combined uncertainties of the standards.

Thus the average of corrected standard readings = 4.955°C.

Corrections to be applied to the test thermometers to obtain the true temperature then are:

T1 °C	T2 °C	T3 °C	T4 °C
-0.040	-0.023	0.000	-0.005

Further readings are made, by two readers, and the results averaged. The complete set of results is shown in Table 6*(b)*. This procedure is repeated at each 5°C (taking readings somewhere between 9.950 and 10.050, 14.950 and 15.050°C, etc).

Table 6. Example of results of calibration procedure B

S1 and S2 graduated to 0.02°C, uncertainty of corrections 0.005°C
T1, T2, T3, and T4 graduated to 0.1°C

(a) First set of readings

	Standard S1 °C	T1 °C	Test thermometers T2 °C	T3 °C	T4 °C	Standard S2 °C
	4.928	4.995	4.975	4.955	4.960	4.949
	31	95	80	55	60	49
Average	4.930	4.995	4.978	4.955	4.960	4.949
Correction from graph	+0.016					-0.002
Ice point correction	+0.010					+0.007
Corrected standard reading	4.956					4.954

(b) Complete set of readings with corrections to be applied to T1, etc. calculated

Observer	Standard S1 °C	T1 °C	Test thermometers T2 °C	T3 °C	T4 °C	Standard S2 °C	Av. Corrected S reading °C
AY	4.928	4.995	4.975	4.955	4.960	4.949	
	31	95	80	55	60	49	
	4.930	4.995	4.978	4.955	4.960	4.949	
	+0.026	-0.040	-0.023	0.000	-0.005	+0.005	
	4.956					4.954	4.955

BX	4.945	5.000	4.995	4.965	4.970	4.963	
	47	05	95	70	75	63	
	4.946	5.003	4.995	4.968	4.973	4.963	
	+0.026	-0.033	-0.025	+0.002	-0.003	+0.005	
	4.972					4.968	4.970
AY	4.974	5.020	5.020	4.990	5.015	4.994	
	78	25	20	95	18	94	
	4.976	5.023	5.020	4.993	5.017	4.994	
	+0.027	-0.022	-0.019	+0.008	-0.016	+0.005	
	5.003					4.999	5.001
BX	4.990	5.050	5.025	5.015	5.025	5.007	
	94	53	30	20	30	07	
	4.992	5.052	5.028	5.018	5.028	5.007	
	+0.027	-0.036	-0.012	-0.002	-0.012	+0.005	
	5.019					5.012	5.016
Average:		-0.033	-0.020	+0.002	-0.009		
Range:		0.018	0.013	0.010	0.013		

When the calibration is complete, an ice point determination is made, after storage at ambient temperature for about 16 hours, in the same way as before, and the results of the calibration are tabulated as follows:

Correction to be applied to T1

Temperature °C	Correction °C
0	+0.025
5	-0.033
10	-0.006
15	+0.002
20	0.000
25	+0.034
30	-0.002
35	-0.016
40	+0.009
45	+0.037
50	+0.002
0	+0.050

In this calibration, the temporary depression of zero is greater than the normal value for a thermometer of this range so the uncertainty is assessed as 0.03°C.

The results are tabulated as in Table 7, rounding the figures off to approximately one-fifth of the uncertainty.

Table 7. Tabulated corrections for T1
Uncertainty 0.03°C

Reading °C	Correction °C	Temperature °C
-0.025	+0.025	0.000
5.000	-0.035	4.965
10.000	-0.005	9.995
15.000	0.000	15.000
20.000	0.000	20.000
25.000	+0.035	25.035
30.000	0.000	30.000
35.000	-0.020	34.980
40.000	+0.010	40.010
45.000	+0.035	45.035
50.000	0.000	50.000
-0.050	+0.050	0.000

7. Steam Point

If a thermometer is not graduated at 0°C but is graduated at 100°C, the steam point may be used as a reference temperature. The thermometer is preconditioned according to method C (Section 4) and is then placed in a hypsometer[19]. After about 10 min, readings are taken of the thermometer and also of the ambient pressure, the latter by means of a mechanism or a Fortin barometer[20] capable of being read to 0.05 mm. The barometer should be fairly close to, and at the same height above sea level as, the hypsometer. Calibration, altitude and temperature corrections are made to the pressure reading and the temperature of the steam is calculated.* The difference between the temperature of the steam and the thermometer reading gives the correction which has to be made to the thermometer reading at 100°C.

8. Partial Immersion

When a thermometer is used at partial immersion its reading depends to some extent on the temperature of the emergent liquid column. The increase δ in the reading of a thermometer due to an increase of t degrees in the mean temperature of a liquid column of length equivalent to n degrees may be expressed by the formula

$$\delta = knt$$

where k = 0.000 16/°C for a mercury-in-glass thermometer and 0.001/°C for a spirit-in-glass thermometer.

* International Practical Temperature Scale of 1968.
The equation of $t^{68} = [100 + 28.0216\ (p/p_o - 1)]$ will introduce an error of less than 0.01°C.

If there are graduations down to the immersion line, the value of n can be read directly off the thermometer scale. If, however, the immersion line is below the lowest graduation line, a rule or calipers can be used to measure the length which is exposed, and this length read off in degrees from a graduated part of the stem.

Also, for a thermometer used at partial immersion, the average temperature of the emergent column must be measured during the calibration so that a correction can be made for different conditions during later use. Two methods of measuring this average value are described below. The results obtained by method *(a)* are closer to those obtained by calculation from total immersion conditions because the Faden thermometers compensate for the effect of conduction up the stem of the main thermometer.

(a) By means of a Faden thermometer

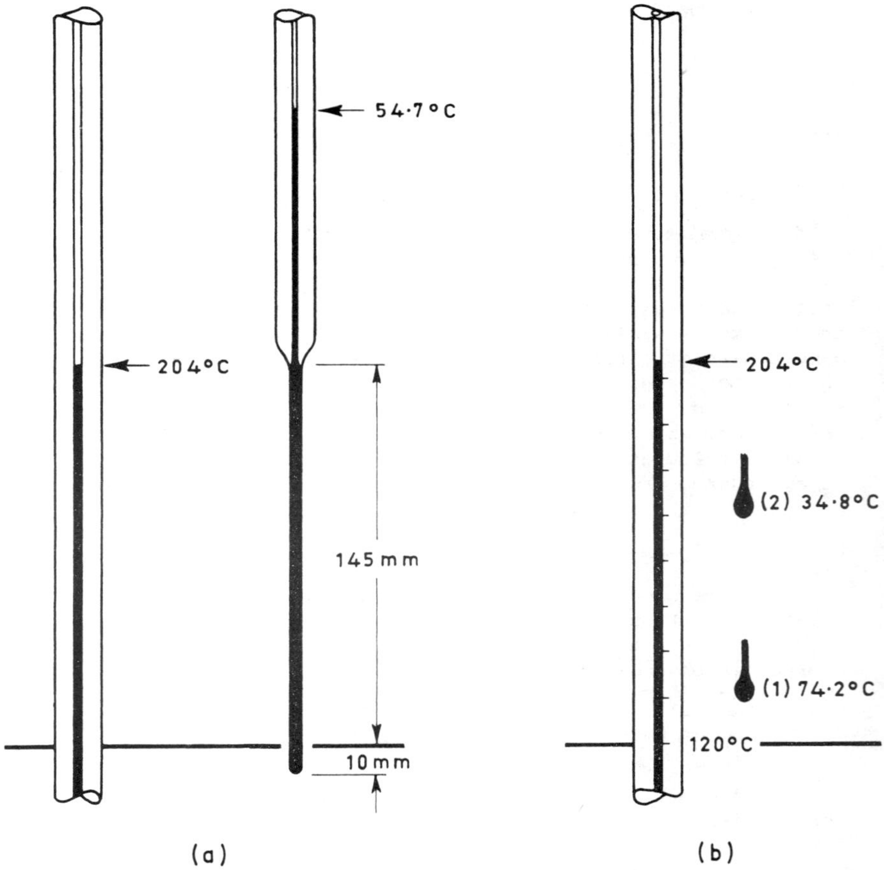

Fig. 9. Mean temperature of an emergent mercury column, measured by means of:
(a) a Faden thermometer, and *(b)* short-stem thermometers.

Faden thermometers are designed to have their bulbs simulate as closely as possible, the length of emergent column whose temperature is to be measured. They can be obtained with bulbs ranging in length from 55 mm to 305 mm (Appendix 1).

It has been found that accurate results are obtained by using these thermometers with at least 10 mm of the bulb immersed in the comparison bath. This ensures that the conditions of the Faden thermometer bulb reproduce those of the emergent stem.

As an example (see Fig. 9*(a)*), the length of the emergent column of the main thermometer is 145 mm so a Faden thermometer of length 155 mm is placed alongside the main thermometer with 10 mm of its bulb immersed in the liquid medium. The Faden thermometer reads, say, 54.7°C while the main thermometer reading is, say, 204°C. The average temperature of the emergent liquid column is therefore

$$\frac{54.7 \times 155 - 204 \times 10}{145} = 44.4°\text{C}.$$

(b) By means of short-stem thermometers placed alongside the emergent stem

The number of short-stem thermometers used depends on the length of the emergent liquid column. The middle of the bulb of the lowest short-stem thermometer should be not more than 25 mm above the top of the comparison bath because there is a large temperature gradient in this region. Other short-term thermometers should be positioned at intervals of about 100 mm.

In Fig. 9*(b)*, the middle of the bulb of short-stem thermometer (1) is placed level with the 130°C graduation line (18 mm above the top of the bath) and that of (2) level with the 172°C graduation line. The readings of these short-stem thermometers are weighted according to the length of liquid column they correspond to, i.e. in the example, from 120 to 140°C is at an average temperature given by the reading of short-stem thermometer (1) and from 140 to 204°C is at an average temperature given by the reading of short-stem thermometer (2). The average temperature is therefore

$$\frac{74.2 \times 20 + 34.8 \times 64}{84} = 44.2°\text{C}.$$

This temperature (measured by either method) is then rounded off in DAP Reports to a figure which would not alter the quoted correction by one-tenth of the quoted uncertainty. The rounded figures are tabulated with the rest of the calibration.

9. Temperature of Expansion Chamber

With gas-filled thermometers, an increase in the temperature of the expansion chamber will cause a decrease in the reading. In most circumstances, the effect will be of the same order in use as it was during calibration and the uncertainty assigned to the calibration should be broad enough to cover any slight variations or any effect resulting from difference in conditions when the secular change is measured.

In two special cases, however, the effect of increasing the temperature of the expansion chamber is measured and quoted in the calibration Report, together with the temperature of the expansion chamber at the time of calibration. This is done for bomb-calorimeter thermometers so that they may be used within an uncertainty of 0.003°C and also for thermometers of type ASTM22 or IP32 which may be used either in oil or in water at 98°C.

10. External Pressure

The effect of an increase in external pressure on the bulb of a thermometer is of the order of 0.0001°C per mm of mercury (133.3 Nm^{-2}). In the case of a thermometer with a reference point, the effect of a major change in ambient pressure would be registered at the reference point as well as at the temperature of measurement and so would be compensated for.

The effect of variations in ambient pressure at the time the secular change is measured is taken into account in assessing the uncertainty of the calibration.

In two cases, however, the effect is measured and given in DAP calibration Reports. These are:

(1) unprotected deep-sea reversing thermometers which are designed to register and measure pressure,

(2) bomb-calorimeter thermometers, designed for differential measurements only, when used for absolute measurements at altitudes different from those at which they were calibrated.

11. Precautions

Care must be taken that the thermometers are vertical and the line of sight at right angles to the thermometer.

At high temperatures, thermometers are examined to see if there is any distillation of mercury. To avoid distillation, a portion of the mercury column is left exposed until a few minutes before readings are taken, but if distillation still occurs, the thermometer is tested with, say, 5 or 10 divisions exposed and this is given in the report as the immersion at this point.

In the case of spirit-in-glass thermometers, sufficient time must be allowed for all the liquid to drain out of the capillary above the liquid column. This is usually of the order of half an hour.

12. Mending of Broken Liquid Column, etc.

If the thermometer has a break in the liquid column or a bubble of gas trapped in the bulb, the bulb is cooled (eg with dry ice) until all the liquid is in the bulb (or contraction chamber). The thermometer is tapped gently to unite the liquid. The thermometer is held vertically and allowed to warm up. On no account should the bottom of the bulb be warmed first as the mercury in the bulb may have frozen (-38°C). If the bottom were to be heated, some of the mercury would melt and expand, cracking the bulb. Sometimes it is best to make a large bubble in the bulb (by tapping the horizontal thermometer gently when all the mercury is in the bulb) and warm the thermometer so that the separated mercury rises up the capillary and picks up any isolated drops. It can then be mended as before. Broken columns can usually be mended by cooling the bulb and it is the best method as it avoids heating the bulb.

If this procedure does not work, it may be possible to mend the broken column by centrifuging or by heating the expansion chamber to distil separated liquid into the main reservoir (particularly with the case of a spirit-in-glass thermometer). With any of these methods it is important not to overheat the bulb as this may:

(1) alter the calibration, or

(2) break the instrument.

13. Identification number

If the thermometer does not already have a serial number, a number can be etched on it as follows.

A thin coating of nail polish or wax (prepared by melting together equal parts of paraffin wax and beeswax) is applied by means of a brush or knife. The number is engraved using a pantograph or a mounted needle. A wall of petroleum jelly is built around the number with a syringe and the inside of the wall is filled with hydrofluoric acid (using an eye dropper).* The acid is allowed to remain for 5-10 min and then the thermometer is cleaned.

The number is blackened using the mixture:

1 drop water glass (sodium silicate),
1 drop water,
MnO_2 or other black powder.

A finger is dipped into the water glass and water, then into the black powder, and is then rubbed across the area to be blackened. Just as the mixture starts to dry (about half a minute), it is wiped off with a dry tissue, leaving the blackening in the engraving. The thermometer is kept dry for a further half hour.

References

Kemp, W.R.G. (1977). "International Practical Temperature Scale of 1968 (Amended Edition of 1975)". Australian version of the official French text. CSIRO National Measurement Laboratory.

Beavis, I.M. (1982). Division of Applied Physics Technical Paper No. 5. "The Melting Point of Ice as a Reference Temperature for Mercury-in-glass Thermometers, Thermocouples and Platinum Resistance Thermometers".

* Hydrofluoric acid is a particularly dangerous acid. Contact of the liquid or fumes with parts of the body must be avoided. Adequate ventilation must be provided.

Appendix 1

General Equipment

The equipment listed is necessary for all but a crude standards laboratory, but is also sufficient for calibrations to the least possible uncertainty over the range covered by glass thermometers.

1. Ice pot (Drawings PD 567, PB 568, P 776); thermometer support (P 882-3).*
2. Ice maker (suggested supplier Crystal Tips, 59 Padstow Road, Padstow, NSW 2211).
3. Distilled or de-ionised water supply.
4. Ice crusher (Crystal Tips or similar).
5. Stirrer for ice point (see Fig. 3*(a)*.
6. Conditioning furnace (Drawings P 417-1, -2).
7. Platinum resistance thermometers.
8. Resistance bridge (see Section 6*(c)*).
9. Water bath (Drawings P 1043-1, -2, -3, -4, -5, -6, -7, -8, -9
Oil bath P 1136-1, -2, -3, -4, -5, -6, -7
Alcohol bath P 1099-1, -2, -3
P 1312-1, -2, -3).
Commercial baths are also obtainable, e.g. Heto, Jones, Tamson.
10. Salt bath (Drawings PC 3044 to 3052, PE 3065, PP 2692, P 328, 510, PC 3171 to 3173, P 598-1, -2).
11. Low temperature block (Drawings P 1114-1, -2, -3).
12. Examination table and thermometer storage cabinet (Drawings PC 2181, PC 7281, PC 5055 to 5058, P 314).
13. Strain viewer (Drawing P 882-2).
Commercial models are available from Hilger & Watts, London; Herbert Arnold, West Germany, Carlisle Gas Burner Corpn., USA.
14. Etching equipment
    (a) pantograph (Drawings PC 2959 to 2961, PC 3038 to 3040)
    (b) hydrofluoric acid
    (c) equal parts paraffin and beeswax
    (d) petroleum jelly.
15. Rubber o-rings (Consolidated Bearing Co., Artarmon, NSW 2064).
16. Metal bushes (samples available on request from DAP).
17. Storage refrigerator.
18. Viewing system (see Guide Book of Japanese Optical Precision Instruments; Guide to Scientific Instruments, special annual issue of Science; The Optical Industry & Systems Purchasing Directory (published by the Optical Publishing Co. Inc., Pittsfield, Ma.))

* Available on request from Division of Applied Physics.

19. Hypsometer (Y 147 + circuit diagram).**
20. Fortin or mechanism barometer.**
21. Report forms (samples available on request).
22. Register cards (samples available on request).
23. Exhaust fan and fume cupboard.
24. Salt (Houghton's Draw Temp 275).
25. Silicone oil (MS 550) or paraffin or vegetable oil.
26. Stainless steel or glass conservatoire.
27. Gloves for use with salt bath.
28. Mercury-in-steel control thermometer.
29. Suggested room layout (Drawing P 882-1).

Reference Thermometers

In view of the individual behaviour of thermometers, it is essential that the reference thermometers be of the best available quality. It has been found that the following manufacturers produce reliable instruments.

Dobbie Instruments, 18-20 George Street, Sandrignham, Vic. 3191. (Branch: Queens Road & Regatta Road, Five Dock, NSW 2046.)

Arno Amarell, Postfach 1280, 6983 Kreuzwertheim, West Germany. (Agent: J M Scientific Pty Ltd, 61-63 Victoria Avenue, Chatswood, NSW 2067.)

G H Zeal, Lombard Road, Merton, London SW19, England (Agent: Arthur Bailey Pty Ltd, 83-88 Chandos Street, St Leonards, NSW 2065.)

Thermometers should be obtained in sets of three each.

(1) Thermometers to be used as secondary reference thermometers (see B.S. 1900: 1976).

(2) -50 to 20°C mercury-thallium in glass, divided to 0.2°C.

(3) Faden thermometers with bulb lengths from 55 mm to 305 mm in 10 mm steps. (Karl Schneider & Sohn, Am Bildaker 14, Postfach 58, 6980 Wertheim 1, West Germany.)

** Only used by DAP for measuring effect of increase in temperature of expansion chamber of thermometers similar to ASTM 22.

Appendix 2

Least uncertainty to be expected from liquid-in-glass thermometers

These figures are based on results from large numbers of thermometers and do not necessarily apply to any given instrument.

Section 6*(k)* on uncertainty <u>must</u> be in conjunction with this table.

Thermometer divided to °C	°C -183 to 50	°C -80 to 50	°C -50 to 50	°C -10 to 50	°C -10 to 100	°C -10 to 200	°C -10 to 300	°C -10 to 450	°C -10 to 500
0.01				0.005 0.005	0.01 0.01				
0.02			0.02 0.02	0.01 0.01	0.02 0.02				
0.05			0.05 0.05	0.01 0.02	0.02 0.05	0.05 0.05			
0.1			0.1 0.1	0.02 0.02	0.02 * 0.05	0.05 0.1	0.1 0.2	0.2 0.5	
0.2			0.1 0.1	0.05 0.05	0.05 0.1	0.05 0.1	0.1 0.2	0.2 0.5	
0.5	0.5 0.5	0.5 0.5	0.2 0.5	0.1 0.2	0.1 0.2	0.1 0.2	0.2 0.2	0.2 0.5	1 1
1	1 1	1 1	0.2 0.5	0.2 0.2	0.2 0.2	0.2 0.5	0.5 0.5	0.5 0.5	1 1
2	±2 ±2	1 1	0.5 1	0.5 0.5	0.5 0.5	0.5 0.5	0.5 1	1 1	1 1

The upper figure in each case applies to thermometers tested immersed to the reading
The lower figure applies to thermometers tested at partial immersion

Immersion < 57 mm, 0 to 50°C, uncertainty not less than 0.03°C
< 57 mm, 50 to 100°C, " " " " ½ division
< 57 mm, > 100°C, " " " " 1 division
76 mm, > 100°C, " " " " ½ division
spirit, " " " " 1 division

* This figure applies only when the thermometer is treated in accordance with Preconditioning Method A or B. In all other cases, 0.05 is applicable.

Appendix 3

Uncertainty factors

The figures tabulated in Appendix 2 are the least uncertainties to be expected from well-made thermometers calibrated under good conditions. The uncertainty of the calibration will be increased unless the following conditions apply.

(1) Temperature variations in the comparison bath are of the order of 0.002°C up to 250°C and 0.1°C above 250°C.

(2) The standard and measuring equipment is known to within 0.002°C.

(3) The thermometer can be read accurately to one tenth of the smallest graduation interval.

(4) The thermometer is kept at the calibration temperature long enough to reduce the effect of further expansion of the glass to about 0.002°C.

(5) The thermometer is always used at the correct immersion and under normal ambient conditions so that the temperature of the expansion chamber will not vary greatly. See Section 9.

(6) To overcome capillarity effects, readings are taken with the temperature rising except at the ice point where the thermometer is dumped before each reading is taken.
Note: Even when these methods are used, the readings of finely divided thermometers may still be affected.

(7) The temporary depression of zero is normally about 0.04°C per 100°C rise in temperature and thermometers for use above 100°C are cooled slowly after each heating to remove this effect.

(8) The error which can be incurred by using linear interpolation between calibration temperatures is due to changes in the capillary diameter. Thermometers are pointed at intervals in order to keep the corrections small but the pointing* can itself be faulty. It has been found that corrections generally vary most about half-way between pointing marks so the thermometer is tested at, and half-way between, pointing marks.

(9) There are no visible anomalies in the bore or graduations of the auxiliary scale.

(10) The thermometer is used under normal ambient pressure. See Section 10.

(11) The thermometer is used under normal ambient conditions - preferable in an air-conditioned room so that there is no change in the temperature of the expansion chamber.

(12) The partial immersion thermometer is used at the correct immersion and under normal ambient conditions so that the temperature of the emergent liquid column will not vary greatly.

*The temperatures at which a thermometer was pointed are often evidenced by marks or lines scratched on the stem. In other cases they can be deduced from differences in the length of, say, 100 divisions.

APPENDIX 4. Summary of preconditioning methods for BS 1900:1976 thermometers

I. Calibration to least possible uncertainty			II. Calibration to greater uncertainty in range 0 to 100°C
SR2/2C to SR6/101C: SR7/51C, SR7/101C for uncertainty ≤0.05°C	SR2/2C to SR6/101C: SR7/51C, SR7/101C for uncertainty ≤0.05°C	SR8/151C to SR12A/505C	SR7/51C, SR7/101C for uncertainty ≥0.05°C
Method A	Method B	Method C	Method D
Store at ≤0°C at least overnight	Heat to top temperature for about half an hour	Heat to top temperature for a few hours	Store at ambient temperature for 1 month (approx.)
Measure ice point correction		Cool over at least 12 hours (b)	
		+	
Repeat these two steps at intervals over several months until ice point correction is constant	Store at 0°C for approximately 16 hours (a)	Measure ice point correction after 5 min in ice	
		+	
		Repeat these 3 steps until ice point correction is constant	
+		+	
Measure ice point correction	Measure ice point correction	Measure ice point correction	Measure ice point correction after 5 min in ice
Calibrate	Calibrate	Calibrate	Calibrate
Store at ≤0°C overnight	Store at ≤0°C for same period as above (a)	Cool from top temperature over same period as above (b)	Store at ambient temperature for 16 hours (approx.)
		+	
Measure ice point correction	Measure ice point correction	Measure ice point correction after 5 min in ice	Measure ice point correction after 5 min in ice
Estimate temporary depression at lowest temperature of working range. Use this figure in the uncertainty assessment	A more negative correction would indicate secular change	A more negative correction would indicate secular change	Estimate temporary depression at lowest temperature of working range. Use this figure in the uncertainty assessment
			Repeat last 3 steps to obtain time needed for recovery
Use for 6-12 months	SR2/2C to SR6/101C - use. SR7/51C, SR7/101C - before each use, heat to top temperature and store overnight at temperature below temperature of use	After each use, cool over same period as above (b)	Use
Measure secular change as described below at intervals specified in Section 3 of text			
Store at ≤0°C at least overnight each time ice point correction is measured until there is no longer any temporary depression effect and ice point correction is constant over at least a month	Heat to top temperature	Cool from top temperature over same period as above (b)	Store at ambient temperature for time needed for recovery
	Store at ≤0°C for same period as above (a)	+	+
	Measure ice point correction	Measure ice point correction after 5 min in ice	Measure ice point correction after 5 min in ice

Platinum Resistance Thermometry

J. J. Connolly
CSIRO Division of Applied Physics, Sydney, Australia 2070

Abstract

The International Practical Temperature Scale of 1968, Amended Edition of 1975, is realized in the range 0 to 630.74°C by the Standard Platinum Resistance Thermometer. However the procedures required in such a realization are time consuming, expensive and the thermometers are very fragile. An alternative approach is to use an Industrial Platinum Resistance Thermometer as a transfer standard. The advantages and limitations of this approach are discussed.

1. Introduction

Resistance thermometers use the change in electrical resistance with temperature to measure temperature, and a wide variety of materials have sufficiently sensitive temperature coefficient of resistance to be used as thermometers. These range from pure metals through alloys to semiconductor materials. However stability, reproducibility and linearity requirements limit the choices to just a few.

Platinum, copper and nickel are the only metals used to any extent and the latter two are rather restricted in temperature range because of corrosion problems. Modern construction techniques have made platinum the most common sensor material for precision resistance thermometers and its use for such sensors is likely to increase. The vast bulk of precise thermometry is in the range 0 to 250°C, and especially 0 to 100°C, and this is best achieved with resistance thermometers. Alloys and semiconductor materials, both with positive and negative temperature coefficients, have many specific applications but stability considerations prevent application in precision temperature measurement except in limited temperature ranges. This treatment of resistance thermometry will be restricted to platinum resistance sensors.

2. Platinum

The obvious question is why is platinum used as a sensor material when its high cost is considered. The answer is simple; the amount of platinum used in thermometers is very small and the overall cost of the instrument is governed more by the problems of construction than by the cost of component materials.

Platinum has many of the desirable properties for a temperature-sensor material. It is inert and corrosion resistance, has a high melting temperature and has a predictable and reasonably linear resistance-versus-temperature characteristic. (The quadratic coefficient is less than 0.02% of the linear coefficient.) Platinum can be obtained in a very pure state, draws fairly easily, has a high electrical resistivity and hence can be fabricated into a reasonably compact sensor. On the debit side, pure platinum recrystallises above 400° C and is subject to severe grain growth at high temperatures. The wire becomes so ductile that there can be large scale movement of a minimally supported coil. There is also a range of temperature where volatile oxides of platinum form and this effect can cause dimensional changes to the wire as well as insulation problems in cooler parts of the thermometer where the deposited oxide reverts to platinum metal.

Platinum resistance thermometers can be categorised roughly into two classes.

(i) standard Platinum Resistance Thermometers (SPRT) are those which are used as the interpolating instrument of the International Practical Temperature Scale of 1968 (Amended Edition of 1975), (IPTS-68), in the temperature range of 13.81 K to 630.74° C.

(ii) Industrial Platinum Resistance Thermometers (IPRT).

The general characteristics and calibration procedures of each type will be discussed separately.

3. Standard Platinum Resistance Thermometers

(a) International Practical Temperature Scale of 1968 (Amended Edition of 1975)

The SPRT is the interpolating instrument of IPTS-68 over the temperature range 13.81 K to 630.74° C. The scale is established by defining the temperatures of a series of reproducible phase transitions in pure substances and setting out the requirements both for the properties of the interpolating SPRT and the method for its calibrations. Below 0° C the interpolating procedure is in the form of a standard reference function, equation 5 of IPTS-68, and an individual SPRT is calibrated by a resistance measurement at each of seven fixed points. From these determinations a series of six deviation equations are obtained, each covering a small temperature region between fixed points. The procedures and equations are set out in IPTS-68.

In the temperature range 0 to 630.74° C the calibration procedure is a little less complicated. Two defined temperatures as well as the water triple point are used. These are the freezing point of pure zinc (419.58° C) and the freezing point of pure tin (231.9681° C), the latter being a more easily obtained alternative to the water boiling point (100° C).

The SPRT's resistance is measured at these points and the constants R_o, α and δ (or R_o, A and B) of the interpolation equations are obtained.

$$t_{68} = t' + 0.045\left(\frac{t'}{100^\circ C}\right)\left(\frac{t'}{100^\circ C} - 1\right)\left(\frac{t'}{419.58^\circ C} - 1\right)\left(\frac{t'}{630.74^\circ C} - 1\right)^\circ C \quad \ldots(10) \quad \text{IPTS-68}$$

where t' is defined by the equation

$$t' = \{ \frac{1}{\alpha}[W(t') - 1] + \delta \left(\frac{t'}{100°C}\right)\left(\frac{t'}{100°C} - 1\right)\}°C \qquad \text{... (11a) IPTS-68}$$

where

$$W = \frac{R(t')}{R(0°C)}.$$

Equation (11a) is equivalent to the equation

$$W = 1 + At' + Bt'^2 \qquad \text{... (11b) IPTS-68}$$

As can be seen these equations amount to a quadratic relationship between resistance ratio and t' with the IPTS-68 temperature, t_{68}, being obtained from equation (10) which is independent of the thermometer.

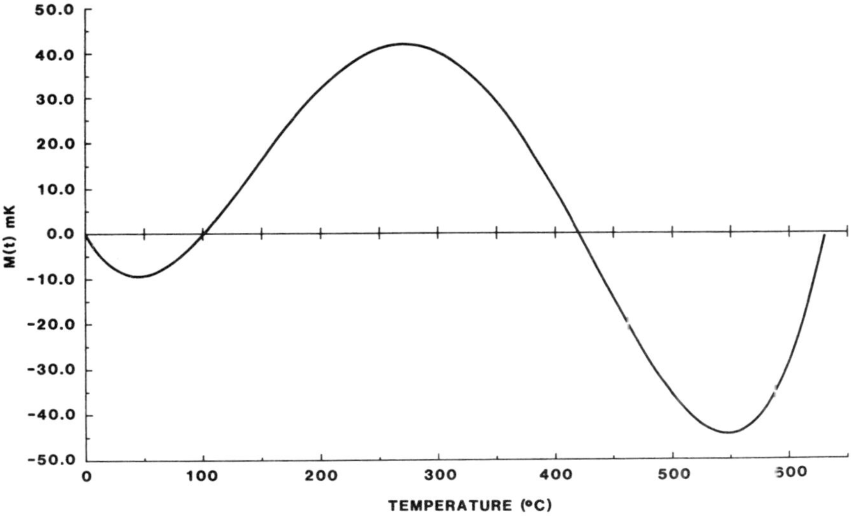

Fig. 1. Plot of M(t') against t_{68} for the range 0 to 630.74°C

Figure 1 is a plot of M(t') with respect to t' where M(t') is defined from equation (10) by

$$M(t') = t_{68} - t'.$$

This is a slowly varying function with extremes of about -0.01 K at 45°C, +0.04 K at 250°C and -0.04 K at 549°C.

(b) SPRT Sensors

The IPTS-68 lays down the general requirements for an SPRT resistor. The platinum must be strain-free, annealed and pure and its resistance ratio W defined by

$$W = R_{t_{68}}/R(0°C)$$

at t_{68} = 100°C must not be less than 1.392 50. This is equivalent to requiring the "α" coefficient of (11a) to be greater than 0.003 925 0 $(°C)^{-1}$. Typical sensor structures are shown in Fig. 2.

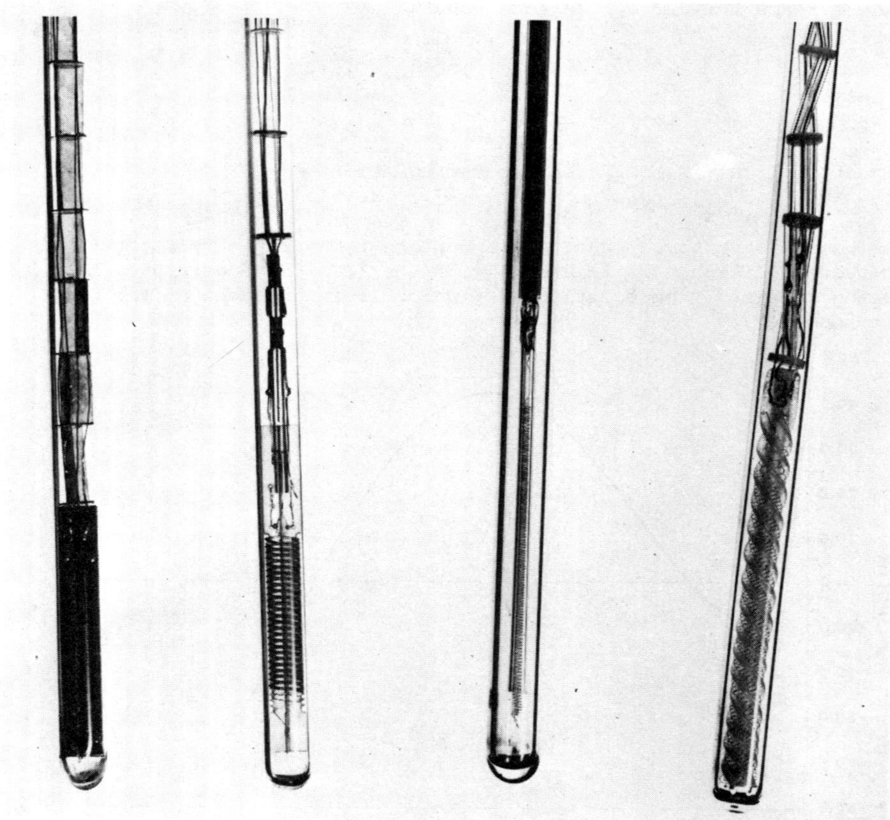

Fig. 2. Typical standard platinum resistance thermometer sensors

Most modern patterns use silica as the mounting structure for the platinum wire. It can be obtained in a very pure state and can be readily fabricated into appropriate shapes using modern technology, for example, laser machining. Older designs used a mica-cross construction. Mica has excellent insulation properties but has disadvantages for higher temperature use in that above 500°C, and sometimes at much lower temperatures, it gives up water both from between leaves and also from its crystal structure. This will cause a lowering of the insulation resistance. Mica also has iron in its crystalline structure and, at high temperatures, this can diffuse into the platinum wire causing a degradation of the wire purity and temperature coefficient of resistance.

(c) SPRT Calibration

(i) Water Triple Point. The water triple point provides the most accurate and easily attained reference temperature. The purified and gas-free water is normally contained in a sealed glass cell with a re-entrant thermometer well. Ice is formed around the well by cooling and the cell is maintained frozen by immersion in an ice bath. Figure 3 shows a typical frozen water triple point cell.

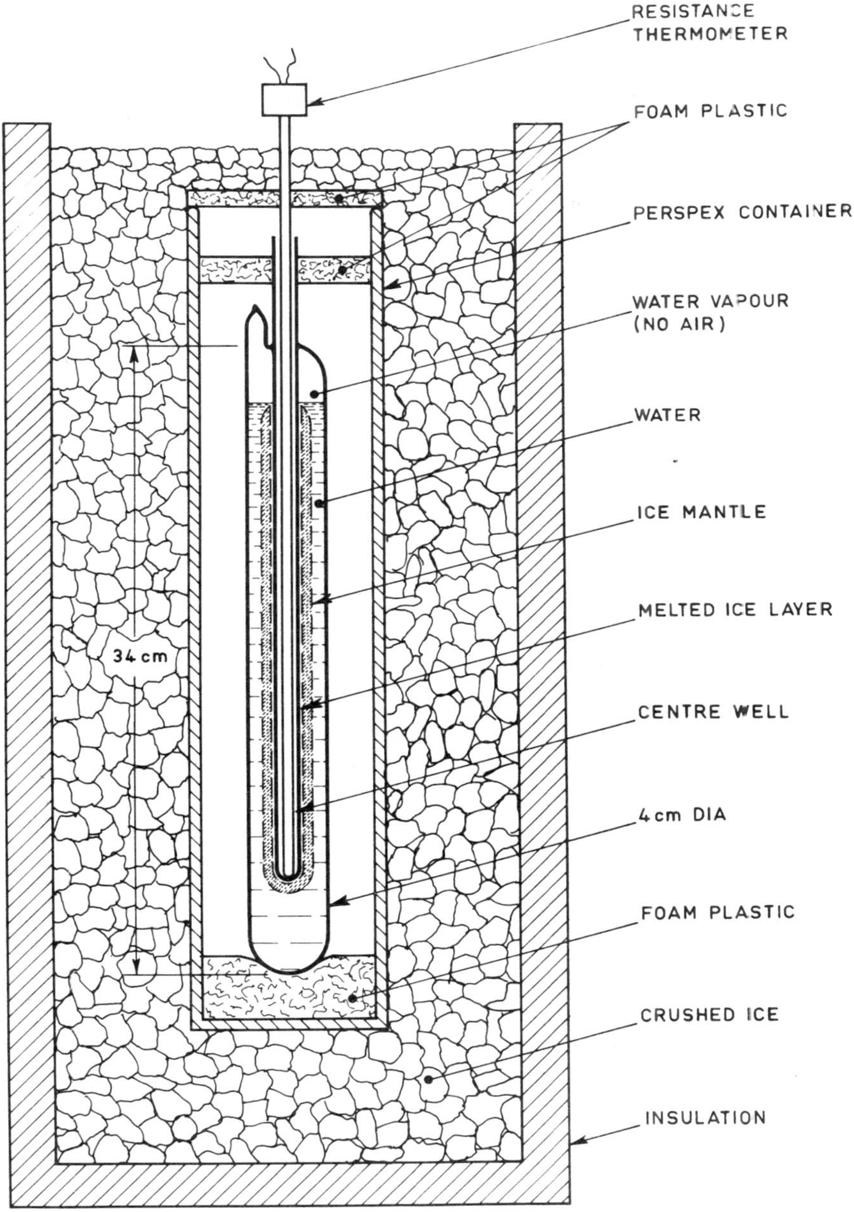

Fig. 3. Water triple point assembly

Although the glassware can be readily constructed by a competent glassblower, the cleaning and procedure for filling with pure degassed water is time-consuming and requires care and patience. commercial cells are available in the USA and in Europe and a description of the freezing procedures and method of use is set out in the NBS monograph 126 p.29 (Riddle *et al.*1973). The procedures need not be as rigid as implied in the reference. A solid CO_2-cooled freon heat pipe (Evans and Sweger 1969) is an effective method of forming an ice mantle without the danger of the ice mantle cracking and the top piece floating free. A usable ice mantle can be maintained for many months when the cell is stored in a polycarbonate tube of diameter a little greater than the cell and immersed in the ice bath.

The so called "inner melted layer" will normally be formed by the insertion of the precooled thermometer. The only care required is the periodic melting off of crystals which have grown onto the outside wall of the cell, mainly across the liquid-vapour interface. It also should be noted that, because of the 0.01°C temperature difference between the water triple point and the ice-bath temperature, the cell is in a freezing situation and ice will form across the liquid-vapour interface. If the ice bath is close to 0°C the build-up of ice in the cell will only occur until the pressure in the liquid is about 1-1/3 atmospheres at which point the equilibrium temperature will have dropped to 0°C and freezing ceases. This pressure will not break the glassware. The equilibrium pressure at solid-CO_2 temperatures is enormous and care must be taken not to allow the ice to freeze solidly to the outer wall during the mantle freezing procedure.

(ii) Metal Freezing Points. The procedures for realizing the metal freezing points are also well documented (Riddle *et al.*1973; McLaren 1957, 1962). The reproducibility of the freezing points of metals with impurity content of less than one part per million can be about 0.2 mK and the sample quality can be monitored using the melting range of the sample. The criterion of 90% of the melt being within 1 mK of the immediately preceding freezing point will ensure that that freezing point was within 0.3 mK of the "true" value for most impurities (McLaren 1969; McAllan 1982). A SPRT generally will reproduce temperature to within 0.3 mK between 0°C and 420°C (McAllan and Connolly 1982) and to within 2.0 mK when the calibration is extrapolated to 630.74°C (Riddle *et al.* 1973). It is a very stable and well behaved instrument if handled carefully but because of the strain-free construction it is very susceptible to vibration and there is always the risk of its calibration being ruined by vibration during transportation. SPRTs are expensive and those with silica sheaths are very fragile. Some constructions have a limited life due to mechanical instabilities brought on by excessive grain growth. In particular the coiled coil type of construction will eventually suffer short circuited turns arising from this grain growth (Ray and King 1978).

(iii) Resistance Measurement Equipment. Traditionally direct-current resistance bridges, adapted for lead compensation, were used for the measurement of the resistance of a SPRT. The calibration of these bridges was recurrent and a time-consuming procedure. The main resistors needed precise temperature control and the calibration history could be lost by a failure of the temperature-control equipment. Precise measurements were very tedious and required skilled and patient operators.

In more recent times a variety of alternating-current bridges have become available mostly based on the inherent stability and linearity

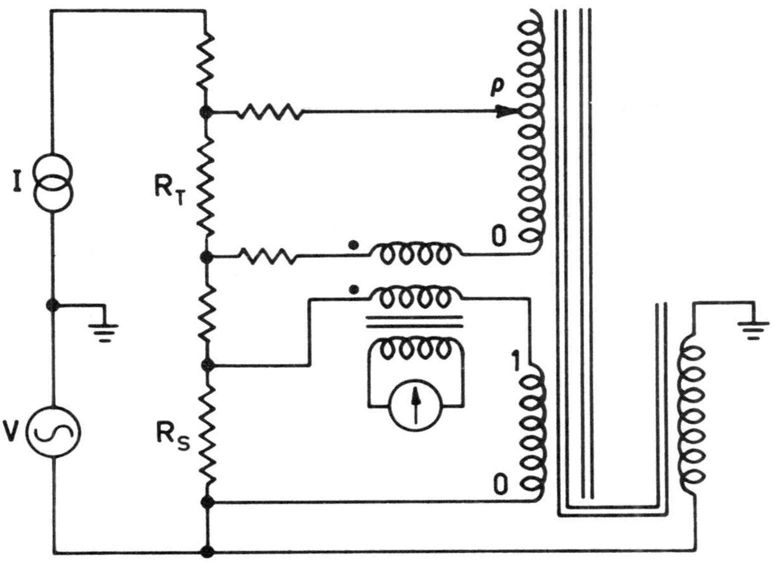

$$\frac{R_T}{R_S} = \rho$$

Fig. 4. Schematic AC ratio transformer resistance bridge

of well-designed ratio-transformer dividers. Figure 4 shows a basic AC ratio transformer bridge circuit. The problems of inter-lead insulation and earth loop currents which occur with a DC resistance bridge have their parallels in AC-resistance-measuring equipment. In particular some perfectly satisfactory DC insulating materials have poor AC insulating properties even at the low frequencies used in these instruments. PVC is one example. Active-lead shields are often employed to reduce inter-lead capacitance had active guarding can minimise other leakage current problems (Cutkosky 1981). With the availability of microprocessors very sophisticated self-balancing AC resistance bridges are being produced but they are quite expensive.

An alternative approach to the bridge method of measurement of resistance has always been to measure the voltage drop across the sensor when a known current flows in it. The potentiometers used for this are as difficult to calibrate and use as the DC resistance bridges and problems of parasitic emfs make it necessary to take a second measurement with the current flowing in the reverse direction. Thus, four voltage measurements, two across a standard resistor to

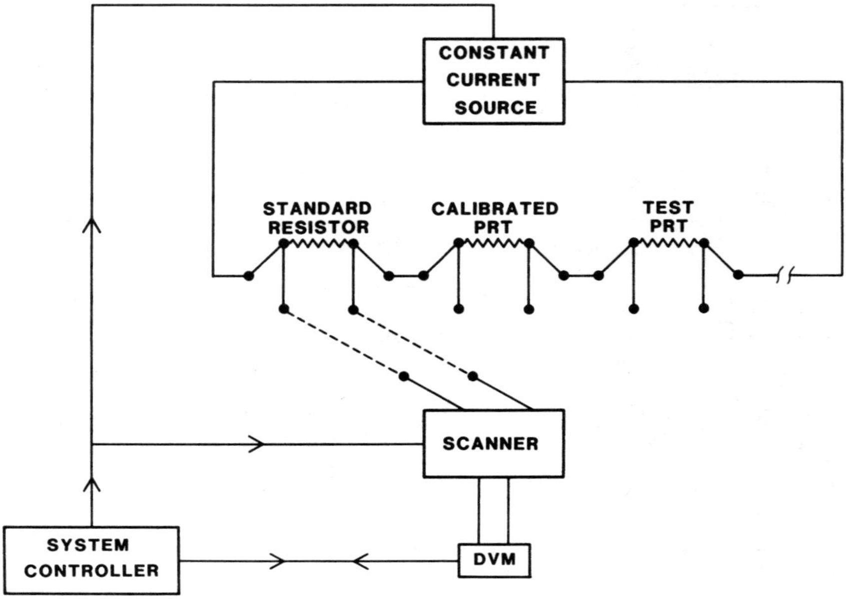

Fig. 5. Schematic of a DVM based resistance data acquisition system.

determine the current and two across the thermometer itself, were required for each temperature determination. Digital voltmeters with very stable current sources have simplified this approach and a computer-controlled data-acquisition system can now be constructed which can rival the AC resistance bridge for speed and accuracy.

4. Industrial Platinum Resistance Thermometers

(a) IPRT Sensors

The sensors used in IPRTs can be divided into two classes, firstly encapsulated wire sensors and, secondly, platinum film sensors. The film sensors are mostly thick film although very recently some thin film types have become commercially available. The thick film sensors, also comparatively new, are not as stable or reproducible as the wire types but it is expected that further development will improve their quality (Bass and Connolly 1980). There is little information on the behaviour of thin platinum film sensors at the moment.

The main requirement of an IPRT as distinct from an SPRT is robustness. The wire is normally held firmly in place on its former either by a ceramic glaze or sintered powder. There is, therefore, a requirement that the thermal expansion of the mounting and fixing materials match that of platinum otherwise effects due to strain will occur and, in extreme cases, the wire will break during thermal cycling. The various materials in contact with the platinum should be sufficiently inert so as to avoid contamination of the wire. These are fairly stringent requirements when it is considered that

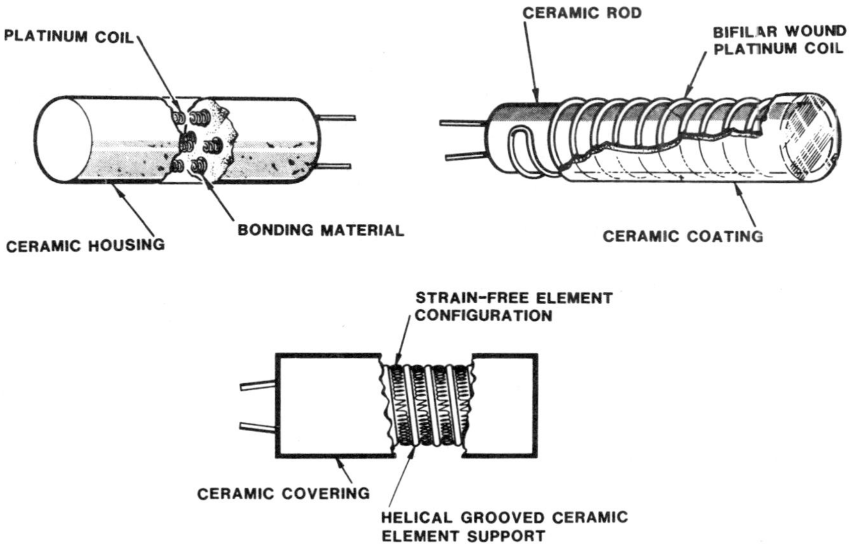

Fig. 6. Typical industrial platinum resistance thermometer sensors

such thermometers are normally only used when the temperature accuracy requirements are less than 0.1 K, that is when a resistance stability of better than 0.04% is required. Figure 6 shows some of the structures used in IPRT sensors.

The "α" coefficients of IPRT sensors are normally lower than those of SPRTs, which implies less-pure wire. In fact the industry standard α of 0.00385$(^\circ C)^{-1}$ is now obtained by deliberately doping the platinum with known impurities to produce this value. This has the advantage of making the wire less susceptible to diffusion impurities and possibly inhibits grain-growth problems. There does not appear to be any difference in the form of the resistance-versus-temperature relation between the wire used in IPRTs and that used in SPRTs. This, of course, means that IPTS-68-type equations must be used to describe the R/T characteristics of these thermometers when precision of better than 0.1K is required. A simple quadratic equation as specified in most standard specification codes is not sufficient, an equation like IPTS-68 equation 10 will be needed.

(b) IPRT

An IPRT sensor is usually mounted in a stainless-steel sheath with the internal leads insulated either individually or with 4-bore tubing. Powder fill is often used to prevent movement. It is very important to expel any moisture before the sheath is sealed as its presence will cause errors due to the effect of electrical leakage. Insulation resistance between the measurement leads and the sheath for 1 100 Ω IPRT must be greater than 10 MΩ.

(c) Calibration of an IPRT

The fixed-point calibration procedure is usually not applicable for IPRTs either because of thermometer-demension limitations or the high fixed-point precision not being needed. Fixed point calibration would also be unduly time-consuming.

Calibrations can be carried out by ice-point measurements and intercomparison measurements with an SPRT at a number of temperatures in a temperature-controlled zone. Ice points made from crushed or shaved-distilled-water ice are very reproducible and make water triple-point cells unnecessary unless accurancies of better than 2 mK are required (Beavis 1981). The main danger with ice points, apart from the obvious one of contamination, is inadequate immersion. When choosing an IPRT consideration should be given to the details of its calibration. Large-diameter, short-sheathed IPRTs will conduct sufficient heat along their length to melt the ice away from the sheath, preventing the sensor becoming isothermal with the melting ice. Lack of adequate immersion can be a serious problem in any temperature-measurement situation.

There are many stirred liquid baths which are sufficiently uniform to be used as intercomparison calibration baths. The fluids can range from ethyl alcohol through water to silicone and other oils. Ethyl alcohol can be used down from about 30°C to -50°C and water from above 0°C to 80°C or higher if precautions are taken to prevent heat loss by surface evaporation. Silicone oils have an upper limit of about 250°C where they begin to vaporise excessively. Above 250°C molten salt mixtures are possible fluids but as temperatures increase any liquid heat-transfer material will become dangerous if there is a chance of volatile components being present. This will cause the liquid to splatter with the resultant danger of burns or fire. Other types of intercomparison equipment such as fluidised particle baths or controlled furnaces are more suitable.

A thermometer should be intercompared with the SPRT at more than the two temperatures (plus an ice point) required to determine the coefficients of the quadratic equation (IPTS equation 11b). This is to enable an assessment to be made of the accuracy of fit of the test data to the equation. Six temperatures in the range would be adequate and allow a least squares regression procedure to be used.

The spatial and temporal uniformity of the calibration bath, the response-time differences between the standard and test thermometers and also the thermometer stabilities will govern the uncertainty of the calibration. The controller should have sufficient response to minimise time-constant-related temperature differences between the SPRT and the test PRT. For a uniform temperature drift the error is the product of bath-temperature rate of change and the difference in the time constants of the standard and test PRTs.

An intercomparison calibration bath can be tested for spatial uniformity by employing two similar thermometers, one in a fixed location to monitor temporal variations and the other moved to plot out the variations around the bath working volume. A typical stirred-oil or water bath designed for temperature or kinematic viscosity calibrations should have temporal stability of better than 0.02°C/minute and spatial uniformity of better than 0.01°C over centrally situated cylindrical volume of say 10 cm diameter and 20 cm length. The stability can be further improved by the insertion of a copper or aluminium temperature smoothing block which has in it thermometer wells blanked at their bottoms. This will be at the expense of settling time and the block should not be so large as to radically alter the bath stirring action.

The fluidised particle baths mentioned above can cover a very wide range of temperatures but their control is difficult and there are problems with dust. Controlled cylindrical furnaces with temperature-smoothing block inserts can also be used over a wide temperature range. However it is difficult to establish and maintain a uniform working volume as the insertion of the thermometers can dramatically alter the temperature distribution. In more recent times heat-pipe furnace liners and gas ballasted heat-pipe furnaces have become available and these offer a practical method of obtaining the constant temperature conditions required to calibrate IPRTs by intercomparison with a SPRT at temperatures above 250°C.

5. Calibration Facilities

(a) Primary Temperature Scale

In theory it is quite easy to set up the IPTS-68 for the temperature range 0 to 630.74°C. A PRT with "α" coefficient greater than 0.003 925 0 $(°C)^{-1}$ is purchased and is calibrated by measurement of its resistance at the water triple point, the freezing point of tin and the freezing point of zinc. It coefficients R_0, A and B (or R_0, α and δ) are calculated and it then defines temperatures between the fixed points and extending to 630.74°C. The practice is not quite as straightforward. The requirements for realization of the standard metal freezing points are exacting. The furnace uniformity needed is of the order of 0.01°C/cm which necessitates the use of "end heaters". Materials have to have less than one part per million impurity and have to be monitored for contamination, usually by measurement of melting ranges (McAllan 1972). The metals require controlled atmospheres and so gas handling and vacuum equipment is required. The temperature as sensed by the thermometer varies with depth of immersion and atmospheric pressure (the Clausius-Clapyron relation) as well as with measurement current (self-heating). Allowances have to be made for all these effects.

Thus, in the setting up of a primary temperature scale, the accuracy of realization must be estimated by analysis of the likely errors arising from the various identified systematic errors and an assessment of random errors could be made by repeated fixed-point measurements. This ignores problems of thermometer stability and the precision of the resistance-measurement equipment.

In this laboratory as with most other large national laboratories a substantial number of SPRTs have been calibrated and an overall assessment of the likely accuracy of IPTS-68 realization has been built up. In particular this laboratory routinely uses the freezing point of pure cadmium to monitor the performance of the SPRT being calibrated (McAllan and Connolly 1982) and it is considered that, certainly within the range 0 to 420°C, the IPTS can be realized by a small group of SPRTs with a non-uniqueness of about 0.3 mK. However it requires constant vigilance to maintain this accuracy.

(b) Transferred Primary Temperature Scale

An alternative approach for the establishment of the IPTS-68 is to "buy" the scale. This involves having an SPRT calibrated at a laboratory which maintains the IPTS-68 fixed points and then using this thermometer as the temperature-defining instrument. The precision with which the scale could be transferred would depend on a number of factors. Firstly, although the primary thermometer would probably reproduce IPTS-68 with uncertainty less than 0.5 mK at the time of its calibration, the stability of the thermometer, in most cases, would limit the obtainable precision to somewhat less than this figure, probably to about 1 mK.

Resistance, in resistance thermometry, always appears as a resistance ratio normally with the 0°C resistance as the reference. The constant R_o is used as a monitor of stability as it is easily measurable. The resistance ratio $R(t_{68})/R(0°C)$ at any set temperature will not change as rapidly as the temperature equivalent of changes in R_o or $R(t_{68})$ so that if the present R_o value is used to calculate the resistance ratio the temperature so obtained will not be influenced by the change of R_o from its calibration value. In other words the other thermometer constants are, to some extent, independent of changes in R_o. To put some figures on this, a report of calibration of an SPRT from this laboratory would normally specify uncertainties of 1 mK to 200°C and 2 mK from 200 to 420°C. It is usually stated that these uncertainties will apply until the value of R_o has changed by more than the equivalent of 0.01 K in temperature from its calibration value. The justification for the assumption is that most of the changes which occur are proportional to the temperature and are presumably due to dimension changes in the platinum wire. Problems will arise after use at higher temperatures and, in some thermometers, even at 300°C where contamination of the platinum will cause an increase in R_o. This effect will be largely temperature independent (Matthiessen's Rule) and hence the α coefficient will degrade. This is the reason for the broader uncertainty normally given for temperatures above 200°C.

Thus by using an SPRT calibrated at a national laboratory it would be possible to establish IPTS-68 with an uncertainty of about 2 mK from 0 to 200 or 250°C and somewhat worse at higher temperatures. This is only a little worse than maintaining a primary IPTS-68 scale, the difference being that when the primary fixed points are maintained one has immediate access to SPRT calibrations. The scale below 0°C is difficult and very expensive to maintain. "Stick" thermometers, because of heat conduction along the thermometer sheaths, are not really suited to most designs of calibration cryostats, however it should be noted that the scale above and below 0°C passes through 0°C smoothly and IPTS-68 equations 10 and 11a will reproduce the scale down to -40°C with error less than +1 mK. The error rises rapidly to about +0.1 K at -100°C. Figure 7 graphs the error against temperature. It is possible to extrapolate down to say -80°C by applying corrections and without having to provide any calibration point in this region. The uncertainty in the correction would be less than 10% of the correction.

The disadvantage of using an SPRT as a transfer standard lies in their cost and fragility. Each thermometer would cost at least $A2500 with the calibration adding another $A340 (CSIRO charge). There is always the worry that the calibration will be lost because of mishandling during shipment from the calibration laboratory to the user.

6. IPRT as a Transfer Standard

The IPRT with an encapsulated platinum wire sensor usually has better stability and reproducibility than the manufacturer or standard specification codes imply. This is because they are mostly interested in interchangeability and this is unimportant when

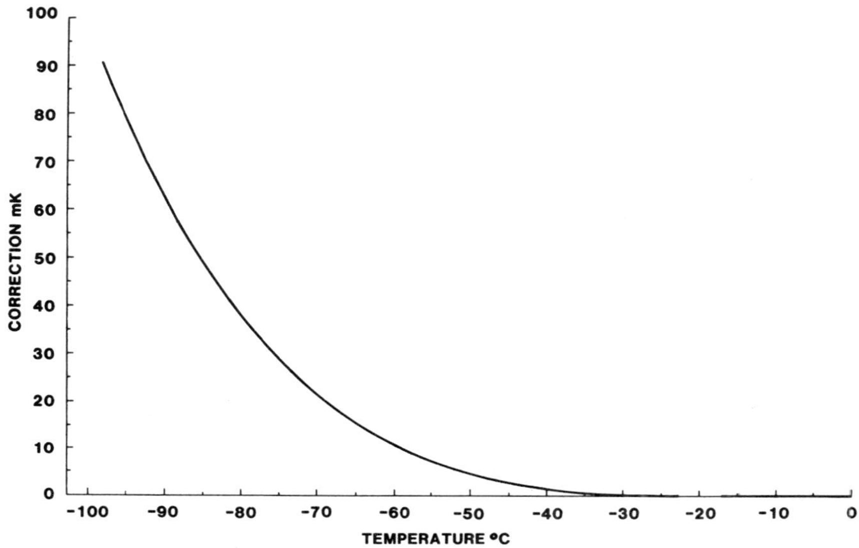

Fig. 7. Plot of the error arising from the use of Equation 10 and 11a in the range 0 to -100°C against temperature.

thermometers are to be individually calibrated. The crucial properties are their stability, reproducibility and predictability.

A substantial number of IPRTs have been calibrated in this laboratory in the last five years and various aspects of their behaviour are discussed elsewhere (Connolly 1982). It was found that over the temperature range -40 to 250°C, which is the range of interest of the bulk of our clients, most IPRTs with platinum wire sensors reproduce IPTS-68 to better than 2.5 mK and have sufficient stability to warrant working at better than 5 mK accuracy level. Selected thermometers are even better than this but some appear to have a higher-order term in their resistance-versus-temperature relation which is characteristic of their actual structure. In the range 250 to 500°C IPRTs are not quite so reliable although still substantially better than manufacturers' claims. For this range ice-point stability is not as good and some thermometers show effects which may be strain hysteresis. Much more work needs to be done to identify the problems. Even with these qualifications, selected IPRTs can still provide a temperature scale with uncertainty of about 5 mK to 400°C and 10 mK to 500°C.

The region above 500°C is much less satisfactory. Stainless-steel-sheathed IPRTs are generally very unstable even at 0.1 K uncertainty. This is most unfortunate as accuracies better than this are often required in power-station, 550°C-steam temperature measurement. Again, there is an investigation underway in this laboratory to identify and characterise suitable IPRTs for this temperature region.

7. Secondary Calibration Facility

Standardising a laboratory temperature scale is only part of the problem. The real aim is transferring it to working standards and other thermometers, be they IPRTs, liquid-in-glass thermometers, thermocouples or semiconductor thermometers.

For temperatures in the range -50 to 250°C temperature-controlled stirred liquid baths, as referred to above in 4(c) have proven most satisfactory. There are many commercial designs specifically made for temperature-probe calibrations and suitable types can also be found in those designed for kinematic viscosity measurements. All that is required is a good stirring action with a well-designed controller. The type of bath, controller and operating volume will be dictated by the particular instrument being calibrated. Electrical thermometers require adequate immersion and the allowable bath spatial and temporal uniformity is set by the accuracy required. As also mentioned above, the inclusion of a temperature-smoothing block in a bath will reduce controller fluctuations substantially at the expense of response time. This has the advantage of removing time-constant-related temperature differences.

A well-designed system can transfer IPTS-68 to a test thermometer with an uncertainty which is limited only by the test thermometer's stability. Experience in this laboratory puts this at about 10 mK and it often approaches the 5 mK mentioned above. It is possible to obtain bath stability approaching 1 mK/minute and temperature differences between thermometer wells in an aluminium or copper block will normally be less than 1 mK.

Intercomparison procedures for IPRT calibrations are set by the resistance measurement equipment available. At this laboratory the SPRT and test IPRT are monitored on separate AC bridges making time synchronisation easy. However if only one bridge is available the SPRT measurement has to be repeated to ensure that the bath temperature has not altered. A potentiometric measurement of resistance based on a digital voltmeter can be employed with the advantages of being able to scan the PRTs quickly without having to wait for the current heating to equilibrate. The required resolution in the DVM makes it rather expensive and the other systematic errors mentioned in 5 (iii) have to be considered.

The problems of intercomparison calibration at temperatures above 250°C have been dealt with. However it is worth noting the promise being shown by controlled furnaces smoothed by heat pipes in meeting the need. Commercial heat-pipe furnace liners with mercury as the working fluid are designed for use to 360°C and with sodium to temperatures from 580 to 950°C. The gap would be well covered by caesium but these are not as yet a commercial product.

8. Conclusion

Although the IPTS-68 is normally established by maintenance of the defined fixed points it is a difficult, expensive and time-consuming procedure. The IPTS-68 can be maintained in a laboratory by purchasing some SPRTs which have been calibrated on IPTS-68 and then using them to transfer the scale to other thermometers using intercomparison calibration procedures. Unfortunately SPRTs are also expensive, are very fragile and there is a real risk of losing the calibration during transportation.

Selected IPRTs can transfer IPTS-68 with an uncertainty which is only a little worse than SPRTs. They are more robust and not nearly as expensive so that a reasonable number can be maintained as a "standard group" and circulated at intervals for recalibration. With such a group of thermometers it would be possible to meet the needs for calibration accuracy of the bulk of likely customers, particularly in the temperature range from -30 to 250° C and especially in the 30-100° C range where the most precise temperature measurement is needed.

References

Bass, N. M. and Connolly, J. J. (1980). *Aust. J. Instrum. Control* 36, (5), 88-90.

Beavis, I. M. (1981). CSIRO Division of Applied Physics Technical Paper No. 5.

Connolly, J. J. (1982). The calibration characteristics of industrial platinum resistance thermometers. In: Temperature, its Measurement and Control in Science and Industry Vol. 5 (American Institute of Physics, Washington, 1982) Ed. J. F. Schooley.

Cutkosky, R. D. (1981). *IEEE Trans. Instrum. Meas.* IM30, 217.

Evans, J. P. and Sweger. D. M. (1969). *Rev. Sci. Instrum.* 40, (1), 376.

IPTS-68 (1976). The International Practical Temperature Scale of 1968 (Amended Edition of 1975). *Metrologia* 12, 7-17.

McAllan, J. V. (1972). *Journal of Crystal Growth* 12, 46-52.

McAllan, J. V. and Connolly, J. J. (1982). The use of the cadmium point to check calibration on the IPTS. In: Temperature, its Measurement and Control in Science and industry, Vol. 5 (American Institute of Physics, Washington, 1982) Ed. J. F. Schooley.

McLaren, E. H. (1957). *Can. J. Phys.* 35, 1086-1106.

McLaren, E. H. (1962). Temperature, Its Measurement and Control in Science and Industry, Vol. 3 (Reinhold Pub. Co., New York, 1962) Ed. C. M. Herzfeld. p.185.

Raj. R. and King, G. W. (1978). *Met. Trans.* 9A, 941.

Riddle, J. L., Furukawa., G. T. and Plumb, H. L. (1973). N.B.S. USA Monograph No. 126.

An Approach to Thermocouple Calibration Based on the Distributed Nature of Thermo-emf

Robin E. Bentley
CSIRO Division of Applied Physics, Sydney, Australia 2070

Abstract

A calibration philosophy and the techniques that follow as a consequence are developed from a consideration of the distributed nature of thermocouple emf. It is suggested that the thermocouple be thought of in terms of its "thermoelectric signature", and that if a thermocouple must be re-calibrated in a laboratory facility after use elsewhere its length should be scanned to assess the inevitable changes to its signature. The effectiveness of calibration can be increased by taking into account the likely changes in signature during a defined period of use subsequent to calibration.

1. Introduction

After considerable developmental time and cost a laboratory may establish the necessary skills and equipment to produce its own primary standards for defining the International Practical Temperature Scale (IPTS) (e.g. Kemp 1977) over any desired temperature range. The laboratory would begin by developing a series of freezing- and boiling-point facilities to reproduce the known fixed-point temperatures. Then with a set of defined interpolation formulae the IPTS is transferred to a number of interpolation devices (primary standards).

A primary-standard thermocouple covers the range 630.74 to 1064.43°C. Because of thermoelectric inhomogeneities and the inadequacy of the IPTS interpolation formula such a thermocouple realises temperature to ± 0.2°C, at best.

As an alternative to using a series of fixed points a laboratory may base its realisation of the scale on secondary-standard thermocouples. Temperatures up to at least 1600°C can then be realised, and the need to establish and maintain a fixed-point calibration system is transferred to the laboratory which calibrated the secondary-standards. For example, temperatures in the range 0 to 1100°C may be established with an uncertainty of ± 0.1°C at 0°C to ± 0.4°C at 1100°C and is little different from that of a primary-standard thermocouple in the temperature range covered by both standards. This uncertainty is more than adequate to support the thermocouple calibration needs in most areas of industry and research.

This paper describes a calibration philosophy and the resultant techniques for establishing a laboratory facility for the calibration of thermocouples by comparison with a secondary-standard thermocouple. The approach makes full use of the knowledge that emf in thermocouples is not produced at junctions but is distributed along the wires (e.g.

Bentley 1982a) and that the thermopower, which dictates the magnitude of emf, will change in a non-uniform manner whenever a thermocouple is used. For details of the conventional approach to calibration and of basic measurement techniques the reader is referred elsewhere (ASTM 1974).

2. Emf in Thermocouples

The emf in a thermocouple is distributed along the wires in response to the temperature gradient and hence coexists with the flow of heat (Bentley 1982a). The emf produced per unit length of thermocouple is the product of temperature gradient and thermopower, and since the latter is much the same throughout the length of a thermocouple the emf distribution is primarily dictated by the longitudinal temperature profile. The nett effect of this distribution is to produce a total measurable emf that depends mainly on the temperatures of both the tip and the cold junction. However the thermopower does have some effect on emf distribution since all thermocouples are to some extent inhomogeneous. As a consequence their emf-temperature relationships have a second-order dependence on temperature profile. For example, if a calibration is repeated at different depths of immersion but with the same tip and cold-junction temperatures, differing values of emf will occur. At each depth of immersion the measured emf represents the average thermopower of that section of wire in the throat-gradient region provided the wire elsewhere (within the furnace and outside at ambient temperature) lies in regions of uniform temperature. The plot of such data (see Figs. 1 and 2) reflects the variation in thermopower along the wires and will be referred to as the "thermoelectric signature" of the thermocouple.

A knowledge of the signature and of its likely changes with use are required for any thermocouple being calibrated. It should be determined by scanning the thermocouple length with a steep temperature gradient, e.g. by progressively inserting it into a relatively long uniform-temperature furnace having a short throat-gradient region. Recent measurements using the apparatus described show that the signature of new type R or S thermocouples, given the annealing sequence detailed in section 7, can be expected to extend over a range of ~ ± 0.02% of developed emf. These results are in agreement with an earlier study using a specialised high-resolution scanning facility (Bentley and Jones 1980). This inhomogeneity range or dependence on depth of immersion could increase by orders of magnitude if care is not taken to avoid contamination, for example. Even with such care a 50-100 h use of a rare-metal thermocouple could increase the inhomogeneity range to ± 0.05%.

Changes in the signature occur because of various effects, some of which are avoidable and some not. If platinum-based thermocouple wires are kept within twin-bore insulation of the best quality re-crystallised alumina (Degussit AL23 or Morganite ΔRR) extending in one continuous length from tip to ambient the problems arising from cold work (in handling), contamination, and vapour transfer between the wires are minimised. However unavoidable impurity pick-up from even the best insulation material, at wire temperatures above ~1400°C, and the reversible effects of hysteresis (above ~700°C) and rhodium oxide (500-900°C) affect the signature and must be carefully considered.

Hysteresis (Bentley 1982b) due possibly to trapped vacancies and short-range order in the platinum-rhodium alloy results in a reversible change in thermopower for those sections of a thermocouple cycled from above ~ 700°C to below ~ 500°C. Under normal circumstances, temperature along the length of a thermocouple is non-uniform and hence a withdrawal from a high-temperature application will lead to non-uniform hysteresis. For example, consider a type R or S thermocouple insulated in 3-4 mm O.D twin-bore insulation and given a 16 h isothermal anneal at 450°C. If it is then used in a furnace above 800°C and withdrawn from it at a practical rate (2-5 s for 500 mm) that portion of the signature corresponding to wire previously at 600°C or below will be unchanged whilst that portion corresponding to a withdrawal from 800°C or above will show a loss in output of up to ~ 3 μV. Further details are given in the studies by McLaren and Murdock (1979) and Bentley and Jones (1980) which also describe the effects of rhodium oxide.

Rhodium oxidises at temperatures from 500 to 900°C, especially near 800°C, and the process is reversed after a frief exposure to 1000°C and above. Oxidation in a type R or S thermocouple, initially given a set of recovery anneals (see below), will cause an *in situ* drift of -0.01% in 3 h and -0.03% in 100 h whilst the thermocouple monitors a temperature of 900°C or above. However the change in calibration of the oxidised portion of the thermocouple would be ~ 0.1% (100 h) and could be evident on moving the thermocouple to a shorter depth of immersion. This local change in thermopower could also be seen in a signature scan.

For oxidation and hysteresis to have maximum effect on the measured output of a thermocouple, that section of wire producing most of the emf must be fully affected by the processes. Usually this implies the earlier use was at a deeper immersion by at least 200 mm. For example, a thermocouple used to monitor a temperature near 800°C would have a relatively long oxidation zone and if moved to a smaller immersion the affected wire could easily extend over the main temperature gradient (emf-producing) region of the furnace. A thermocouple used at a fixed depth of immersion is not as sensitive to those processes since the affected zones span a smaller temperature interval.

If a thermocouple's signature reflects a nominally homogeneous state it does not follow that its calibration is unique. For example, a thermocouple uniformly oxidised by an exposure to 800°C and then uniformly quenched by a withdrawal to ambient will be in a state that is just as homogeneous as that achieved with a brief anneal at 1100°C and an overnight anneal at 450°C. However the latter state is characterised by a ~ 0.15% greater emf output and is the preferred condition for platinum-based thermocouples. Any perturbations to the signature that subsequently develop through oxidation and hysteresis are easily reversed for any critical application by the following recovery anneals. Kept in their insulator the wires are first annealed at 1100°C for 0.5 h and withdrawn quickly ($<$ 5 s) from the furnace to minimise oxidation while cooling through the 500-900°C zone. This is followed by a 16 h (overnight) anneal at 450°C.

The processes in base-metal alloys are different from those in the platinum/rhodium alloys but there are similarities and the basic principle of using a knowledge of the signature to describe a thermocouple's behaviour is valid. The effect of inherent inhomogeneity in 'as-new' base-metal wire tends to be ~ ± 0.1% and in nickel-based thermocouples, for example, hysteresis has the additional effect on thermopower of up to ~ 0.5% (Bentley 1982b). However oxidation in these alloys is not reversible. It is accumulative and its effect rapidly increases with temperature above 500°C. The *in situ* drift of such thermocouples (Chromel-Alumel* and Nicrosil-Nisil) would be ~ 0.1% in 10 h at 600 to 1100°C for 0.5 mm diameter wires but the *in situ* calibration may undergo a far greater change if the thermocouple is subsequently moved. An increase of 3% (~ 30°C) at 1100°C can occur on reducing the immersion after a 10 h use.

3. Calibration Philosophy

This paper treats thermocouple calibration in terms of intercomparison, in a laboratory calibration facility, with another thermocouple whose calibration with respect to the IPTS is known. Other techniques are available with advantages in special situations.

For example, intercomparisons with platinum resistance thermometers in stirred liquid baths below ~ 400°C are potentially the most accurate but require an overnight anneal at 350-450°C, depending on the thermocouple, to stabilise the processes of hysteresis. A reliable measure of the signature is also required for each thermocouple and is obtained by scanning in the bath over that section of the thermocouple likely to produce significant emf in the intended use. A calibration conducted *in situ* also has special merit. It is useful for fixed-immersion applications particularly for base-metal types above 600°C where significant irreversible changes to the signature are to be expected. An *in situ* calibration, which involves exactly that temperature profile present in normal use, would be directly relevant whilst the calibration in a special calibration facility would not. The temperature profile in the latter case would be different and so the various features of the signature would have different contributions to the total emf. These special calibration techniques are not given further consideration in this paper.

All thermocouples have some degree of inhomogeneity, and so the calibration or emf-temperature relationship obtained at the particular depth of immersion used in a calibration facility can at best be regarded as representative only. The fact that the calibration applicable to a future use will differ from this may be covered in the calibration uncertainty with a suitable component. Indeed it is the initial inhomogeneity and the subsequent changes in signature with use that govern the accuracies achievable with thermocouples.

An initial calibration supplies a representative emf-temperature relationship and an associated uncertainty. To a first order the calibration will continue to apply indefinitely. What changes with use is the signature and hence, possibly, the calibration uncertainty. Localised changes occur which increase the dependence of the thermocouple's performance on depth of immersion and so the prime function of re-calibration must be to assess this growth in signature. With rare-metal thermocouples the re-calibration procedure serves other functions as well if it includes adequate cleaning and annealing (section 7). It would then reduce potential contamination and reverse changes in the signature due to cold work, rhodium oxide and hysteresis. For base-metal thermocouples it is far more economical to replace the used thermocouple than submit it for re-calibration.

* Trade name of Hoskins Manufacturing Co. (USA).

Indeed a re-calibration attempt is likely to lead to a rejection of the thermocouple on the grounds of an excessive growth in signature expecially if its use involved temperatures in excess of ~ 600°C.

The procedures adopted in a calibration facility depend on three questions:

(a) whether the thermocouple wires are new or used,

(b) whether the wires are rare-metals or base-metals, and

(c) whether the calibration results are to apply to a reasonable period of use after calibration or merely to the condition of the thermocouple at the time of calibration only.

If the wries are new a measure of their inhomogeneity can be taken as ± 0.02% for rare-metal and ± 0.1% for base-metal thermocouples and a scan is not necessary. Used wires require a scan because their inhomogeneity could easily be an order of magnitude or more worse than this even after an apparently careful use.

Different accuracy levels apply to the calibration of rare-metal and base-metal thermocouples. There is an order of magnitude difference in their degrees of inhomogeneity and under clean oxidising conditions the rare-metals suffer no irreversible deterioration in signature (to 1300°C) whereas the base-metals do. Cleanliness and care in handling are required for rare-metals and cleaning and annealing procedures are applied with advantage. Base-metals are calibrated in the 'as received' condition with one exception. Those thermocouples limited in use to temperatures below 400°C and requiring an improved accuracy should be given an overnight anneal at 350-450°C, depending on the alloys, to stabilise the processes of hysteresis.

The uncertainty of calibration can be evaluated on the assumption that the calibration results apply only to the condition of the thermocouple at that time. There are applications for which this option is useful. One is the use of a rare-metal thermocouple whose 'as-calibrated' state is continually maintained with recovery anneals (section 2) as required. A useful approach for a National Standards laboratory but not for the users of its calibration service since they have neither the expertise to apply these anneals nor the need for the slight improvement in accuracy that results. However there is a second, far more practical, application of calibrations expressed in terms of the 'as-calibrated' condition. This relates to thermocouples intended for a fixed-immersion use in which the (*in situ*) drift is either insignificant because of the brevity of the use or is monitored by regular *in situ* calibrations. An example of the former would be base-metal thermocouples used for a one-off measurement in assessing the uniformity of a heat-treatment furnace (Bentley 1982c), and of the latter would be a thermocouple used for the long-term monitoring or control of a critical heat-treatment process.

Thermocouples may be selected as samples to represent a batch of wire and calibrated to test for conformity with the manufacturers' tolerances, to ASTM E230 or to some other specification. This common situation also requires an 'as-calibrated' calibration because the results are not directly applied to a subsequent use. The calibration served its purpose once the wires are accepted or rejected.

If the period of intended use is not brief changes in signature will affect the behaviour of the thermocouple. There are changes, such as due to contamination, which are avoidable if the user successfully follows 'good practice' and there are those which are inevitable because they arise from inherent properties of thermocouple alloys.

It is therefore advisable to assume and advocate 'good practice' and to cover the latter changes by an appropriate component in the calibration uncertainty (section 8). The component will be least if a fixed-immersion application is intended and considerably greater if no such restriction is assumed. The former, conditional, form of calibration is relevant to base-metal alloys because of their poentially large changes in signature and the unrestricted form is appropriate to the rare-metals.

4. Equipment

Thermocouple calibrations are usually expressed as an emf relative to the internationally accepted reference tables (Powell *et al.* 1974) based on a 0° C cold junction. This cold-junction temperature is preferred for calibration and may be achieved with wet crushed ice made from distilled water or with commercial units that automatically control thermocouple wells at 0° C to ~ ± 0.02° C using Peltier cooling. Ideally the calibration standard and the thermocouple being calibrated should extend in one piece from tip to cold junction. If they do not additional uncertainties are introduced because of the need for extension leads (same alloys as thermocouple) or compensation leads (different alloys). Estimates of these uncertainties are given in Table 1 in the form ± (x μV + y%).

Table 1. Uncertainties due to the use of leads for extending thermocouples to 0° C

Thermocouple Type	Extension Lead	Compensation Lead
B	± 1 μV	± 3 μV
R or S	± (1 μV + 0.015%)	± (15 μV + 0.2%)
Base-metal	± (30 μV + 0.1%)	N.A.

The first component, x, refers to the 0° C to ambient section of the lead and is independent of tip temperature. The second, y% which refers to the total emf, was calculated assuming the connections between thermocouple and lead have a temperature which rises linearly with tip temperature to a value 40° C above ambient when the tip is at 1000° C.

During calibration or scanning, thermocouple emfs should be measured with a single emf-measuring instrument and low-thermal switching. The switch should be in the copper circuit between the 0° C cold junction and instrument. Otherwise errors arise because of differences in temperature between its input and output terminals. The switch should include a short circuit so that the circuit zero can be measured and corrected for as needed.

A digital voltmeter (DVM) with a least count of 1 μV is adequate for calibration but 0.1 μV is preferred for the scanning function. Its long-term drifts due to instabilities in gain and in zero and to its temperature coefficients are not as important as might at first appear if all thermocouples involved in the calibration and the circuit zero are measured on the same instrument range. The effects of zero and gain errors on measurements taken at similar times are not independent. Consider a span (gain) error of ε%, a measured emf (after zero correction) for the calibration standard of E_o and for the thermocouple being calibrated of E and respective thermopowers of θ_o and θ. The aim of a calibration is to determine the difference (E-REF) between the thermocouple emf and the reference tables value (REF) for the same tip temperature. The temperature is obtained from the calibration standard. The resultant error in (E-REF) expressed as a percentage of E is

$$(1 - E_o\theta/E\theta_o)\varepsilon \text{ \%}.$$

Clearly, the effect of span errors is negligible if the thermocouple being calibrated and the standard are of the same type ($\theta \sim \theta_o$, $E \sim E_o$). If one is a type R and the other a type S the error is less than $\sim (\varepsilon/20)$% and if either type is used to calibrate base-metal thermocouples the error varies from $(\varepsilon/10)$% for Nicrosil-Nisil to $(\varepsilon/4)$% for Chromel-Alumel. These values should be compared to $(\sqrt{2}\varepsilon)$% the uncertainty that applies if the thermocouples are measured using different (and independent) instrument ranges. It is only the short-term instabilities and the non-linearity (usually negligible) of a DVM that can have a direct bearing on thermocouple intercomparisons if the above steps are taken. Hence it is relatively easy to select a DVM which contributes insignificantly ($\lesssim$ 0.01%) to the calibration uncertainty.

A tubular laboratory furnace with a relatively long uniform-temperature zone is required if the intention is to re-calibrate used thermocouples. It is needed for scanning and to apply recovery anneals to rare-metal thermocouples. It may also be used for the intercomparisons as well. The one in use at this Division is suitable for temperatures up to 1200°C, is based on a 1200 mm long 15 mm I.D. mullite former and was constructed in a similar manner to that outlined by Jones, Meszaros and Morgan (1982). The axial temperatures are uniform to ± 1% over the central 900 mm and 90% of the temperature step from ambient to furnace termperature occurs in 150 mm of the throat-gradient region. For the scanning function it is important that this emf-producing zone be no longer than will apply during calibration or in subsequent use. This ensures that local inhomogeneity peaks will be observed with at least as much sensitivity. The 150 mm zone of the above furnace is considered adequate for most applications. If necessary it may be reduced using an air jet across the entry port.

Within the mullite former a silica tube is used to contain platinum-based thermocouples and so avoid contamination. It is wound with an earthed nichrome wire to prevent ac leakage from the heater affecting the DVM. To facilitate measurements during a calibration or a scan, temperature drifts should be kept below ~ 0.1°C in the short term (10-20 s) and so a proportional controller with derivative and automatic-reset actions could be required. The effect of rapid switching of the heater power in time-proportioning controllers may be reduced by locating the control thermocouple between the former and its winding.

Calibration by thermocouple intercomparison may be undertaken in commercial laboratory furnace assemblies which are available for different temperature ranges. However to the author's knowledge they are too short to allow adequate scanning.

5. Inhomogeneity Scans

Whilst the level of inhomogeneity of new thermocouple wires can safely be assumed, that of previously used ones cannot. Both the calibration uncertainty of a used thermocouple and the usefulness of its calibration to future use needs to be assessed with a signature scan. And it follows that such a calibration can refer only to that length of thermocouple scanned. In a future use no more than 50°C of the temperature step spanned by the thermocouple (tip to cold junction) should lie outside this length.

For maximum coverage in scanning, the emf-producing throat-gradient zone (~ 150 mm for the furnace described) should be adjacent to the thermocouple tip at minimum immersion. At maximum immersion it should just lie within the length annealed. For example, the furnace described allows annealing in its central zone for up to ~ 1000 mm of wire (temperature within 5% of that at the centre). Since its emf-producing zone extends ~ 50 mm outside the furnace end plate a maximum depth of immersion of 950 mm, relative to this end plate, can be achieved for any scan. The minimum covenient depth is 150 mm.

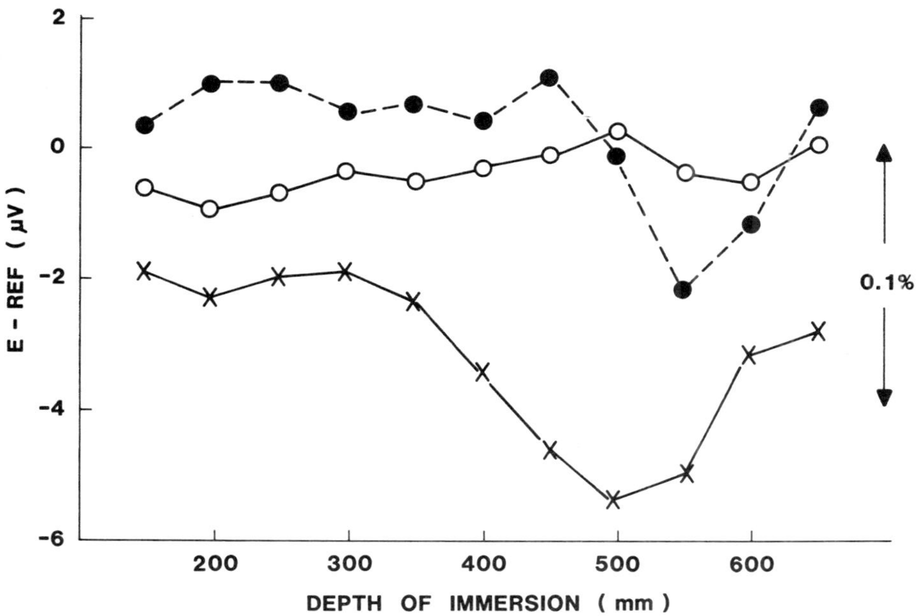

Fig. 1. Signatures of rare-metal thermocouples limited in use to 1100°C: full lines, used thermocouple before (x) and after (o) annealing; dashed line, thermocouple slightly contaminated near the end of its 500 mm insulator.

The temperature at which the scan is conducted is important. When used as a diagnostic tool the furnace can be set at any appropriate temperature but routine signature assessment should be standardised. At present the Division uses 450°C. This has advantages over the use of a higher temperature, such as 1000°C, but it requires a smaller measurement least count (~ 0.1 μV) and a better short-term stability. At 450°C there is far less ac-leakage, the thermocouple signatures are not affected by the thermal experience received in a scan, and hence the reference thermocouple (see below) remains in a stable 'homogeneous' state. It therefore does not require repeated anneals to recover its 'homogeneity'. The depth of immersion can be changed in either direction or by any amount and the measurement at any one depth or even the entire scan can be repeated at will.

The signature obtained at one temperature can be related to that at another by assuming the inhomogeneity effect is proportional to thermocouple emf. This is true for inhomogeneities produced by rhodium oxide (Bentley and Jones 1980) and by impurity pick up from alumina insulation at temperatures up to 1700°C. Recent measurements in this Division on discarded standards (inhomogeneity $< \pm 0.05\%$) and other rare-metal thermocouples after various 'clean' applications ($\pm$ 0.08 to $\pm$ 0.2%) revealed a similar dependence. The signatures of some of these are given in Figs. 1 and 2.

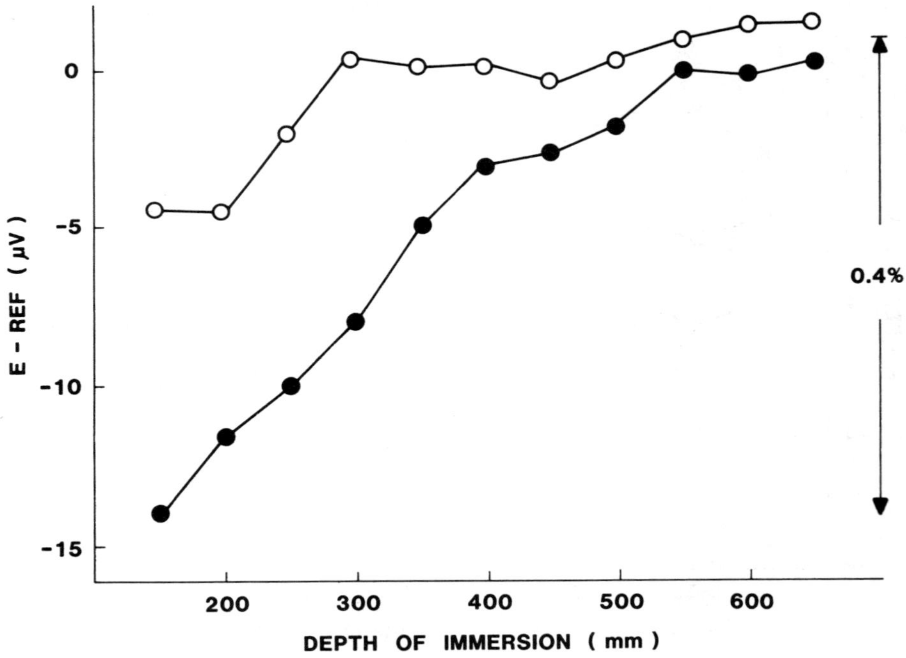

Fig. 2. Signatures of rare-metal thermocouples used to 1550°C.

All thermocouples should be in the fully annealed state (section 7) before scanning and the reference thermocouple should be mounted in a twin-bore insulator long enough to allow use at the maximum depth envisaged. Its tip should be welded or platinum-wire wrapped to the thermocouples being studied. If need be extension leads can be used for any thermocouple not long enough to reach the 0°C cold junction chamber at maximum immersion. To prevent distortion of the measured signature due to this lead having a different calibration from that of the thermocouple their interconnection should be kept at the same temperature (eg ambient) throughout the scan.

Sufficient time should be allowed during scanning for the establishment of the temperature profile at each position, eg 5-10 min. for 50 mm steps. From the reference thermocouple the common tip temperature is derived and thence the relevant reference tables value (REF). For each thermocouple (E) the signature (E-REF plotted against depth) is thereby obtained relative to that of the reference thermocouple. To avoid the temperature dependence of E-REF affecting the measured signatures the values should be corrected to the same nominal temperature using calibration data for the thermocouples being assessed. Alternatively the measurements should be limited to tip positions within the uniform-temperature zone of the furnace so that the tip temperature, for all immersion depths considered, is held within a range of ± 20°C. The effect of differing tip temperatures on E-REF is thus kept to within ± 0.5 μV (rare-metals at 450°C) and may be ignored in most cases.

6. Calibration Standards

The calibration standard should be a Pt10%Rh or Pt13%Rh versus Pt thermocouple traceable to the IPTS by a suitably accurate calibration. The calibrations available from this Division are issued with uncertainties that depend on the temperature range and on whether recovery anneals are used. For 'as-new' wires the calibration uncertainties are as follows:

(a) assuming the standard will be given 1100°C and 450°C recovery anneals by the user when required:

± (1 μV + 0.04%) from 0 to 1100°C

or

± (1 μV + 0.05%) from 0 to 1300°C

(b) assuming the standard will not be annealed and the calibration applies to a 100 h period of use above 500°C at any practical depth of immersion:

± (1 μV + 0.06%) from 0 to 1100°C

or

± (1 μV + 0.06%) from 0 to 1300°C

or

± (1 μV + 0.1%) from 0 to 1550°C

The use of recovery anneals on calibration standards results in a significant improvement in their accuracy. However as seen later this improvement does not fully transfer to the thermocouples calibrated by them. Clearly a laboratory setting up a calibration facility should carefully assess the accuracy needs of its end-users before committing itself to the use of recovery anneals. Calibration standards having this requirement need careful administration to decide when anneals are necessary. To simplify the problem and to minimise the need for recovery annealing the use of standards should be regulated.

To achieve this, each calibration run should have a similar time-temperature sequence and the standard should always be at the same depth of immersion. Consider the example of a standard initially in an annealed state and used at progressively increasing temperatures until the upper limit is reached where it is withdrawn. Hysteresis will have the following effect. On the first run the standard will operate in accordance with its calibration. On withdrawal that section of wire that exceeded 600°C will suffer hysteresis on cooling to ambient temperature. In the early stages of the next run, while the tip is less than 600°C, the section will produce less emf than it did before. The total thermocouple output will then be low by ~ 0.02%. As the tip temperature approaches its upper limit the error will reduce to zero. Thereafter the standard will reproduce the hysteresis behaviour of the second run.

The effect of rhodium oxidation is similarly estimated by looking at three sections of the standard. For that section never exceeding 500°C the wire remains unaffected. The section which exceeds 900°C at the end of each run will not accumulate a significant level of oxide because oxide produced during the heating-up stage tends to be removed at the end of the run. It is the intermediate section, which reaches 500-900°C, that accumulates significant oxide and causes a progressive loss in calibration. The loss will become significant (~ 0.02%) after a total use of 10-20 h with the furnace above ~ 1000°C and recovery anneals will then be required. In practice the period of use between anneals should be assessed by the occasional intercomparison with another standard (see below).

It is recommended that at least three standards be acquired with a calibration uncertainty appropriate to the laboratory's needs. Each should be 1000-1200 mm in length and be installed in a single 500-600 mm long twin-bore insulator at the tip end and insulated in PVC tubing at the other. The insulator should be of the best quality re-crystallised alumina (Degussit AL23 or Morganite ΔRR). The standards are used to calibrate clean platinum-based thermocouples only and should be re-calibrated at regular but staggered intervals. The standard with the earliest calibration is used routinely in the calibration facility and every 5-10 runs it is intercompared with a second standard to detect significant drift. The third, with the most recent calibration, is kept as a spare and as a check in cases of doubt. While the most-used thermocouple is being calibrated two standards remain to support the facility.

For the calibration of base-metal thermocouples the standard should be placed in a clean silica or alumina sheath to isolate it from the thermocouples being tested. This may require an equaliser block within the furnace to reduce the temperature difference between their tips. In any case, separation of the tips represents an additional source of error to be considered when assessing the calibration uncertainty. Alternatively a lower grade type R or S standard could be used in direct contact with the base-metal thermocouples. It is first calibrated by intercomparison with the more accurate standards described above. However such intercomparisons should not be repeated because of possible cross contamination. Instead the lower-grade standard is checked occasionally against a second kept for the purpose and discarded when significant drift is noted. If possible the standard should be used at the same depth every time.

7. Calibration Procedure

The following description relates to the calibration of rare-metal thermocouples. The approach required for base-metal types differs only in the cleaning, annealing and insulating requirements.

The wires first should be cleaned by placing them for 15 min. in each of the following boiling liquids: nitric acid (20% of full strength), distilled water, hydrochloric acid (20%), distilled water and again in distilled water. After each acid wash the wires are rinsed in distilled water. From the moment cleaning begins the wires should be handled as little as possible and then only with clean tweezers or disposable plastic gloves and on a clean glass-topped bench.

After cleaning, the wires are fed into a clean twin-bore alumina insulator of a length to suit scanning (if required) or calibration and annealed for 2-3 h at 1100°C and then overnight (~ 16 h) at 450°C. Both anneals should extend over all that length of thermocouple to which the calibration will apply. If the rare-metal thermocouples are not annealed but are simply calibrated in the 'as-received' state the potential growth in inhomogeneity is greater and the calibration uncertainty must be increased to suit. New wires received from the manufacturer will be reasonably homogeneous but lower in output by ~ 0.05 to 0.1% compared to the fully annealed state. The signature of an un-annealed but used thermocouple would not only show the effects of hysteresis and acquired oxide but it would indicate increases in thermopower in that section of wire closest to the tip due to the non-uniform annealing effect of its use.

The wires not within the alumina should be insulated with flexible PVC tubing. The thermocouple is then scanned, if required, and calibrated. The scan may require a longer length of twin-bore insulator than can be managed in calibration since the latter should be done without extension leads. In such cases, the change to a shorter insulator could involve inadvertent cold work during handling and hence necessitate a further 450°C anneal.

The possible differences between the tip temperatures of the standard and of the thermocouples being calibrated is a potentially large and variable source of error. If the tips must be kept apart it may be necessary to incorporate a thermal-equalising block in the furnace and in either case it is essential that the likely error be assessed. This error (~0.1%) is avoided if the thermocouple tips are welded or platinum-wire-wrapped together.

Intercomparisons should be made, in increments of temperature no greater than 150°C, from the lowest to the highest temperature of interest. At each temperature the circuit zero should be measured and the effect of any drift in furnace temperature should be overcome, for example, by averaging the pair of readings obtained for each thermocouple on switching through them all and back to the first in equal time intervals.

To minimise changes in signature a minimum of time should be spent in the calibration furnace and once results are obtained at the upper temperature the thermocouples should be withdrawn from the furnace rather than allow them to cool with it.

Once the thermocouple emfs have been corrected for furnace drift and circuit zero they are used to calculate values of E-REF as shown in Table 2. All values are expressed to 0.1 μV, the least count of the DVM, in all stages of calculation. For this example the standard is assumed to be a type S and the other thermocouple, type R. From the emf (E_o) of the standard and its associated calibration value (E_o - REF_o) the reference tables value equivalent to tip temperature is obtained:

$$REF_o = E_o - (E_o - REF_o)$$

The temperature is then converted to REF suitable for the thermocouple type being calibrated and hence to E-REF given in column 6 of Table 2. computation would have been simpler if both thermocouples had been of the same type ($REF = REF_o$) for then the result could have been obtained directly from the measurements E and E_o.

$$(E-REF) = E - E_o + (E_o - REF_o)$$

The values of E-REF should be plotted (see Fig. 3) to provide a means of interpolation to other temperatures.

Table 2. Values of (E-REF) determined from measurements E and E_o

Measured EMFs Thermocouple (E)	Standard (E_o)	REF_o	T	REF	E-REF
118.3 μV	112.0 μV	119.9 μV	21.2°C	118.2 μV	0.1 μV
1 112.6	1 098.1	1 097.6	158.6	1 112.0	0.6
2 557.5	2 463.2	2 463.1	315.3	2 554.9	2.6
4 248.9	4 026.1	4 025.5	478.9	4 242.0	6.9
5 878.9	5 499.6	5 495.1	625.2	5 869.6	9.3
7 681.2	7 106.7	7 098.9	777.3	7 670.5	10.7
9 494.6	8 703.2	8 693.2	921.8	9 482.9	11.7
11 483.7	10 493.6	10 428.0	1072.4	11 471.5	12.2
13 253.6	11 973.9	11 961.5	1201.2	13 240.6	13.0

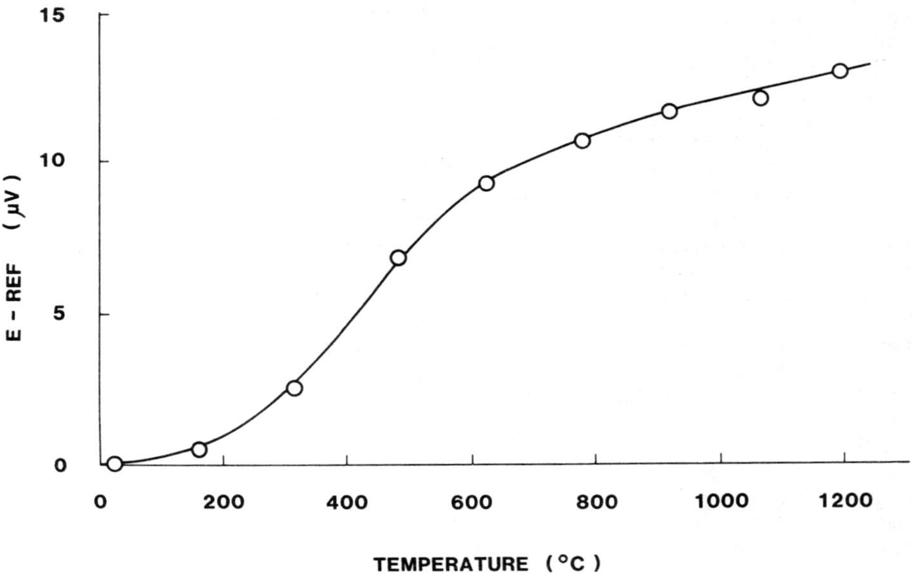

Fig. 3. Plot of E-REF against temperature

This is a relatively uncritical technique since the extent of the plot (< 25 μV at 1000°C) is not large when compared to the uncertainty given for the calibrations (at least 7 μV at 1000°C - Table 6). The simplest possible curve should then be drawn consistent with the requirement that it should deviate from the data points by no more than the expected scatter due to measurement (short-term instabilities of DVM, inadequacies of zero and drift correction, etc.). Alternatively if the data suggest a higher-order polynomial than cubic the deviates that result from using a cubic-like fit are used to revise the estimate of measurement scatter, a quantity needed when calculating the calibration uncertainty.

The temperature at which the calibration is reported should be selected at a sufficiently small spacing to allow linear interpolation without significant error. If only temperature and emf values are reported the number of values required is far greater than if E-REF is also given. In the latter case it should be assumed that interpolation is to be accomplished using E-REF, which varies with temperature by less than ~ 5 μV/100°C for types B, R and S, rather than values of emf (~ 1000 μV/100°C). In the case illustrated by Fig. 3 a temperature interval of at least 200°C would be adequate for a report. Table 3 indicates the method used in obtaining thermocouple emf at the chosen temperatures from interpolated values of E-REF. A 100 h period of anneal-free use after calibration is assumed and so a -0.04% bias is applied (section 8). Otherwise the thermocouple would have a systematically lower output in use than indicated by its calibration report. Column 6 of Table 3 is the sum of columns 2 and 5.

The calibration report could then supply the results (columns 1, 5 and 6 of Table 3) in a form similar to Table 4 and the calibration uncertainty could be enumerated in a fourth column or separately expressed as a function of emf as is done throughout this paper. In Table 4 the emf data has been rounded to the nearest 1 μV.

Table 3. Calculations of emf and E-REF at selected temperatures

Temperature	REF	E-REF	Bias (-0.04%)	E-REF With Bias	Emf
0°C	0.0 μV	0.0 μV	0 μV	0 μV	0 μV
200	1 486.1	1.1	-0.6	0.5	1 468.6
400	3 407.2	4.8	-1.4	3.4	3 410.6
600	5 582.3	8.7	-2.2	6.5	5 588.8
800	7 948.8	10.8	-3.2	7.6	7 956.4
1000	10 503.2	12.1	-4.2	7.9	10 511.1
1200	13 223.9	12.9	-5.3	7.6	13 231.5

8. Calibration Uncertainty

A calibration facility based on standards calibrated by another laboratory cannot, with advantage, issue calibrations conditional on the user applying recovery anneals. The potential improvement in accuracy to the user is marginal.

Table 4. Suggested presentation for calibration report

Temperature	Emf (E)	Difference from Reference Tables (E-REF)
0°C	0 μV	0 μV
200	1 469	1
400	3 411	3
600	5 589	7
800	7 956	8
1000	10 511	8
1200	13 232	8

The calibration uncertainty must be determined for each report issued. Its magnitude depends on the type of thermocouple calibrated, on its inhomogeneity and on the amount of future use covered by the calibration. It also depends on the calibration facility, i.e. the standard, the equipment and the technique.

Under clean, stress-free, oxidising conditions the changes in signature of rare-metal thermocouples are relatively insignificant for most practical applications. For this reason their calibrations can be given on the basis that in use, subsequent to calibration, any practical depth of immersion is permitted and the depth may be changed at will. Table 5 indicates the likely effect of such use for a 100 h accumulated time with a tip temperature above 500°C. The tabulated values are expressed in the form A ± B% to separate the bias (A) from the uncertainty (± B). The bias should be incorporated in the calibration report, as shown in Table 3, and reflects the fact that the relevant sources of error affect the signature in one direction only. For example it was assumed for Table 5 that during use oxide equivalent to 20-30 h exposure at 800°C accumulated in at least one section of the alloy wire before exposure to a higher temperature reversed the process. For this section the signature would show a potential loss in output of 0.06%. Hence for the 100 h period the thermocouple emf would be affected by oxide to the extent of 0 to -0.06% or -0.03 ± 0.03%.

Table 5. The effect of 100 h use on the signatures of new type R or S thermocouples

Temperature Range	If Annealed Before Calibration	If not Annealed
0 - 1300°C	-0.04 ± 0.04%	0 ± 0.08%
0 - 1550°C	-0.08 ± 0.08%	-0.04 ± 0.12%

A similar treatment could be given to other specified forms of use. For example the calibration could be given on the condition that the thermocouple be used at a fixed depth of immersion in one furnace only. The changes in calibration that will then take place are much less than indicated above and would tend to be progressive. As a guide it is worth noting that freshly annealed type S (and hence type R) thermocouples held at fixed immersion will drift -0.03% in 100 h at 1000°C (Bentley and Jones 1980) and that nickel-based thermocouples of 0.5 mm wire diameter will experience an *in situ* drift of 0.1% in 10 h at temperatures between 500 and 1100°C (Bentley 1982b).

As an example in calculating the total calibration uncertainty and as an indication of the quality of service that a calibration facility can offer, Table 6 was constructed on the basis of an optimum use of the facility. In other words, an 'as-new' type R thermocouple was first given the 1100 and 450°C anneals and was calibrated while welded to the standard, without the use of extension leads and with the cold junction of both thermocouples at 0°C. For comparison purposes two accuracy levels are given for the standard depending on the temperature range of its calibration and on its need for recovery anneals (section 6). As can be seen from Table 6, the extra work involved in maintaining the standard with recovery anneals is not warranted. The uncertainty is reduced only slightly, for example, from ±0.7 to ±0.6°C (type R at 1000°C).

A rare-metal thermocouple calibrated in the facility and intended as the standard for calibrating base-metal thermocouples could have an uncertainty of ±(1.5 μV + 0.075%). This is adequate since it is an order of magnitude smaller than the manufacturer's tolerance (±0.75%) for base-metal types and is less than the effect of their inherent inhomogeneity.

Table 6. Minimal calibration uncertainty for a new type R or S thermocouple

Error Source	Uncertainty Component
Standard	±(1 μV + 0.04%) or ±(1 μV + 0.06%)
Measurement	±(1 μV + 0.01%)
Inhomogeneity	± 0.02%
100 h use	± 0.04%
Rounding error in Report	±0.5 μV
Total	±(1.5 μV + 0.06%) or ±(1.5 μV + 0.075%)

9. Conclusion

A calibration philosophy has been developed from an understanding of the distributed nature of emf in thermocouples. It involves the view that the effect of a thermocouple's signature in calibration and the subsequent changes in signature, during that use to which the calibration report will be applied, must be fully taken into account.

This led to a calibration technique with a number of novel features. For example, there is a need to scan thermocouples after any significant use if re-calibration away from its position of use is required. Also, the emfs given in a calibration report may need a bias if they are to correctly represent the thermocouple after calibration.

References

ASTM (1974). American Soc. for Testing and Materials Special Tech. Publ. 470 A.

Bentley, R.E. (1982a). *Aust. J. Instrum. Control* 38, (oct.).

Bentley, R.E. (1982b). "Short-term instabilities in thermocouples containing nickel-based alloys." Submitted to *High Temp. - High Pressure*.

Bentley, R.E. (1982c). Proc. 35th Annual Conf. Aust. Inst. Metals, 152-155.

Bentley, R.E. and Jones, T.P. (1980). *High Temp. - High Pressure* 12, 33-45.

Jones, T.P., Meszaros, S.M. and Morgan, T.L. (1982). *Aust. J. Instrum. Control* 38, (Oct.).

Kemp, W.R.G. (1977). "International Practical Temperature Scale of 1968 (Amended Edition of 1975)". Australian version of the official French text. CSIRO.

McLaren, E.H. and Murdock, E.G. (1979). Nat. Res. Council of Canada publ. APW 2213/NRCC 17408.

Powell, R.L., Hall, W.J., Hyink, C.H. Jr., Sparks, L.L., Burns, G. W., Scroger, M.G. and Plumb, H.H. (1974). Nat. Bur. Stand. (US) Monogr. 125.

Calibration of Visual Optical Pyrometers

J. Tapping
CSIRO Division of Applied Physics, Sydney, Australia 2070

Abstract

The paper describes the apparatus and procedures for calibrating visual optical pyrometers using tungsten-strip lamps. The discussion includes criteria for the selection of suitable equipment, correct pyrometer techniques and the computation of calibration results.

Introduction

The calibration of visual optical pyrometers is usually carried out by using calibrated tungsten-strip lamps as sources of known spectral radiance temperature (often called blackbody temperature). The first part of this paper describes the apparatus necessary for this process and the criteria to be used in selecting items of equipment. The second part gives a recommended procedure for calibrations using this apparatus which closely follows the method used in this Division for the calibration of secondary standard instruments. Techniques for using pyrometers are also given. Appendices illustrate the kind of information which should be kept on record, the computation of results, and the format of a report.

Aspects of pyrometer calibration other than apparatus and procedure are not discussed in detail. The theory of optical pyrometry can be found in Kostkowski and Lee (1962), and information on tungsten-strip lamps and their calibration in Kostkowski and Lee (1962) and Coates (1981a, 1981b).

Part 1: Apparatus

1. The Principle of Calibration

Figure 1 shows the optical arrangement of the most common type of visual optical pyrometer. The much less common "variable wedge" type, although different in principle, is calibrated using exactly the same apparatus. These pyrometers measure spectral-radiance temperature (commonly referred to as blackbody temperature); that is, they measure the brightness of an object in a narrow wavelength band and indicate a temperature on the basis that this radiation is being emitted by a perfect radiator (ie a blackbody). To calibrate a pyrometer we therefore need to be able to supply it with radiation while knowing the temperature of a blackbody which would provide the same amount of power to the pyrometer. The most accurate and convenient way to do this over a range of temperatures is by means of tungsten-strip lamps. These lamps are calibrated by determining the relationship between the electric current passed through them and the spectral-radiance temperature of an area at the middle of the tungsten strip. A pyrometer is calibrated by sighting it on to the strip and comparing its readings with the known temperatures for a series of lamp

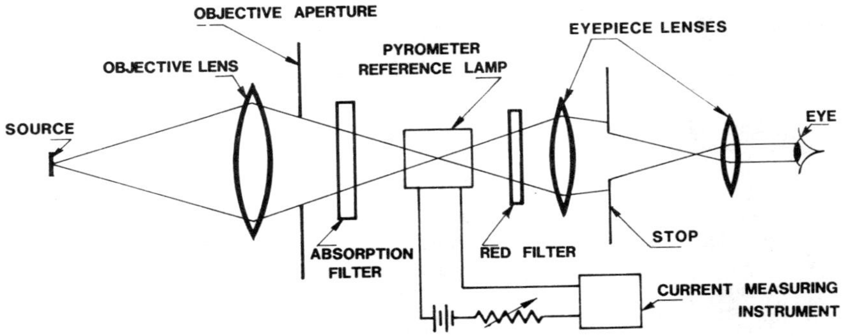

Fig. 1. Schematic diagram of a disappearing filament optical pyrometer

currents. The essential apparatus for this consists of a calibrated strip lamp, a power supply and a current measurement circuit for the lamp (see Fig. 2).

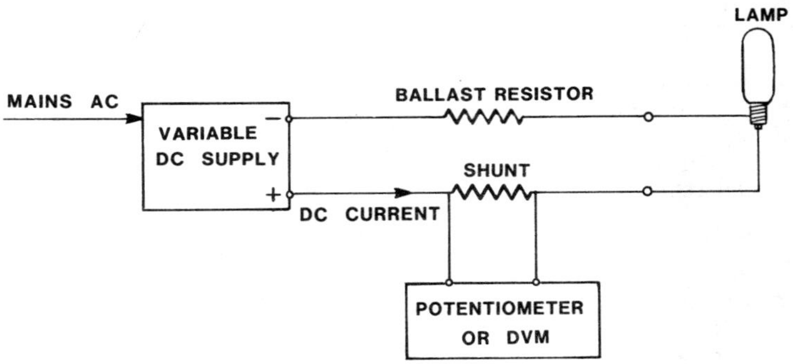

Fig. 2. Tungsten strip lamp current supply and measurement circuit.

To obtain the optimum accuracy for pyrometer calibrations the contribution by the strip lamp to the total uncertainty should be small compared to the pyrometer errors. The uncertainty in the radiance temperature is comprised of the lamp calibration uncertainty and contributions from the components of the strip lamp current supply and measurement circuit. Only limited control can be exercised over the lamp uncertainty, but the uncertainty due to the other factors can be easily made small by using components of sufficient quality. Where practicable the individual contributions should therefore be made small, say five to ten times smaller than the lamp calibration uncertainty.

2. Tungsten Strip Lamps

Strip lamps are of two kinds, vacuum lamps which are used for temperatures below about 1600°C, and gas-filled lamps which are used above about 1300°C. They are available in either ordinary or high-stability grades (Quinn and Lee 1972). It is recommended that a calibration laboratory have a minimum of two each of the vacuum and gas-filled types so that a pair (one vacuum and one gas-filled lamp) is available for use while a pair is being recalibrated. A national laboratory should have at least three of each type, so that one pair can be retained as a rarely-used cross-check for the working standards, and also so that the minimum requirement of two of each type is available if one lamp is damaged or becomes faulty. The vacuum lamps should be calibrated over the range 800 to 1500°C, although for higher accuracy separate lamps can be used for the upper and lower sections of this range. The gas-filled lamps should be calibrated from 1300 to about 2200°C, so that the overlap with the vacuum lamp range can be used for cross-checks between the two types.

The ordinary quality lamps are made with threaded brass end-caps which are intended to be screwed into sockets. It is recommended that sockets are not used mainly because of the risk of damage during insertion and removal, and the unreliability of the electrical contacts. In this Division these lamps are fixed into bases of the design shown in Fig. 3. This design allows the lamps to be safely stored in the upright position, enables firm attachment to the lamp support, and provides electrical connections which are safe, reliable and convenient.

High-stability lamps are fixed into bases of a different design (Jones and Tapping 1971) which allow the temperature of the lamp pins to be controlled and measured. For these lamps the base is essential because it also prevents relative movement between the pins, which can cause calibration changes or damage to the lamps.

An important consideration in deciding between normal and high-stability lamps should be the present and future calibration accuracy needs of the laboratory. The uncertainty of a strip lamp calibration, which can be a

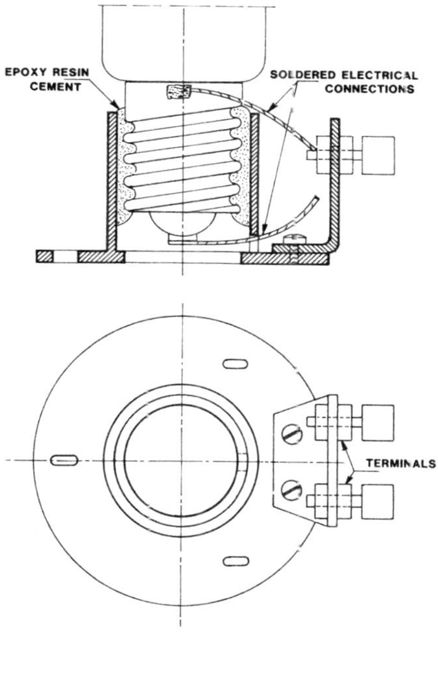

Fig. 3. Lamp base used at DAP, cross-section and top views.

major component of the uncertainty for pyrometers calibrated with it, is influenced by a number of lamp characteristics. Changes in the lamp calibration take place with use, and these depend on the conditions of operation (Barber 1946, Jones and Tapping 1979). There are also short-term changes which can be reversible (eg hysteresis effects which result in different temperatures depending on whether the set current is approached from a higher or lower value), or non-reversible (eg changes in the resistance of the strip due to annealing). Other characteristics of concern are the temperature gradients across the strip and variations in apparent radiance with angle of view. These properties vary from lamp to lamp and it is not usually practicable to investigate them in detail in each case, so estimates are made of their effects from limited tests during lamp calibration. These estimates usually form a major part of the calibration uncertainty of vacuum lamps, but for gas-filled lamps they are generally small compared to the variations caused by gas convection currents in the envelope.

The overall uncertainties assigned to lamps vary according to the type of lamp, the method of calibration, the radiance temperature and the conditions of use, as well as the characteristics already mentioned. Typical uncertainties are 1 to 5°C for ordinary vacuum lamps, 2 to 7°C for ordinary gas-filled lamps, 0.5 to 2°C for high-stability vacuum lamps and 1 to 5°C for high-stability gas-filled lamps.

Although most ordinary quality lamps are satisfactory for the calibration of visual pyrometers there is a risk that some will not be acceptable. Consideration should be given to having some high-stability lamps because their reliability can be assumed with confidence. By gradually instituting the procedures necessary for their optimum use (Jones and Tapping 1979), their calibration uncertainties can be improved with the need and capabilities of the laboratory. On the other hand their cost is high, about five times that of the ordinary lamps. If funds are limited it is better to have three ordinary lamps than one high-stability grade.

3. Power Supply

Strip lamps are usually supplied with direct current because dc is more easily and accurately controlled and measured. Electronic power supplies which convert mains ac to stable and variable dc are readily available commercially at reasonable cost, and these are the best form of lamp supply in most circumstances. Care should be taken in selecting a supply to ensure that it is adequate not only in terms of the maximum current and voltage it produces, but also its other specifications such as stability, ripple and control resolution. The maximum current and voltage needed will depend upon the types of lamps which are used. Table 1 gives typical requirements for some lamps, but a margin of at least 10% should be allowed above any specified maximum because of variations from lamp to lamp. But while it is important that an excess of power is available, a supply should be chosen which does not greatly exceed requirements because specifications in absolute numbers become worse for higher maximum power output. Many models can be used in series or parallel to increase voltage or current capability, so it may be better to have two smaller supplied of this kind than one large one.

Table 1. Typical currents and incremental rates for tungsten strip lamps. Lowest and highest currents are for 800°C and 1550°C for vacuum lamps, and 1300°C and 2200°C for gas-filled lamps. GEC is the General Electric Co, England, and GE is General Electric, USA.

Type	vacuum		gas-filled		
Manufacturer	GEC	Philips	GEC	Philips	GE
Lowest current (amperes)	4	2.8	9	6	18
Incremental rate (mA/°C)	5	3	8	6	18
Highest current	11	7.5	19	14	42
Incremental rate	13	8	13	10	30

An important consideration is whether the ripple and noise in the current from the supply is sufficiently small, because dc meters measure the average current but the lamp temperature is a function of the power dissipated, and hence of the root mean square of the current. The effect of ripple is to dissipate a quantity of extra power which is equal to that which would be produced by the ac component acting alone. For a dc current I, a superimposed ac current i is therefore equivalent to an increment in I given by $\delta I = i^2/2I$. The magnitude of ripple is expressed in a variety of ways but if the amplitude given is assumed to be that of a sine wave a sufficiently good estimate can be made of the effect.

It should be noted that many of the specifications given for power supplies are expressed as worst case situations which are not applicable to pyrometry applications. For example stability may be given as the change in output voltage when the current is changed from zero to maximum. It may take considerable effort to determine the behaviour of the supply in your application.

4. Current Shunts

To measure the lamp current it is necessary to pass it through a resistor of known value and then measure the voltage across the resistor. Resistors used in this way are known as current shunts. Shunts of the four-terminal type (that is, possessing separate voltage measurement and current connections), are the most accurate and should be used in this application.

The resistance value of the shunt must be high enough for the uncertainty in current measurement to be less than the current resolution required, but not so high that the power which will be dissipated in it will be excessive. The shunt should have a power rating at least twice the maximum required, and since size and cost increase rapidly as the power rating increases, the resistance value should be kept as low as practicable. Table 2 shows a typical example of the relevant figures needed to choose the shunt resistance. Note that in this table the resistances used are exact power of ten. These are very convenient because they allow the conversion of measured voltage to current by simply shifting the decimal point. (In practice the conversion will not be exact because the resistance will not be exactly equal to the nominal value and a correction may be necessary.)

Table 2. Current measurement parameters for some shunt resistance values, for a current range of 4 to 20 A, an incremental rate of 5 mA/°C and a resolution of 0.1°C

Shunt resistance value (Ω)	0.001	0.01	0.1
Maximum voltage produced (V)	0.02	0.2	2
Voltage resolution required (μV)	0.5	5	50
Maximum power dissipated (W)	0.4	4	40

The most important parameters to consider when choosing between different shunts with the same power rating and resistance, are the temperature coefficient (that is the change in resistance with change in ambient temperature), the power coefficient (that is the change in resistance due to the self-heating of the shunt), and stability (ie the expected rate of change of resistance with time). The values of these parameters may be different for different samples of the same model, and manufacturers will usually quote only typical values or upper limits. For a good quality shunt the drift rate will decrease with time so that the intervals between calibrations should be increased accordingly. The temperature and load coefficients will not change, and need be measured only once at the time of the first calibration. Oil filling improves these two coefficients considerably, but also adds significantly to the cost.

5. Voltage Measurements

The specifications for the instrument used to measure the voltage across the shunt are determined by a number of factors, the most important of which are the range of strip lamp currents, the resistance of the shunt and the temperature resolution required. As an example, suppose that the currents for GEC lamps are to be set to the equivalent of 0.1°C at 800°C and 0.5°C at 2200°C, with a current shunt of 0.01 Ω. From Table 1 it can be seen that 40 mV would need to be measured to within 5 μV, and 190 mV to within 65 μV. The voltmeter must therefore be able to measure to five digits, and with an uncertainty not larger than 5 μV.

The voltage should be measured with either a digital voltmeter (dvm) or a potentiometer. Digital voltmeters are particularly convenient to use and they can often be interfaced to computers for automatic control and data collection. However it is important that the specifications of a dvm are examined carefully and interpreted correctly to ensure that it is adequate because its accuracy may not be immediately obvious. For example, uncertainties specified by manufacturers are generally the sum of two or more factors, and the resultant uncertainty is often up to ten times the resolution of the dvm. Digital voltmeters can also have significant calibration errors in small parts of their range which may not be detected by checks at widely spaced voltages. It is therefore advisable to frequently check a dvm over the whole of the voltage range used. Partly because of their electronic nature and partly because of their high input impedances, dvmsare subject to a number of measurement problems, for example common mode, series mode and earth loop errors. These errors are often difficult to detect so measurements should be checked with an instrument not as prone to them, such as a potentiometer. Their complex circuitry also makes it difficult to carry out repairs, and when choosing a dvm substantial weight should be placed on the availability of a local repair service.

Potentiometers have a number of disadvantages when compared to dvms. They are less convenient to use because they require manual adjustment, they are less mobile and they need a more controlled environment. They also require a stable voltage supply and a sensitive galvanometer. If wet-cells are used for the supply these will need frequent attention. Electronic galvanometers are subject to some of the same electrical errors as dvms, and mechanical galvanometers are more reliable but more fragile. Potentiometers are however usually more accurate, stable and reliable, and in standards work these factors are of over-riding importance. Digital voltmeters should therefore be used only if there is some means of ensuring that their measurements are free of significant errors.

6. Lamp Supply Circuit

The arrangement of the circuit is shown in Fig. 2. The cables which carry the lamp current should be rated at a current higher than the maximum requirement which can be foreseen. They should preferably be permanently fixed in position for both electrical and mechanical safety. The polarity of the supply should be clearly marked to ensure that lamps are not connected incorrectly.

The electrical resistance of a strip lamp varies with the temperature of the strip, and may be typically 0.2 Ω at 800°C and 0.4 Ω at 1500°C. If in the circuit of Fig. 2, the ballast resistor was not included the circuit resistance would be dominated by the lamp, so that for a constant voltage supply the current would vary as the lamp temperature stabilized. If current control is not used, the inclusion of a ballast resistor will help to stabilize the current so that fewer adjustments will be necessary. The extra total circuit resistance will also increase the necessary supply voltage, and hence will improve the precision with which the current can be set with the voltage control of a particular power supply. It will also reduce the risk of inadvertently raising the lamp current to an excessive level. The resistance value of the ballast is not important, but should be a compromise between a high value desirable for stability and safety, and a low value to limit the maximum voltage and power dissipation. A value equal to one to two times the maximum lamp resistance is reasonable. The ballast can be adjustable to different values for different circumstances, but then care must be exercised in switching values while current is flowing.

7. Lamp and Pyrometer Supports

The lamp support must be adjustable to allow for lamps with different viewing heights and directions, and it must hold lamps firmly in position after adjustment. Figure 4(a) shows a simple support suitable for holding lamps fitted with the recommended base. The lamp is placed on the plate and held by screws inserted through the holes in the lamp base into the threaded holes in the plate. The pitch adjustment (shown in detail in Fig. 4(b) has its total travel limited by the bevels for safety reasons. The lamp height is adjusted by moving the shaft in the tube. The safety ring is a collar which is then clamped to the shaft at the top of the tube to allow rotation about the vertical axis without changing the height. The main clamp knob holds the shaft firmly in the tube. The base of the stand is heavy (eg thick plate or cast metal) to provide stability.

Pyrometers can be supported on a stand which is identical to that used for lamps but with a plate such as that shown in Fig. 4(b). This plate has an offset hole to enable pyrometers to be held by means of the threaded hole which is usually provided in their bases. An alternative to constructing a support is to adapt a commercially available device such as a camera stand or a radiation pyrometer support.

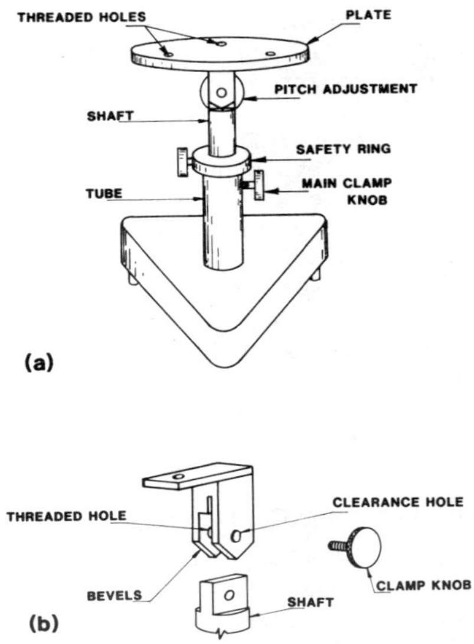

Fig. 4. (a) Strip lamp support, (b) bracket used for the pyrometer support, also showing detail of the pitch adjustment.

8. Pyrometer Supply

Visual pyrometers are usually powered by dry cells or rechargeable cells, and are designed for intermittent use only. In calibration, and particularly for a stability test, the pyrometer must be operated for extended periods, and it may be convenient or necessary to use an external supply. The supply voltage should be adjustable, and although exact voltages are not necessary they should have good stability over a few tens of seconds. Typical pyrometer requirements are for 100 mA at 6 V with less than 50 mV ripple, but needs can be up to 12 V or 250 mA. The adequacy of the ripple for a particular reference lamp can be checked by calculations similar to those used for strip lamps. Some instruments require two voltages, so two supplies should be available.

For pyrometers which use dry cells the supply connection can be made by removing the cells and attaching wires by means of clips, but where rechargeable cells are used these must be disconnected by unsoldering their leads before the supply connection is made. The supply should not be connected across the cells because of the risk of damage to them.

9. Reference Lamp Current Measurement

Since each temperature on the scale of a pyrometer actually represents a particular reference-lamp current, a number of sources of errors can be eliminated by calibrating directly in terms of this current instead of the scale readings. Many pyrometers have some means of facilitating the current measurement, such as an internal high-stability resistor or a removable wire bridge in the current circuit. The requirements for the instruments measuring this current are similar to those for the strip lamp current measurement. Typical currents are 30 to 80 mA with corresponding incremental rates of 20 and 80 μA/°C, for which a shunt of one to ten ohms is suitable. If the shunt is built into the pyrometer its precise value is not important because this becomes part of the pyrometer calibration. By using some form of switching, the voltage across the shunt may be measured with the same voltmeter as is used to measure the strip lamp current.

10. Automatic Data Collection

If the voltmeter which measures the reference-lamp current has a digital electronic output, pyrometer readings could be recorded on a suitable printer or transmitted to a computer or calculator for recording and processing. Besides the obvious advantages of increased convenience, efficiency and precision, this method enables readings to be taken objectively, thus preventing unconscious bias being applied by the observer. Calibrations which might otherwise require a second person to read and record the results can then be done by a single observer. If required, reference-lamp currents can be converted to pyrometer scale readings by determining a mathematical relationship between the two quantities. This can be done for example by least-squares polynomial curve fitting to a number of corresponding currents and scale readings.

11. Miscellaneous

Lamps are calibrated for a particular area of the tungsten strip and for a particular direction of view (Barber 1946, Jones 1963). The appropriate area is defined either by a wire pointer or a small notch in the edge of the strip, and the direction by a mark on the rear of the glass envelope. The correct alignment is such that the pyrometer views the middle of the strip adjacent to the pointer or notch while the mark appears to be behind the target area. This can be achieved with a collimated beam of light projected along the viewing axis from behind the lamp. The light source should be adjustable in position and direction to allow for different strip lamps, and switchable from the observer's position so that alignment can be checked frequently. It is convenient to have the brightness of the light adjustable so that the lamp markers are comfortably seen on all ranges of the pyrometers.

The strip lamp should be shielded from direct illumination by the sun or by room lights because at lamp temperatures below about 900°C reflected energy can give significant errors in pyrometer readings.

At lamp temperatures greater than about 1400°C the brightness of the lamp is distracting and annoying to the observers. This can be prevented with a black screen, made for example from cloth or painted aluminium sheet, with a hole through which the pyrometer is sighted. The position and height of the screen should be adjustable, and the screen should be removable for pyrometer alignment.

The table or bench on which the pyrometer and the lamp rest must be stable enough to prevent vibration. It must be sufficiently low to allow ample vertical adjustment of the pyrometer support to accommodate pyrometers of different sizes.

To minimize the risk of damage to strip lamps they should be stored in a cupborad or box in individual compartments, and placed so that there is clear access to the location at which they are used. Each lamp and compartment should be clearly marked with the identity and type of the lamp to avoid confusion and unnecessary handling.

It is vital to provide good protection for lamps when they are being transported to and from the laboratory at which they are calibrated. A design for a suitable transportation box is shown in Fig. 5. The box can be made from plywood or particle board, and if it is to hold only one lamp should have internal dimensions of about 350 × 350 × 350 mm. The lid should be screwed in place so that it is easily removed and replaced. The battens hold the screws, provide hand-grips, and help to keep the box upright. The lamp should be encased in a compressible material (such as foam-plastic), inside a smaller box made of wood or cardboard. The small box should fit inside the main box with at least 50 mm clearance all round which is filled with a shock-absorbing material such as polystyrene balls or sheet, foam-plastic sheet or shredded paper. The lamp should be placed so that the strip is vertical to reduce the possibility of damage if the box is dropped. The box should be marked to indicate that it should be kept upright and that the content is fragile.

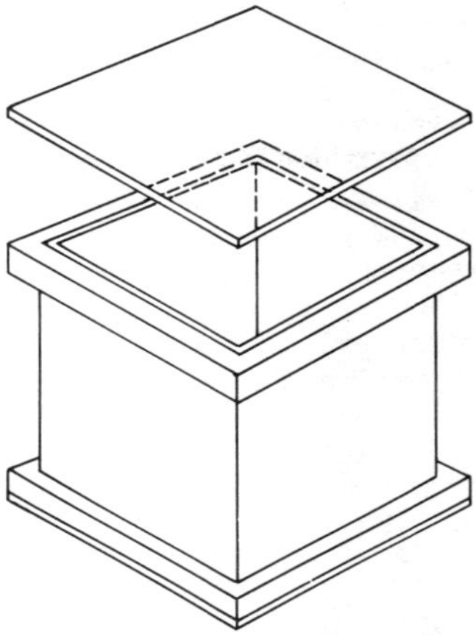

Fig. 5. Strip lamp transportation box.

Part II: Calibration

12. Preliminary

As soon as notification is received of a calibration, a record should be established of any relevant information which is available. An example of such a record is shown in Appendix A. The record may be kept in the same book as the results or elsewhere, but in any case should be permanent and accessible.

13. Inspection and Cleaning

A pyrometer should not be calibrated if it is not in good working order. It should first be examined externally for obvious faults, for example in the adjusting knob, meters or reference lamp. Unless the calibration has been requested for the "as received" condition, the instrument should then be examined internally. The mechanical components, and in particular the slidewire which adjusts the reference lamp current, should be inspected and cleaned if necessary. The reference lamp should be inspected, its serial number noted, and if any other parameters are marked on the lamp these should also be recorded. The lenses and other optical components which are accessible should be cleaned with particular attention to the surfaces between the target and the reference lamp. Surfaces after the reference lamp are less important because attenuation here will affect the radiances of the reference and the target equally. Special care should be exercised when cleaning the reference lamp to avoid static electric charges on the glass, because these can distort the tungsten filament. Tests made in this Division (Jones and Tapping 1979) have shown that commercial window cleaning fluid is satisfactory for cleaning optical surfaces, and has constituents which eliminate static charges.

An important reason for cleaning the optical components is so that the pyrometer is calibrated in a reproducible condition. For the same reason, any adjustments in the instrument should be set according to the manufacturer's directions (unless there is a good reason for doing otherwise), and the settings noted. A copy of the manufacturer's manual for the pyrometer can be very useful if it contains directions for dismantling, cleaning and adjusting the instrument.

14. Positioning

The strip lamp should be secured on its support and adjusted to approximately the correct position. Gas-filled lamps must be set to within one to two angular degrees of their orientation at the time of their calibration, generally with the tungsten strip vertical (Barber 1946), but vacuum lamps are not affected by orientation. Properly designed pyrometers do not vary their readings with distance from the target (Kostkowski and Lee 1962), and since it is easier to take readings on larger target images the strip lamp should be placed close to the minimum focusing distance of the pyrometer. To focus, the eyepiece should be first adjusted until the reference lamp image is sharp, then the target (in this case the strip lamp), focused using the objective lens adjustment. This places the target image in the same plane as the reference lamp.

The final adjustment of the direction of view should be made by moving the pyrometer. Most pyrometers have an indication of the position on the reference lamp which is to be used, such as a wire pointer or a bend in the tungsten filament. This position should be made to appear to coincide with the strip lamp target area and with the direction mark on the lamp envelope. It may be necessary to slightly de-focus the pyrometer so that both the mark and the target area can be seen simultaneously. The position of the mark is chosen so that there is only a small rate of change in radiance temperature with angle, and it is usually such that adjustment within 5 angular degrees gives an error no greater than 0.2°C.

15. Pyrometer Readings

Each reading of a pyrometer contains both random and systematic errors. The uncertainty due to the random errors should be reduced by taking a number of readings but there is little advantage in decreasing it to a value much less than that due to the systematic errors in the measurement. Systematic errors should be minimised where practicable, and the probable magnitude of them known so that the most efficient number of readings can be taken.

In matching the brightnesses of the strip and reference lamps an observer has a finite sensitivity, so there is a range of readings over which the two lamps appear to be matched. This range, commonly called the dead-band, can be from 1°C for experienced observers to more than 10°C for inexperienced observers, but will also vary with time for a particular person (Tapping 1981). Each reading should therefore be obtained by approaching the match from opposite directions and taking the mean of the two results to cancel the systematic errors due to the dead-band. It is important to take the two observations together to exclude errors due to time variations. The dead-band is usually larger for the first few observations, so some practice readings should be taken at the beginning of each session.

A pyrometer ideally provides only a single wavelength of light to the eye, but in practice a band of wavelengths is transmitted. Because each person has an individual spectral response characteristic, this gives rise to small systematic differences between observers, which are up to 1°C at 1000°C and up to 5°C at 2200°C (Lovejoy 1958). The finite bandpass may also result in the reference and strip lamps appearing to be slightly different in colour. This last effect, together with another in which the reference lamp appears to be non-uniform in brightness across its width (Kostkowski and Lee 1962), make it necessary for the observer to form a mental criterion of a match, which can vary from person to person and from time to time. There is no published information on the variations in criterion but experience in this Division indicates that for skilled observers using typical industrial pyrometers they are probably about 1 to 2°C at 1000°C and 3 to 5°C at 2200°C.

The strip lamp calibrations contribute systematic errors and the possible magnitudes of these are expressed by the uncertainties given in section 2. There are sundry contributions from other sources such as the pyrometer-lamp alignment and the strip-lamp current measurement, but these will usually be negligible. Taking the overall uncertainty as the square root of the sum of the squares of the components, the figures given above for observer variations and lamp uncertainties give the overall uncertainty due to systematic errors as typically about 2 to 5°C at 1000°C and 6 to 10°C at 2200°C.

16. Stability Test

The pyrometer reference lamp should be tested for stability by checking the calibration at a convenient temperature before and after running the lamp at its maximum operating temperature for an extended period. Two types of instability are of interest. The first is the gross type which would quickly invalidate any calibration, and such faults give calibration changes greater than 10°C after a running time of one hour. They are usually accompanied by large corrections.

The second type of instability is the slower long-term drift, whose rate may be required to determine whether the calibration can be expected to be valid for a certain period. To illustrate the difficulty in testing for this, suppose that the calibration at 1000°C should not change by more than 4°C in 50 hours of normal usage. The rate of drift of lamps increases rapidly with increasing temperature, and from the curve given by Coates (1981a) for strip lamps it has been calculated that the rate at 1000°C is approximately one-fifth of that at 1200°C. If "normal usage" is taken to be equivalent to operation at 1000°C, then 4°C per 50 hours is equivalent to 20°C for 50 hours at 1200°C. Therefore operation at 1200°C for 20 hours should change the pyrometer calibration by more than 8°C. If the uncertainty in each set of check readings is ±2°C then a difference of at least 12°C must be measured to be sure that a change greater than 8°C has occurred, and a difference no greater than 4°C is needed to be sure that the actual change has been no greater than 8°C. Differences between 4°C and 12°C are inconclusive. Because of these difficulties in making reliable measurements and because of the uncertain validity of the assumptions made, the test for long-term drift should be carried out only if it is particularly necessary and the results should be treated with caution.

17. The Calibration

It is desirable to have two observers, each taking half the observations while the other reads the pyrometer scale and records the results. There are a number of advantages of this. Firstly, if the observer is not aware of the individual results he is able to concentrate solely on the readings thereby making them more objective. Secondly, it tends to reduce the effects of differences between observers. Thirdly, the results are obtained more quickly with less fatigue and better uncertainty.

Appendix B is a specimen of the type of record made in this Division of pyrometer calibrations, and demonstrates the recommended procedures for calibration and processing of results. In the example the two observers (SM and JT) are shown to have taken two pairs of observations each, and this is the recommended minimum. As discussed above, a greater number may be necessary to reduce the random uncertainty but in general no more than four pairs for each observer are needed.

After the strip lamp current has been adjusted for a new temperature and before readings commence the lamp temperature must be allowed to stabilize. The rate of stabilization depends on the temperatures before and after adjustment, and on the design of the lamp. Typical stabilization times for ordinary quality GEC vacuum lamps are given in Table 3, and those for most other vacuum lamps are similar. After raising the current from zero to the equivalent of 1300°C, a GEC ordinary quality gas-filled lamp will be within 1°C of its equilibrium temperature in about eight minutes, and within 0.1°C in about 16 minutes. For changes of 100°C to temperatures above those mentioned here, both types of lamp will be sufficiently close to equilibrium in less than one minute. As the lamp is stabilizing at each new temperature the pyrometer alignment and the lamp current should be rechecked.

It is usually not practicable to calibrate a pyrometer at a large number of temperatures but fortunately most instruments have correction curves which are continuous and smooth. Only a few calibration points are therefore necessary to define the curve, and the temperatures used in Appendix B have been found to be adequate in most cases. The lower limit has been chosen to be 850°C because below this temperature brightness-matching is very difficult and the stabilization times of strip lamps are large. Note that both the vacuum and gas-filled lamps have been used at 1500°C to provide a check of the calibrations of the lamps.

Pyrometers can be calibrated at temperatures higher than the maximum for strip lamps (2300°C) from the low range calibration and the absorption of the range filter. Details of this process are given by Kostkowski and Lee (1962).

Table 3. Times after current adjustment for GEC ordinary quality strip lamps to reach within 1°C and 0.1°C of their final temperatures

Initial temperature (°C)	20	850	900	1000	1100
Final temperature	850	900	1000	1100	1200
Time to 1°C of final (min)	22	5	2	< 1	$\ll 1$
Time to 0.1°C of final	38	18	15	8	2

18. Results

The calibration of a strip lamp strictly applies only for a particular wavelength, and this should correspond to the effective wavelength (λ_e) of the pyrometer. If it does not then corrections can be applied to the lamp temperature using the factors given in Table 4, provided that λ_e is known. Note that increasing λ_e decreases the lamp radiance temperature. The λ_e for visual pyrometers is usually about 655 nm, but even for instruments of the same model it may be up to 5 nm from the nominal value given by the manufacturer (Lovejoy 1958). Although λ_e for a pyrometer can be measured (Kostkowski and Lee 1962) in most cases this is not justified because assuming that λ_e is equal to the nominal value will give an error no greater than other errors in calibration.

Table 4. Factors for correcting strip lamp calibrations to alternative effective wavelengths

Radiance temp. (°C)	800	1000	1200	1400	1600	1800	2000	2200
Factor (°C/nm)	0.09	0.13	0.17	0.23	0.29	0.36	0.44	0.53

As shown in Appendix B the corrections given in reports issued by this Division are obtained from smooth curves drawn through plots of the measured corrections versus pyrometer readings. The signs of the corrections are such that algebraically adding the correction to the pyrometer reading gives the corrected temperature. The curves are drawn in the simplest form which passes within a reasonable distance (about one standard deviation of the set of readings), of the points. If a more complicated curve than those shown in Appendix B is necessary then consideration is given to checking points which deviate from a simple shape. Account is also taken of the shapes of correction curves found in any previous calibrations. The corrections given in the report are rounded to the nearest 2°C, and although this is not essential it gives a better indication of the resolution of the pyrometer.

Appendix C shows the form of calibration reports issued by this Division. The uncertainties given in the tables are a combination of a number of factors, namely the random uncertainties of the readings, the possible systematic errors discussed in Section 15, the effective wavelength uncertainty and the reproducibility of the pyrometer. The last factor is derived from previous calibrations of the instrument and experience with others of the same type. The uncertainties shown in Appendix C are those usually given for good quality industrial pyrometers, but values up to twice these are quoted for lesser quality instruments. If the reproducibility of a pyrometer is doubtful, the largest reasonable uncertainties are given to ensure that an unrealistic estimate of the accuracy of the instrument is not conveyed to users.

References

Barber, C.R. (1946). *J. Sci. Instr.* 23, 238-242.

Coates, P.B. (1981a). NPL Report Qu61, National Physical Laboratory, UK.

Coates, P.B. (1981b). NPL Report Qu62, National Physical Laboratory, UK.

Jones, T.P. (1963). *J. Sci. Instr.* 40, 101-104.

Jones, T.P. and Tapping, J. (1971). *Metrologia* 7, 45-49.

Jones, T.P. and Tapping, J. (1979). *Metrologia* 15, 135-141.

Kostkowski, H.J. and Lee, R.D. (1962) in Temperature, Its Measurement and Control in Science and Industry. C.M. Herzfeld (ed.) Vol 3, part 1, 449-481, Reinhold Publishing Co. New York. (Reprinted as NBS Monograph 41, National Bureau of Standards, USA.)

Lovejoy, D.R. (1958). *Can. J. Phys.* 36, 1397-1408.

Quinn, T.J. and Lee, R.D. (1972) in Temperature, Its Measurement and Control in Science and Industry. H.H. Plumb (ed.) Vol 4, part 1, 395-410, Instrument Society of America, Pittsburgh.

Tapping, J. (1981). Proceedings of the Australian Institute of Physics Second Applied Physics Conference, Vol 1, 144-147, AIP, Melbourne.

Appendix A:

Example of a record kept of information relevant to a pyrometer submitted for calibration.

Calibrations Co. Ltd.
Bradfield Rd
West Lindfield N.S.W.
P.O. Box 218, Lindfield

Leeds and Northrup optical pyrometer
Type 8632C
Serial no. 97151

Received :	16/8/82	File	GB30/CAL-1/20
Examined :	18/8/82	Reference	RS 99123
Calibrated :	19/8/82		
Report :	20/8/82	Order no.	X 4738
Returned :	20/8/82	Dated	July 17
Report sent :	24/8/82	Contact	Mr. Keen
		Phone no.	467-6211

Calibration requested: Low and high ranges.

Pyrometer details

Accessories :	lens cap, shoulder strap
Last calibrated :	July 1980
Standard cell :	1.01859 v at 20°c
Lamp no. :	A 123
Resistor L :	131 Ω
Resistor B :	193 Ω
Drum setting :	9.5

Note : Packaging not adequate. Will request company to provide a stronger box.

Appendix B:

Specimen record of a pyrometer calibration showing the results of observations and the computation of calibration corrections.

Calibrations Co. Ltd. August 82

Leeds and Northrup pyrometer type 8632C
Serial no. 97151

Stability check
tested at 1000°C using strip lamp E74

SM	JT			
1001	1001	2 hours	1000	1003
1002	1003	at 1225°C	1001	1000
999	1002		1002	1001
1002	1002		1000	1001
1002			1001	

Low range – using lamp E74 (vacuum)

	SM	JT
850°C	848	853
lamp current	850	851
	848	852
3·008A	850	852
	Mean 850	

900°C	900	904
	904	903
3·183A	903	903
	901	904
	903	

1000°C	996	1001
	1002	1002
3·609A	1003	1003
	1000	1002
	1001	

1100°C	1098	1100
	1103	1101
4·131A	1101	1099
	1098	1101
	1100	

1200°C	1191	1197
	1194	1196
4·740A	1195	1196
	1195	1196
	1195	

Appendix B (continued)

Calibrations Co. Ltd. 19/8/82
pyrometer no. 97151

High Range
Lamp E 74

1100°c	1105	1102
	1105	1102
4·131 A	1101	1103
	1101	1102
	1103	

1200°c	1199	1206
	1208	1204
4·740 A	1203	1202
	1205	1207
	1204	

1300°c	1311	1309
	1303	1310
5·423 A	1309	1308
	1308	1309
	1308	

1400°c	1405	1407
	1404	1407
6·168 A	1409	1409
	1408	1406
	1407	

1500°c	1494	1507
	1503	1504
6·959 A	1505	1506
	1501	1507
	1503	

Lamp E48 (gas-filled)

1500°C	1503	1501
	1501	1497
7·740 A	1502	1503
	1501	1498
	1501	

1600°c	1595	1598
	1594	1595
8·425 A	1591	1595
	1594	1593
	1594	

Appendix B (continued)

Calibrations Co. Ltd. 19/8/82
pyrometer no. 97151

1700°c	1681	1695
	1690	1690
9.186A	1688	1694
	1687	1689
	1689	

Summary of Results:

Range	Lamp	Lamp Temp.	Pyrom. Reading	Correction
Low	E74	850°c	850°c	0 °c
	"	900	903	−3
	"	1000	1001	−1
	"	1100	1100	0
	"	1200	1195	5
High	E74	1100°c	1103°c	−3 °c
	"	1200	1204	−4
	"	1300	1308	−8
	"	1400	1407	−7
	"	1500	1503	−3
	E48	1500	1501	−1
		1600	1594	6
		1700	1689	11

Appendix B (continued)

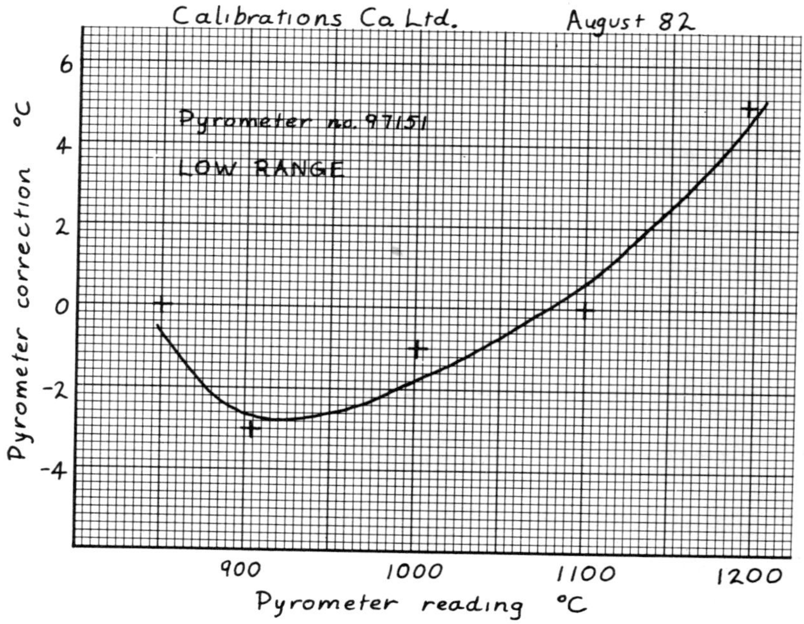

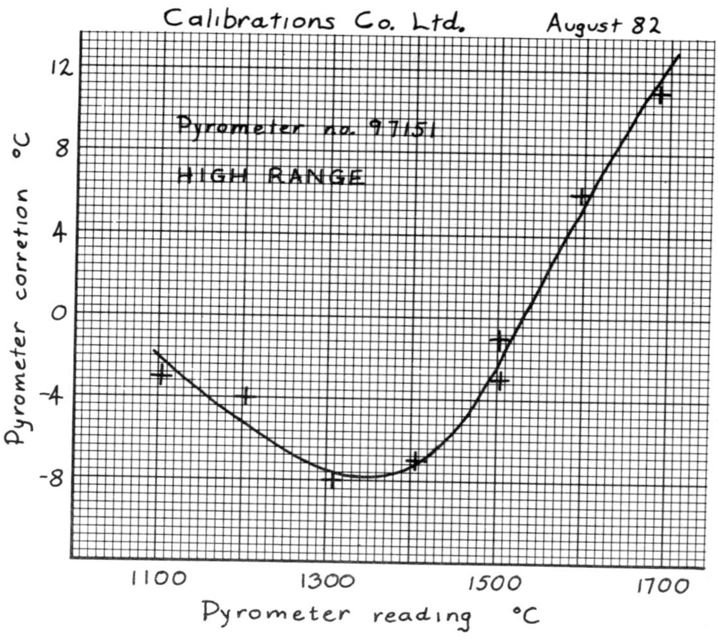

Appendix B (continued)

Calibrations Co. Ltd. 19/8/82
pyrometer no. 97151

Correction table from smoothed curves.

Range	Pyrom. Reading	Correction	Corrected Reading
Low	850°C	0°C	850°C
	900	-2	898
	950	-2	948
	1000	-2	998
	1050	0	1050
	1100	0	1100
	1150	2	1152
	1200	4	1204
High	1100°C	-2°C	1098
	1150	-4	1146
	1200	-6	1194
	1250	-6	1244
	1300	-8	1292
	1350	-8	1342
	1400	-8	1392
	1450	-6	1444
	1500	-2	1498
	1550	2	1552
	1600	6	1606
	1650	10	1660
	1700	12	1712

Appendix C:

Specimen of a report on the calibration of a visual optical pyrometer.

CSIRO

Division of Applied Physics

REPORT

ON

CALIBRATION OF OPTICAL PYROMETER No. 97151

Submitted by:	**Calibrations Co Ltd** **P.O. Box 218** **Lindfield NSW 2070**
Instrument:	**Leeds & Northrup optical pyrometer type 8632C**
Lamp No.:	**A 123**
Calibration Range:	**850°C to 1200°C on the Low Range** **1100°C to 1700°C on the High Range**

The instrument was calibrated in terms of the Commonwealth Legal Units of Measurement.

The uncertainty of calibration as given in the following tables, has been expressed on the basis that there is not more than one chance in one hundred that the true value differs from the stated value by more than the stated uncertainty.

The corrections are given to the nearest 2°C.

/Tables

Ref. **RS 99123** *File:* **GB30/CAL-1/20** *Checked:* [signature] *Date:* **20 August 1982**

Commonwealth Scientific and Industrial Research Organization, Australia

Appendix C (continued)

Continuation of report **CALIBRATION OF OPTICAL PYROMETER NO. 97151** Page 2

TABLE I
LOW RANGE

Pyrometer Reading	Correction	Blackbody Temperature	Calibration Uncertainty
850°C	0°C	850°C	± 4°C
900	− 2	898	"
950	− 2	948	"
1000	− 2	998	"
1050	0	1050	"
1100	0	1100	"
1150	2	1152	"
1200	4	1204	"

TABLE II
HIGH RANGE

Pyrometer Reading	Correction	Blackbody Temperature	Calibration Uncertainty
1100°C	− 2°C	1098°C	± 4°C
1150	− 4	1146	"
1200	− 6	1194	"
1250	− 6	1244	"
1300	− 8	1292	"
1350	− 8	1342	"
1400	− 8	1392	"
1450	− 6	1444	"
1500	− 2	1498	"
1550	2	1552	± 5°C
1600	6	1606	"
1650	10	1660	"
1700	12	1712	"

T P Jones

T P Jones
for John J Lowke
Chief of Division

Ref. RS 99123 File: GB30/CAL-1/20 Checked: Date: 20 August 1982

Commonwealth Scientific and Industrial Research Organization, Australia

Applied Physics and the Electrical Units

J.C. Macfarlane
CSIRO Division of Applied Physics, Sydney, Australia 2070

Abstract

Although the ampere is regarded as one of the base units of the International System of units (SI), the ohm and the volt are maintained in practice by most standards laboratories, and the ampere is derived from Ohm's law. The maintenance of the ohm by means of the calculable capacitor, and of the volt by means of the Josephson effect, are described. Some very recent developments which may lead to improved standards of resistance, voltage and current are also mentioned.

1. Introduction

The units of electrical measurement are defined, and as far as possible realised in practice, by relating them to the S.I. base units of length, mass, and time interval or frequency. Inglis (1982) has described in detail the hierarchy of standards on which the electrical units are based.

This paper illustrates how developments in physics have been applied to the field of electrical measurement, with the aim of providing improved techniques for the realization and maintenance of the electrical units.

The ampere is defined as 'that constant current which if maintained in two straight parallel conductors of infinite length, of negligible circular cross-section, and placed 1 metre apart in vacuum, would produce between these conductors a force equal to 2×10^{-7} newtons per metre of length'. This definition formally allows an absolute measurement of current to be made in terms of the mechanical units of force (i.e. mass) and length, but there are severe practical difficulties in carrying out any experiment with infinitely long, negligibly thin conductors. Measurements based on a current balance using coils of finite dimensions are more tractable, although it is still a difficult and tedious process to achieve parts-per-million accuracies in absolute determinations of the ampere.

It has consequently been the practice for many years in standards laboratories to maintain the ohm and the volt as working standards. This practice was reinforced by the advent firstly of the calculable capacitor in the late 1950s, and of the Josephson voltage standard in the 1970s. In this paper, these developments will be examined together with several other innovations that have taken place, or are in the process of evaluation , in the realization of the electrical units.

2. The Unit of Resistance

It is not many years since the ohm was defined in physics textbooks as the resistance of a column of mercury of specified length and cross-section. In the 1960s A.M. Thompson and D.G. Lampard of this laboratory achieved international recognition for their discovery of a new theorem in electrostatics which proved that the capacitance of a 'cross-capacitor' (Fig. 1) could be precisely defined in terms of a single dimension. The calculable capacitor (Thompson 1968) now provides the definition of the farad, and the absolute standards of capacitance and resistance. A measurement of linear displacement, by optical interferometry, allows the farad to be directly related to the metre:

$$C = \frac{\varepsilon_0}{\pi} \ln 2 \quad F/m.$$

The absolute ohm is determined relative to the metre with an accuracy of 1 in 10^7.

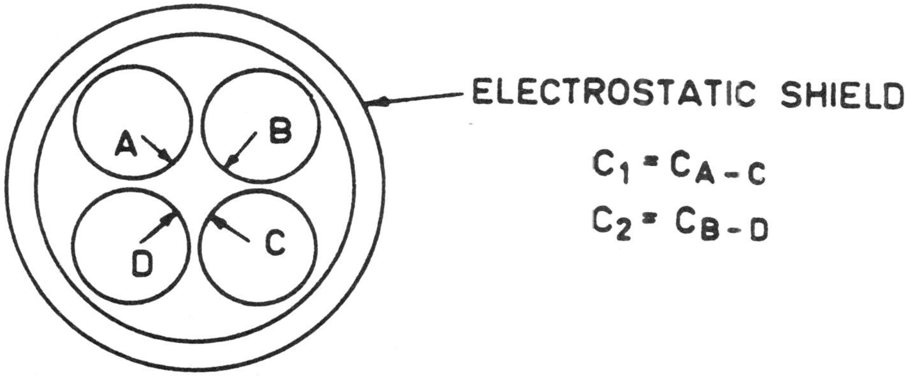

Fig. 1. Cross-section of cylindrical cross-capacitor. There is a simple relation between C_1 and C_2. If they are equal, then they have the value $(\varepsilon_0/\pi)\ln 2$ F/m.

It has been shown very recently (von Klitzing et al 1980) that certain high-mobility semiconductors, when operated at liquid helium temperatures and in high magnetic fields, exhibit a 'quantized Hall resistance'. If this technique is proved to be sufficiently accurate, it will be possible for national laboratories to determine their standard of resistance, independently of any other units, in terms of the quantity h/e^2 (which has the approximate value 25 813 ohms). The results of this experiment, when compared with measurements on the calculable capacitor, may lead to a more accurate value for the fundamental physical constant h/e^2. (h is Planck's constant and e is the charge on the electron.)

3. The Unit of Electric Potential

On the assumption that Ohm's law is valid, any one of the units ampere, ohm and volt may be defined in terms of the other two. It was, however, discovered by Josephson (1962) that extremely non-ohmic behaviour (which has subsequently led to a re-definition of the volt) would take place between two superconducting bodies when they were brought very close together (but not into contact). He calculated that if the 'junction' between the superconductors was stimulated by the application of electromagnetic radiation of frequency f, the potential difference V across the junction would exhibit a number of discrete, quantized levels:

$$V = n(h/2e)f. \quad n = 0,1,2\ldots$$

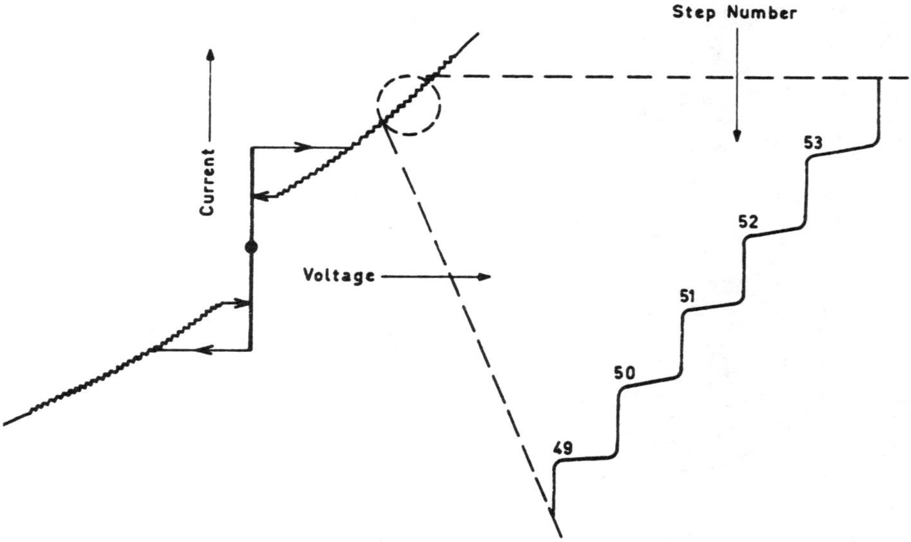

Fig. 2. Current-voltage characteristic of a Josephson junction irradiated with high-frequency electromagnetic waves.

Several national laboratories set up experiments to test the accuracy of Josephson's theoretical predictions, and the first results were reported in 1970 from this Division (Harvey et al, 1970), the National Bureau of Standards (USA) (Finnegan et al, 1970), and the National Physical Laboratory (UK) (Petley & Morris, 1970). It soon became clear that these laboratories, and others that followed, were in sufficiently close agreement that the working standard of electric potential should be re-defined in terms of the Josephson-effect standard, and in 1973, 2e/h was assigned the value 483.594 THz/V by international agreement.

The long-term behaviour of standard cells, which of course are still very widely used, can now be determined (Fig. 3) with a great deal of confidence by reference to the Josephson standard (Harvey et al, 1972).

A detailed account of the maintenance and dissemination of the volt is given by Frenkel (1982).

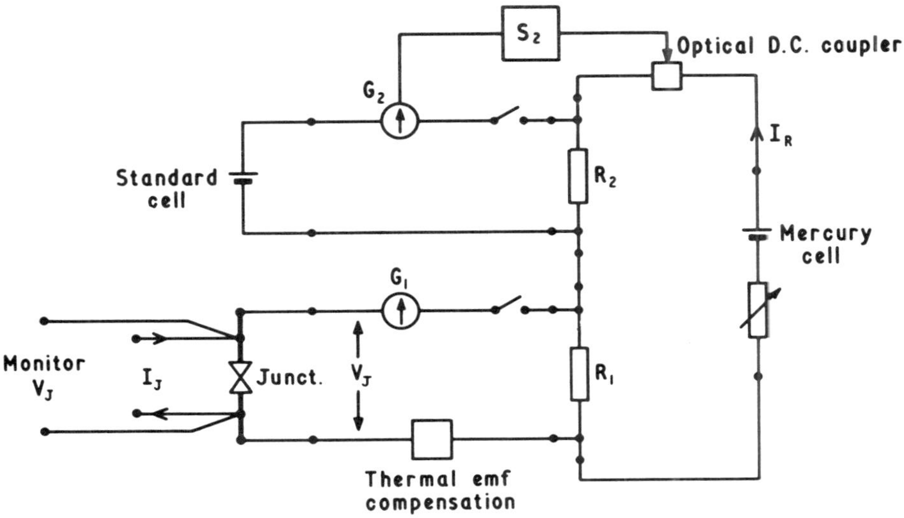

Fig. 3. Calibration of a standard cell against the Josephson junction: simplified circuit diagram.

It has been shown (Macfarlane 1973) that the uncertainty in the Josephson-junction voltage is no more than 2 in 10^9, and the standard cell intercomparison can be done to better than 1 in 10^7. However, the value of 2e/h quoted above is subject to an uncertainty (relative to the SI base units) of about 3 in 10^6. The absolute accuracy of 2e/h may be improved with the CSIRO absolute electrometer (Sloggett 1981). This instrument for the absolute determination of the volt utilizes the upward force of attraction exerted by a high-voltage electrode on the horizontal surface of a pool of mercury. By this means, the volt can be measured in terms of the units of mass and length, so that a comparison of the result with the Josephson standard will yield an absolute measurement of 2e/h.

4. The Unit of Current

Early experiments to determine the absolute ampere in terms of the mechanical units were based on the principle of the stationary current balance. The force due to a steady current flowing in two mutually attracting coils was 'weighed' on a highly sensitive balance. The latest of these determinations (Driscoll & Olsen 1968) had a probable error of ± 6.6 in 10^6, most of which was due to uncertainties in measuring the dimensions of the coils.

Experiments now in progress at the NPL and the NBS make use of an ingenious moving-coil technique to reduce the effect of dimensional errors. It is essentially proposed to measure the emf generated in a coil which moves at a steady speed through a magnetic field, and then to measure the force exerted on the same coil at rest when it carries a known current in the same magnetic field. This will lead to a determination of the ampere in terms of mass, length and time. The measurement of the ratio between currents can be done with extremely high precision, both with conventional apparatus, and particularly with the superconducting direct-current comparator developed in this Division (Harvey 1972). Where liquid helium is in use, as in the Josephson voltage standard, it is often advantageous to make use of the special properties of superconducting devices, such as the current comparator and the SQUID (Superconducting Quantum Interference Device) magnetometer.

5. Conclusion

The realization of the absolute electrical units is not a static technology, nor is it always possible to predict when a radical new development may occur. Advances in physics are constantly taking place which can be applied in the quest for more accurate, more convenient, and more readily accessible standards. Conversely, the techniques developed in establishing these standards may open the way to better measurements and a better understanding of physics.

References

Driscoll, R.L. Olsen, P.T. (1968). Report to Comite Consultatif d'Electricite, Comite International des Poids et Mesures, Sevres.
Finnegan, T.F., Denenstein, A., Langenberg, D.N. (1970). *Phys. Rev. Lett.* 24, 738.
Frenkel, R.B. (1982). These Proceedings.
Harvey, I.K., Macfarlane, J.C., Frenkel R.B. (1970). *Phys. Rev. Lett.* 25, 283.
Harvey, I.K. (1972). *Rev. Sci. Instrum.* 44, 1700.
Harvey, I.K., Macfarlane, J.C., Frenkel, R.B. (1972). *Metrologia* 8, 114.
Inglis, B.D. (1982). These Proceedings.
Josephson, B.D. (1962). *Phys. Lett.* 1, 251.
Macfarlane, J.C. (1973). *Appl. Phys. Lett.* 22, 549.
Petley, B.W., Morris, K. (1970). *Metrologia* 6, 646.
Sloggett, G.J. (1981). 2nd Intl. Conf. on Precision Measurements and Fund. Consts. - in the press.
Thompson, A.M. (1968). *Metrologia* 4, 1.
von Klitzing, K., Dorda, G., Pepper, M. (1980). *Phys. Rev. Lett.* 45,494.

Precise DC Resistance Measurements

J.D. Thane
CSIRO Division of Applied Physics, Sydney, Australia 2070

Abstract

Measurement systems and techiques are given for the precise measurement of DC resistance in the range 10^{-5} to 10^{14} ohms. The effects of temperature, power dissipation, applied voltage and thermal emfs are discussed.

1. Introduction

Measurement of electrical resistance involves magnitudes ranging from 10^{-5} ohms to 10^{14} ohms. Nineteen decades of resistance require the application of three different approaches to the measurement problem. In each approach, appropriate consideration must be given to the effects of temperature, power dissipation, applied voltage, thermal emfs and electrostatic charges. Techniques relevant to each approach are briefly outlined, and some methods of defining or minimising the environmental and operating conditions are given.

The three main approaches (a,b,c), and three other popular approaches (d,e,f) to precise electrical resistance measurement are:

(a) two-terminal testing of resistors between 10^2 ohms and 10^8 ohms using a conventional Wheatstone bridge;

(b) two-terminal testing of resistors of 10^8 ohms or more in a self-calibrating modified Wheatstone bridge circuit;

(c) four-terminal testing of resistors of 10^{-5} ohms to 10^2 ohms using a Kelvin double bridge;

(d) determining the ratio of a test resistor and a standard using a potentiometer for the range 10^{-3} ohms to 10^3 ohms;

(e) using a Universal Ratio Set for the range 10^{-3} ohms and 10^8 ohms;

(f) comparing a test resistor and a standard with the DC Comparator Bridge for values of less than 1 ohm to 10^4 ohms.

In each of the cases, a choice of methods of using the equipment is generally available. Full discussions of these methods are found in most measurement text books (Harris 1957a, Luppold 1969a), but the intention here is to concisely survey the available options and their limitations.

2. Measurement Systems

(a) Wheatstone Bridge

The most common resistance measuring system found in laboratories is the Wheatstone bridge, which is shown in Fig. 1 and provides two-terminal measurements up to about 10^8 ohms. The low limit for two-terminal measurements is governed by the repeatability of the instrument switch resistances, which should be less than 1 milliohm if the bridge switches are periodically cleaned and lubricated. A 100 ohm two terminal resistor can therefore only be measured to an accuracy of about 10 parts per million on such a bridge.

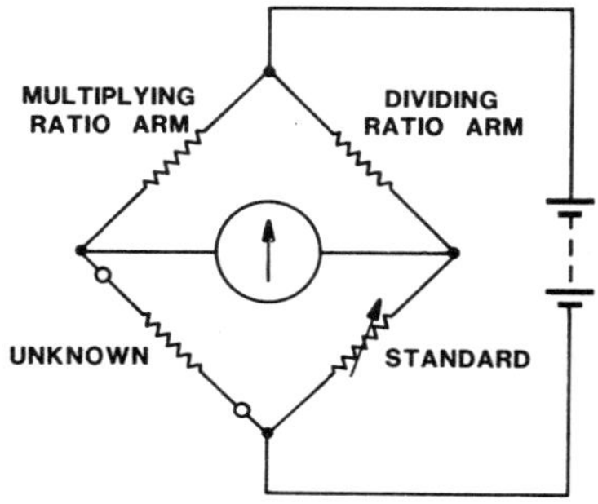

Fig. 1. Wheatstone Bridge.

(b) Self-calibrating Wheatstone Bridge

For values higher than 10^8 ohms, where a wide range of accurate standards is not normally available, the overall accuracy of measurement is limited by the direct reading accuracy of the measuring system. Therefore, it is desirable to use a self-calibrating instrument such as a Wheatstone bridge modified to provide progressive adjustment of each measuring range from an internal standard. A commercial system (Keithley Instruments Inc.) constructed on this principle provides resistance measurements from 10^8 ohms to 10^{14} ohms with an accuracy considerably better than 0.5%.

(c) Kelvin Bridge

Precision resistors of 100 ohms or less need to be constructed with separate current and potential terminals to overcome lead and contact resistances. The traditional measuring system for four-terminal resistors is the Kelvin double bridge, shown in Fig. 2, which can provide a measurement resolution on small resistances of 10^{-9} ohms or better. This degree of resolution is possible because lead, terminal and connection resistances can be virtually eliminated from the measurement after a lead-balancing procedure (Harris 1957b, Luppold 1969b) has been performed.

(d) Potentiometric System

A potentiometer can be used to measure the voltages across unknown and standard resistors connected in series to a stable current source, as illustrated in Fig. 3. The potentiometer readings provide the ratio of the unknown to the standard and, if this ratio is close to unity, an accuracy approaching 1 ppm is achievable. The method suffers mainly from the difficulty in maintaining a constant current for the time required for the two measurements.

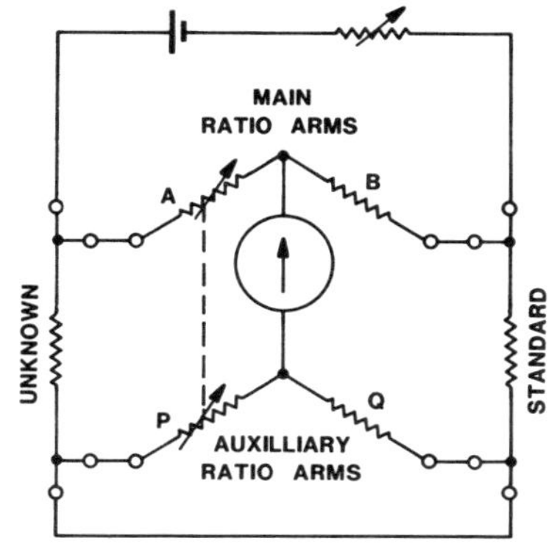

Fig. 2. Kelvin Bridge.

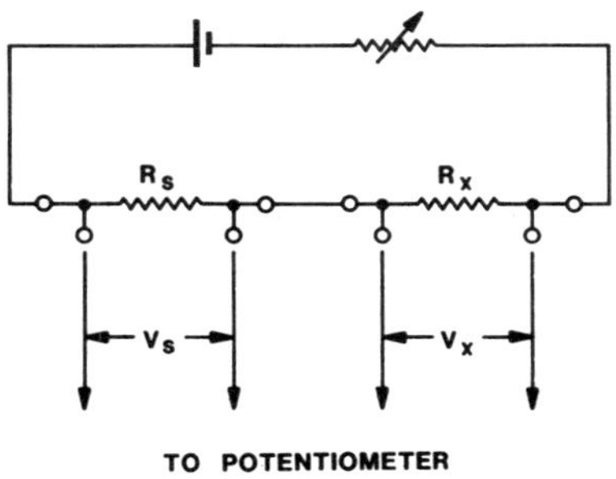

Fig. 3. Using a potentiometer to compare resistors.

(e) Universal Ratio Set

A Universal Ratio Set (URS), (Thomas 1948, Luppold 1969c) may be employed for the measurement of resistance. The URS is a specialised version of the ratio arms of a bridge in which the accuracy is a function of ratio rather than resistance, and which can be used with two or four terminal resistances. The direct-measurement, shown in Fig. 4, or substitution or interchange methods of making resistance measurements can be employed with a URS.

(f) DC Current Comparator Bridge

The most elegant of the measuring systems available is the Current Comparator Bridge (MacMartin and Kusters 1966). This bridge employs decade transformer ratios to obtain an ampere-turns balance when the voltage drops across the standard and unknown resistances are equal, as illustrated in Fig. 5. Under these conditions, the ratio of the standard to the unknown is given by the turns ratio which is inherently accurate to a small fraction of 1 ppm. Other important advantages of this

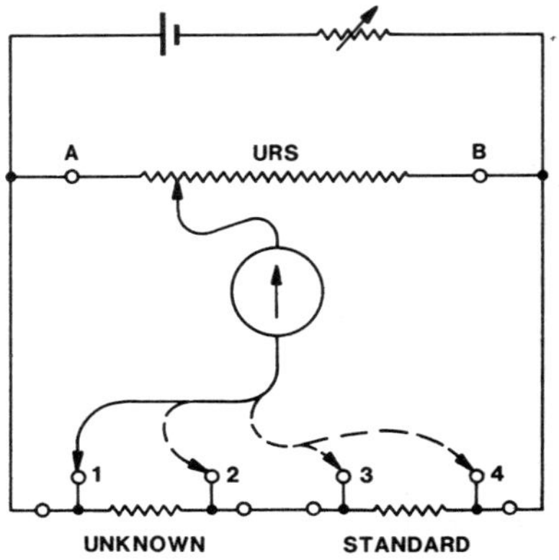

Fig. 4. Resistance comparisons using URS.

instrument are that odd-value resistors may be measured directly with nearly the same accuracy as decade values, and that standard-to-unknown ratios of up to 1000 to 1 can be made where, unlike the equal currents in a Kelvin bridge, only 1/1000 of the current flows in the higher value resistor.

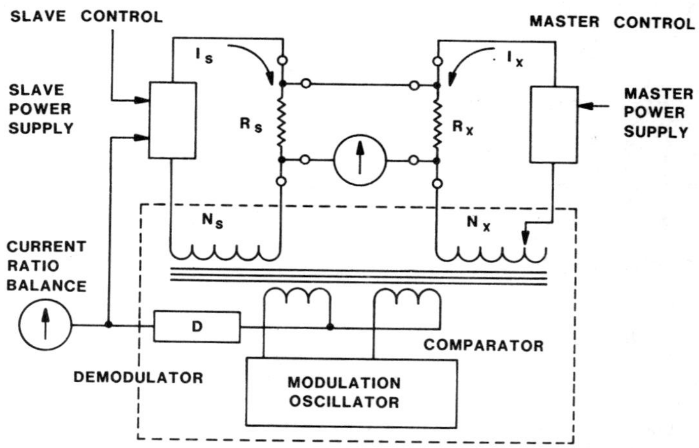

Fig. 5. D.C. Comparator Bridge.

To obtain the best possible results from these six measurement systems, consideration must be given to both the power sources and the detectors. Power sources must be adequate to achieve the desired detector sensitivity, and should also include some power-limiting provision to protect the test resistor and the bridge components from excessive heating. The detectors must not only provide the best possible voltage or current sensitivity, but the input impedance should be of the same order as that of the measuring circuit as seen by the detector. (Luppold 1969d)

3. Conventional Methods of Resistance Measurement

The three principal methods for precise measurement of resistors with a bridge are:

(a) direct measurement using bridge corrections;

(b) substitution techniques;

(c) the interchange method.

Each method has its merits and limitations according to the magnitude of the resistors, availability of appropriate standards, and bridge features such as an incremental ratio dial.

The direct-measurement method requires either a full set of ratio and standard corrections, or some means of adjusting the errors to negligible values. Conventional Wheatstone bridges have no such adjustments. Direct measurement involves not only the need to apply corrections, but also to calibrate the bridge at frequent intervals to keep the corrections up-dated and meaningful.

Where a standard resistor is available with the same nominal value as the test resistor, the substitution technique (Luppold 1969e), illustrated in Fig. 6, can be used.

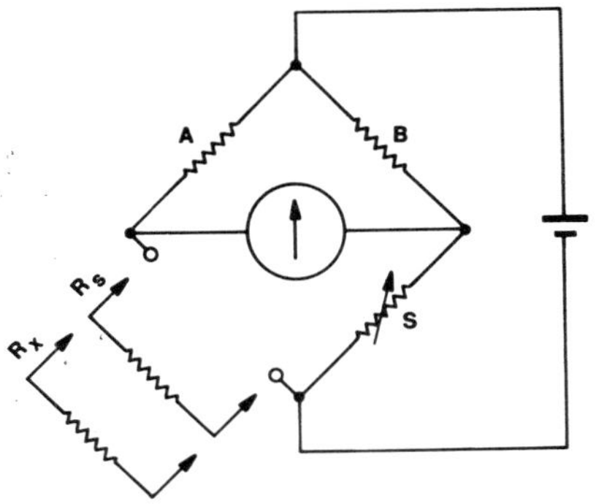

Fig. 6. Substitution technique of comparing resistors.

In this method, the test resistor is substituted for the standard and a new bridge balance is made without disturbing either the ratio settings or the major decades of the bridge. The difference reading obtained by this method may well have an accuracy of the order of 1 ppm if the difference involved is less than 0.1 per cent of the nominal value. Assuming that the bridge resistors are stable for the time taken to make the test, the accuracy of the result is independent of the accuracy of the bridge ratios or the major decades of the bridge, and relies only on the lower dials being within about 0.1 per cent of their nominal value. Therefore the routine calibration requirements for the bridge are considerably less than if the bridge is used in the direct measurement method.

The interchange method (Luppold 1969f) allows the comparison of resistor with a standard of the same nominal value by two separate measurements as in the case of the substitution technique. However, in this method the first test is made with the standard resistor in the standard arm and the test resistor in the unknown arm as before, but the second test is made with the test and standard positions reversed as shown in Fig. 7. Once again, the settings of ratio or major decade dials must not be altered. This method of comparing resistors is not possible on most Wheatstone bridges because firstly, separate 'standard' and 'unknown' terminals are not usually available, and secondly, without a deviation dial to vary the bridge ratios over a small range, no means are available to achieve a bridge balance. However, the Kelvin bridge is generally suitable. The main advantage of the interchange method is that the decade arm which in most bridges is sensitive to temperature changes, is not used and the two interchanged resistors can be placed in an oil or air bath to provide good reproducibility. It is of interest that, apart from determining the difference between the test and standard resistors, this method also allows calculation of the bridge ratio.

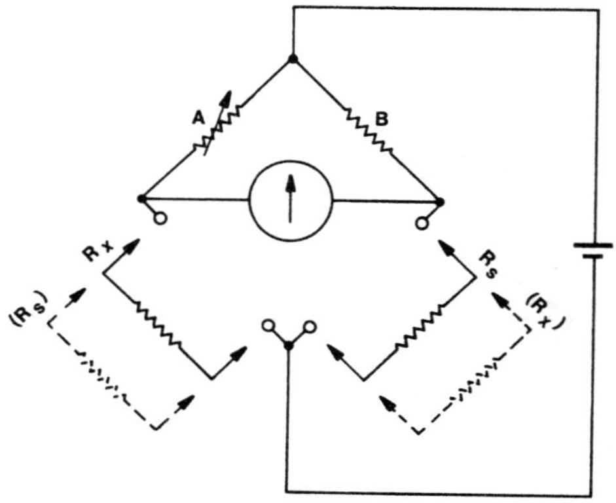

Fig. 7. Comparing resistors by the interchange method.

4. Environmental and Loading Effects

Changes in the temperature of resistors result in changes in resistance of typically several ppm per degree C. Fig. 8A shows a resistance versus temperature curve for a manganin resistor. The particular temperature-coefficients for each standard and test resistor must be accurately known by a calibration laboratory. Temperature changes are usually due to changes in ambient temperature and self-heating of the resistor caused by the test current. To minimise these changes, resistors may be placed in oil or air baths, and the test currents may be reduced to a point where measurement resolution is approximately equal to the change in resistance due to self-heating.

Loading effects are usually associated with resistors with values below one ohm. With loading, self-heating occurs and the resistance element, and oil if used, rise in temperature to a final value where the heat dissipated approaches the heat produced. However, significant temperature gradients can develop, not only between the resistance element and the oil, but also across the cross-section of the element itself. Mathematical constants defining the true value of resistance under load may be determined from measurements made under carefully con-

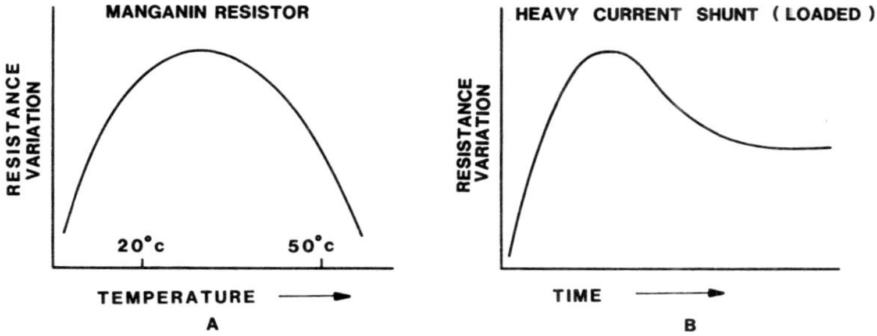

Fig. 8. Resistance variation with temperature and load.

trolled conditions of both temperature and load current. Loaded resistance values as a function of time may increase, decrease, or initially increase and then decrease (see Fig. 8B) depending on the temperature range and the temperature at which the resistance is maximum. A paper describing an investigation on the effects of loading (Briggs 1969) is recommended for further information.

Changes in resistance as a function of voltage in resistors of less than 1 Megohm are usually due to self-heating from the test current. However, with very high-valued resistors between 10^8 ohms and 10^{14} ohms, voltage coefficients of up to -0.1% per volt may be found with no significant self-heating, i.e. power dissipations are considerably less than 1 milliwatt. Since many of these resistors are used in calibration work on megohm-meters and insulation testers, 500 volts or more may be applied resulting in changes of up to 50% from the nominal resistance value. In order to provide customers with useful calibration data, laboratories must obtain resistance values at several voltage levels. Typically, values of resistance may be given at 100, 200, 300, 400 and 500 volts.

Other environmental conditions generally produce smaller effects on the value of resistance, but humidity can be a significant factor where moisture absorbent materials are used in the resistor construction. An early investigation (Dike 1936) showed resistance variations exceeding 100 ppm when the relative humidity was changed from 5% to 75%. These resistance changes were found to be due to stresses placed on the resistance wire by shrinkage in textile insulants. Although modern standard resistors have been constructed so as to have negligible dependence on humidity, atmospheric pressure, gravity and transportation, tests are needed to ensure that a resistance standard does not suffer changes from these factors.

6. Conclusions

Using a Wheatstone bridge and a Kelvin bridge with a range of calibrated resistance standards, substitution techniques provide an accurate and convenient means of making precise resistance measurements with minimum calibration of laboratory equipment. Other methods described are often convenient for particular testing requirements or where the bridges are either not available, have inadequate resolution or are unsuitable for the value being tested. In this regard, special self-calibrating bridges are very satisfactory for high-valued resistors.

In all resistance measurements, attention must be paid to the effects of temperature, humidity, test currents and test voltages. Temperature controlled oil or air baths are useful and often necessary to provide environmental control. Lastly, care is required in the selection of a suitable detector or galvanometer for the particular test - no single galvanometer is ideal for all levels of precise resistance measurement.

References

Briggs, G.H. (1969). *Metrologia*, 5, 61-67.
Dike, P.H. (1936). *Rev. Sci. Instrum.* 7, 278-287.
Harris, F.K. (1957a). Electrical Measurements (John Wiley and Sons: London).
Harris, F.K. (1957b). op. cit. 286-287.
Keithley Instruments Inc. (USA). Megohm Bridge Model 515A.
Luppold, D.S. (1969a). Precision Measurements and Standards, (Addison-Wesley: Reading, Mass. USA).
Luppold, D.S. (1969b). op. cit. 164.
Luppold, D.S. (1969c). op. cit. 99-101.
Luppold, D.S. (1969d). op. cit. 109-115.
Luppold, D.S. (1969e). op. cit. 94-95.
Luppold, D.S. (1969f). op. cit. 96-97.
MacMartin, M.P. and Kusters, N.L. (1966). *IEEE Trans. Instrum. Meas.* IM-15, 212-220.
Thomas, J.L. (1948). Nat. Bur. Stand. (US), Circ. 470, 27.

Secondary Standards of Electromotive Force in a Standards Laboratory

R.B. Frenkel
CSIRO Division of Applied Physics, Sydney, Australia 2070

Abstract

This paper reviews the characteristics, maintenance and use of saturated standard cells as secondary standards of electromotive force. The following topics, probably of special interest to standards laboratories, are covered in some detail: the equipment used at the Division of Applied Physics for maintaining cells and for comparing their emfs; two experimental designs with numerical examples for these comparisons; the long-term behaviour of cells as observed in the Division; and the medium-term behaviour of a particular brand of electronic 'standard cell'. The descriptions of equipment and behaviour have largely not been previously published. The experimental designs are not original, but because they should be better known, are presented here in tutorial form.

1. Introduction

Almost a century after its invention, the Weston cadmium-mercury saturated cell remains in general use throughout the world as a secondary standard of electromotive force (emf) (Vinal 1950; Hamer 1967; Hamer 1972). This paper presents a summary of its characteristics, a description of equipment used at the Division of Applied Physics for maintaining them and for comparing one with another, and an economical experimental design for these comparisons. A design may be considered economical if acceptably precise results are obtained with relatively few data, and if a computer is available for calculations. Next, we describe the long-term variation in emf of saturated standard cells as observed in the Division, these cells including clients' cells submitted for calibration as well as the Division's own cells. These variations can be stated with high confidence, since the Division has since 1970 enjoyed the advantage of a primary monitoring standard based on the ac Josephson effect (Harvey *et al.* 1976). Finally, we discuss the electronic 'standard cells' now emerging as serious competitors to chemical cells, with output voltage and resistance simulating those of chemical cells.

2. Characteristics of Saturated Standard Cells

(a) Their emf is about 1.0186 V at 20 °C and 1.0181 V at 30 °C.

(b) Almost one-half of all long-term changes in cell emfs observed at the DAP fall between 0 and -1 μV/year. Long-term changes are discussed more fully in 6. The long-term stability is determined largely by the pH value of the electrolyte (Froehlich 1974). A reasonable lifetime for a cell in frequent use is 10 to 15 years. When a rapid fall of emf and an increase in resistance are observed, the cause may be gas-bubble formation in the negative limb (Denton 1959). Although shaking or striking the cell may free the bubble, its occurrence normally implies the end of the useful life of the cell.

(c) They have an emf-temperature coefficient of approximately -40 μV/°C at 20°C and -57 μV/°C at 30°C. These coefficients are to be understood as applying only to 'steady-state' emfs, measured when all hysteresis effects have disappeared (see (d) below). Tight control of temperature is evidently essential if a cell's stability is to be fully exploited. At the DAP cells are kept immersed in baths of stirred oil whose temperature is kept constant to within a few mK of 20°C (see part 4 below), except when they are already mounted inside temperature-controlled enclosures, for which 30°C is a popular control temperature. Both environments are intended not only to fix the temperature but also to minimise variations in temperature gradients between each cell's positive and negative limbs. This last requirement arises from the fact that their sensitivity to such variations is roughly ten times greater than to variations in their overall temperature.

(d) When a cell is subjected to a sudden change in temperature, its emf may at first change by a very different amount, in sign and magnitude, from that consistent with the temperature-coefficient. This effect, known as hysteresis, means that a day or more may be required for the emf to settle at the correct value corresponding to the new temperature. As a rough guide, at least one recovery day should be allowed for each degree of change. Cells permanently inside continuously temperature-controlled enclosures are preferred over 'bare' cells as travelling standards, for this reason.

(e) Over a large range of temperature, say from 15°C to 40°C, the variation of emf with temperature is described adequately by a cubic equation. There are several choices of coefficients of this equation. The coefficients of the so-called 'international' formula are quoted in part 5 below. This formula was obtained (Wolff 1908) for neutral cells, i.e., to which no sulphuric acid had been deliberately added. The modern practice is to acidify the solution slightly, but measurements made on the resulting emf-temperature variation (Froehlich et al 1971; Hamer et al 1972) show no clear effect of the addition of acid. Although the 'international' formula is quite old there appears to be as yet no consensus on an improved set of coefficients. Since a standards laboratory would as a rule calibrate its clients' cells at the nominal temperature at which the clients use them, the effective disagreement between any two sets of coefficients is unlikely to exceed a few hundredths of a microvolt.

A modified Weston cell, incorporating a small amount of lead, has been developed and has a smaller temperature coefficient than the normal Weston cell (Hirayama et al 1972). A series combination of two of the lead-containing cells and one normal cell has an emf of 1.0206 V with only 5.1 μV/°C coefficient at 30°C (Shida et al 1980).

(f) When a cell is in circuit as a standard with submicrovolt precision, any direct current through it must be reduced to less than a few tenths of a nA, since its internal resistance is normally of the order 500 to 1000 ohms. After passage of dc, a cell will need time to recover its open-circuit emf, the length of this recovery time depending on the magnitude and duration of the current (Hartshorn et al 1954). Charging currents generally impose a longer recovery time than do discharging currents. A discharge of 1 μA for ten seconds will generally impose several hours of recovery time to within 0.1 μV.

(g) Movement temporarily changes a cell's emf. A laboratory's main group of cells is normally kept in a fixed place, and is disturbed only when a cell needs to be removed or replaced. When other groups of cells are brought for calibration, at least a week of settling (and often more) is needed before comparisons at the submicrovolt level may begin.

(h) When mains-powered equipment is used together with standard cells, it must be borne in mind that any alternating current while passing through a cell may change its emf (Jennings et al 1964). The change is proportional to the square of the amplitude and to the inverse of the frequency. For 10 μA rms the change may be of the order of 1 μV at 50 Hz, but there is much variation from one cell to another. The constant of proportionality is related to cell dc resistance, increasing in magnitude with increase in resistance. Cells that show these effects also show second-harmonic emfs when ac is passed, evidently a related phenomenon (Sloggett 1977).

(i) During manufacture of standard cells, small amounts of foreign materials within the chemicals will affect the emf stability. Investigations of the effect of various contaminants have been made (Froehlich 1967; Klein 1978).

3. Equipment for Maintaining Standard Cells

Control of room temperature within a range of about ± 1°C is generally considered a minimal requirement for maintaining standard cells. Even when this control is provided, a cell cannot be used in precise comparisons with other cells until it has been allowed to settle in a temperature-controlled enclosure or oil-bath, whose temperature is stable over long periods to within a few mK. Its stability is usually ensured by a dc or ac bridge network in which one arm contains the sensing thermistor, an adjacent arm is a precision resistor equal to the thermistor resistance at the desired control temperature, and the other two arms form a 1:1 resistance ratio or ratio transformer. The bridge output voltage is fed to an amplifier and, in the ac case, to a phase-sensitive detector. The resulting dc voltage is normally used to drive the heating current in the enclosure or oil-bath. The DAP uses two baths of stirred oil controlled at 20°C with a long-term stability of a few mK. Since the ambient temperature is normally slightly above 20°C, the oil must be cooled as well as heated. For one of the oil-baths, the cooling is provided by circulated chilled water. For the other oil-bath, which has better stability, no heating is provided other than that due to the action of the oil-stirrers, and the cooling is varied by varying the current through a Peltier cooler in proportion to the output of a bridge network. The cooler consists of a set of Peltier elements in close thermal contact with the pipe through which the oil is circulated

(Richardson). The DAP also uses air-enclosures of its own and of commercial design, with comparable long-term stability. Enclosures with submillikelvin long-term stability have been developed elsewhere (Cutkosky et al 1974; Hill et al 1976).

4. Equipment for Comparing Standard Cell Emfs

Two saturated cells at the same nominal temperature will have emfs that are very probably less than 100 μV apart, and when the cells are at opposite extremes of the normally encountered range in temperature, i.e. $20^{\circ}C$ and $37^{\circ}C$ their emfs will generally be less than 1000 μV apart. In this 'worst' case, therefore, if the emf of one of the cells is known to within 1 pp 10^7, the emf of the other can be determined to the same precision if the 1000 μV difference is measured to within 0.01%. This level of precision is not difficult to attain, and the comparison of standard cells to the submicrovolt level is quite feasible with relatively cheap equipment.

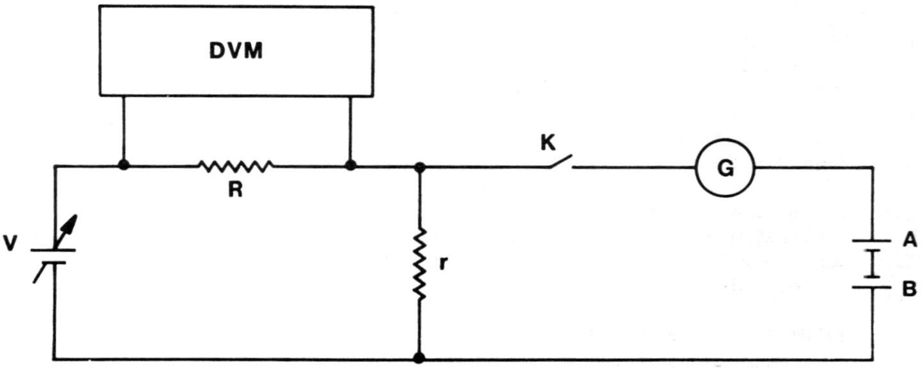

Fig. 1. Principle of the 'Lindeck' potentiometer or comparator, for comparing standard cells A and B. (For initial balance, protective resistors should be in series with A and B.)

Figure 1 shows the principle of the potentiometer, commonly known as the 'Lindeck' potentiometer or comparator, used for comparing cells A and B. A galvanometer G (or other null detector) is connected in series to A and B. When the key K is closed, the galvanometer will show a deflection which is nulled by adjusting the voltage of the variable voltage source V. When the galvanometer deflection is steady at zero, the current through the cells is also zero, and if a digital voltmeter (DVM) is connected across R as shown and reads a voltage V_m when the cell current is zero, then the emf difference E between the cells is given by $E = V_m r/R$. (In older forms of the circuit, an ammeter measured the current I in the resistance R, so that when the cell current was zero, E was then given by $E = Ir$). Since E is normally much less than a millivolt, the precision with which V_m is measured and r and R are calibrated need only be moderate. R is much larger than r, so that V_m is much larger than E by the same ratio. Thus, if $E = 10$ μV, $r = 0.1\,\Omega$

and R = 1000 Ω, then V_m = 0.1 V, easily measured to moderate precision using a cheap DVM.

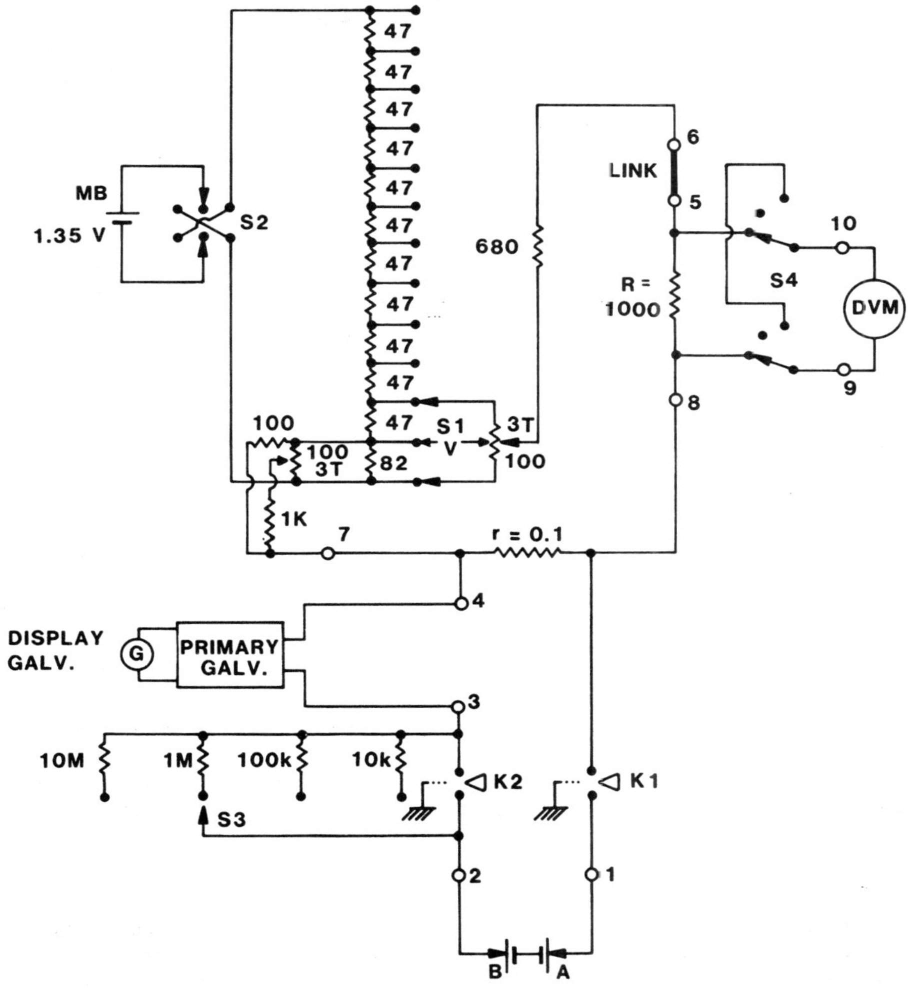

Fig. 2. Practical comparator, with R/r = 10^4. (Values in ohms unless otherwise stated.)

Figure 2 shows a comparator, based on the above principle, with r = 0.1 Ω and R = 1000 Ω, developed at the DAP (Harvey) and, when used with the double galvanometer system described below, capable of measuring emf differences with a precision of a few hundredths of a μV. Its range, 75 μV, encompasses almost all the emf differences encountered between cells at the same nominal temperature. The cells A and B are connected to terminals 1 and 2. A built-in mercury battery MB with reversing switch S2 supplies a constant current of about 2.4 mA along a chain of twelve nominal 47 Ω resistors. One of these resistors is actually 82 Ω in parallel with 100 Ω. The voltage along this chain is tapped by a 2-pole 11-position switch S1, which is the main dial, so that the current through r and R in series is varied by varying its tapping points. The second dial is a 100 Ω 3-turn slidewire potentiometer connected across the two tapping points of the main dial, and has a range nominally one-tenth that of the main dial. The output current through r and R is driven by the voltage V between the slidewire tap and the upper end of the paralleled 82 Ω and 100 Ω. The convenient result of connecting this upper end to the r, R circuit is that a small negative range is obtainable without actuating the reversing switch S2. Thus when S1 is in its first position and the slidewire tap is below the midpoint on the 100 Ω resistor, V is negative, reaching a nominal minimum of minus one-twentieth of the full range of the main dial. The third dial is a 100 Ω 3-turn slidewire potentiometer whose tapped output is then divided by a 100 Ω : 1 K divider to provide a range nominally one-tenth that of the second dial.

To obtain galvanometer balance, the low-thermal key K1 is closed and the switch S3 brings successively smaller protective resistances into series with the galvanometer while the main and second dials are adjusted. The low-thermal key K2 (identical to K1) is then closed and final precise balance is obtained by adjusting the third dial. Both keys are then opened and the DVM is switched across the R = 1000 Ω resistor using switch S4. The reason for using two keys, instead of one, is that since key K1 is connected to one of the cells and key K2 to the other, when both K1 and K2 are open neither cell is exposed to any possible hum or noise between either DVM terminal and ground.

The circuitry in Fig. 2 is oil-immersed, except for the DVM, galvanometers and external terminals (marked with open circles).

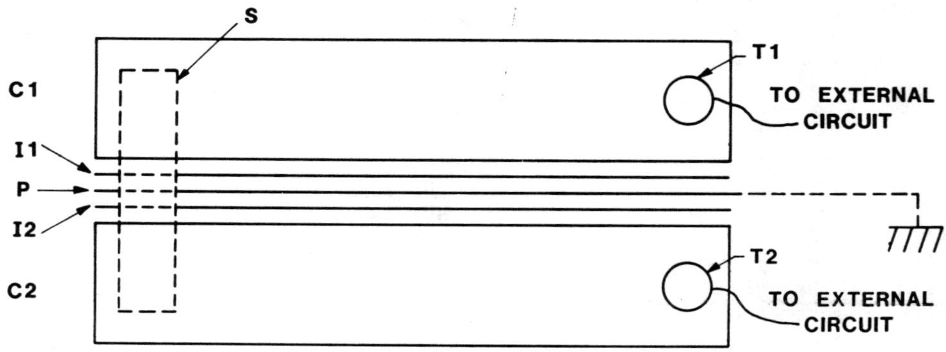

Fig. 3. Plan view of low-thermal key. (Plunger not shown, and separation of bars C1, C2 not to scale.)

The essential features of each key, giving it its low thermals (less than 0.5 μV) are depicted in Fig.3. Two identical copper bars

C1 and C2 (7.5 cm × 1.5 cm × 1.5 cm) are separated by two twin insulating kapton sheets I1 and I2 between which a ground plane P is interposed. The bars are less than 1 mm apart, and so are in good thermal contact, while their mutual capacitance is minimised by the presence of P. Each bar is gold-plated at one end and provided with a copper terminal (T1, T2) at the other end. To close the key, a plunger connects a gold-plated shorting bar S across the plated ends of C1 and C2. By symmetry the thermals generated at the junctions C1-S and C2-S are approximately equal and therefore mutually cancel as seen from the external circuit. To reduce still further the effect of thermals, readings are taken both on 'forward' and 'reverse' polarity, as is standard practice in precise dc measurements. On forward polarity, the leads from terminals 1 and 2 to cells A and B are connected as shown in Fig. 2; on reverse polarity, the lead from terminal 1 is connected to cell B and from 2 to Cell A. Each lead terminates at the cell end in a perspex-insulated copper tong with claws that open when a spring-loaded perspex-insulated plunger is depressed and close to grip the cell terminal tightly when the plunger is released. A constant thermal between the cell positive terminals and the circuit will be eliminated when the mean of the forward and reverse readings is taken. It may be shown that this lead reversal also eliminates another but normally lesser systematic error, namely that due to a combination of low leakage resistance to ground from the commoned cell negative terminals and of unequal leakages to ground from the leads. This effect becomes significant if leakages of the order of 10^{10} Ω are allowed to develop.

The link joining terminals 5 and 6 is removed for periodic calibration of R and r. Terminals 5 and 8 are the current terminals of R, and 10 and 9 are its potential terminals. Terminals 7 and 8 are the current terminals of r, and 2 and 1 are its potential terminals when 3 and 4 are connected by a shorting bar and both keys K1 and K2 are closed.

To measure larger emf differences than 75 μV, the DAP uses a precision dc potentiometer (McMartin et al). Since even the maximum of about 1000 μV emf difference is still much less than the typical range of precision dc potentiometers (2 V), the main potentiometer current need only be standardised to moderate precision. Alternatively, resistor R in Fig. 1 may be switched to one of lower value. A third possibility is to use a source of constant offset voltage switchable in steps of (say) 100 μV, up to 1000 μV.

As null-detector for manual standard-cell comparisons (as indeed for much of its precision dc measurements) the DAP uses a double galvanometer system (Preston 1946) in which the primary galvanometer, connected to the source of current, has no display screen but contains a pair of photocells which are illuminated by different amounts when its coil is deflected. A portion of the differential current of the photocells then deflects the coil of the secondary galvanometer which has a display screen. The current-sensitivity of this double system is typically better than 0.1 nA per mm deflection of light spot on screen. This figure exceeds by a factor of at least 10 the sensitivity of the secondary galvanometer when connected directly to the source of current. More valuable still is the fact that the system uses a feedback arrangement to give effective near-critical damping resistance, even though the secondary galvanometer may have a critical damping resistance around 10^5 Ω. Because of the high sensitivity the galvanometers should be connected together and to the rest of the system by shielded leads, with inner conductors of untinned copper to minimise thermal emfs, and because of the sensitivity to vibrations of the primary galvanometer it should be placed on a heavy base or isolating table.

Automatic comparison systems have been developed at the DAP and elsewhere where there is a large workload of cell comparisons (Arri et al 1979; Dunn 1974; Harkness et al 1977; Brandaway et al 1974; Frenkel).

5. Experimental Designs for Comparing Standard Cell Emfs

In general, when U quantities are to be calculated from P measurement data, then of necessity $P \geqslant U$. As a simple example, when $U = 1$ the calculation may be simply that of the mean of P observations. When $U = P$, there are no 'redundant' data, and the results are very sensitive to small measurement errors. This is a highly undesirable situation, as for example would be the calculation of the mean when only one observation was made, or the calculation of the slope and intercept of a line when only two observations were made. The difference $P - U$ is the number of 'degrees of freedom' of the errors, and should be as high as is permitted by the time and labour needed to collect the data and by the inherent stability of the unknown quantities. If there are redundant data, their internal consistency can be calculated and is a measure of the observational precision. The precision with which any of the unknown quantities has been determined, often called its 'standard error', is proportional to the observational precision and very roughly increases as the square root of P.

When, as with standard cell emfs, the data are provided by the mutual comparison of n objects in pairs, the multitude of experimental designs available arises essentially from the fact that there are many ways of reducing the number of measured permutations from the maximum $n(n-1)$, which may be impracticably high, to a more manageable number. In good designs, the measured permutations will permit each cell to appear in circuit an equal number of times. A large reduction is afforded by 'cyclic' designs (John et al 1972; Wilbur-Ham 1973). Non-cyclic designs (Dunn 1971) generally omit a smaller number of permutations, and are suitable for smaller groups of standard cells. In the following, we describe two types of non-cyclic design, here called respectively 'intercomparison' and 'cross-comparison', of which the latter is the more economical and has been used at the DAP for many years. In this as in most examples of experimental design, the unknown quantities are evaluated by the technique of least-squares, and the random errors are assumed to be distributed normally. A computer is assumed available for the calculations.

(a) Definition of Intercomparison and Cross-comparison

These terms are defined as follows:

'Intercomparison' is a procedure applied to only one group of n cells. The mean emf M of the n cells at a particular temperature must be known in advance. The n cells are 'intercompared' one with another in pairs, so that the data consist of P = n(n-1) differences in emf (obtained in a pattern described below). Calculations on these n(n-1) emf differences yield: the n unknown emfs, of which n-1 are independent since the sum of the n emfs must be nM, and an additional independent small 'bias' emf. Therefore U = n-1+1 = n.

'Cross-comparison' is a procedure applied to two groups of cells, one group containing n cells, the other m cells. One group (e.g. the group with n cells) must have a higher status than the other, and is called the 'reference group' (RG). The other group (with m cells) is called the 'test group' (TG). The n cells in the RG must have a known mean emf M at a particular temperature. Cells in RG are compared with cells in TG, but cells in the same group are not compared with one another. The experimental data consist of mn differences of emf (obtained in a pattern described below). Calculations on these mn differences yield: the n emfs in RG (which are constrained to sum to nM), the m emfs in TG, and again a bias emf. Evidently P = mn and U = n+m.

Normally cross-comparisons are more efficient than intercomparisons, since the amount of observational work is about the same in both cases, yet gives information about two groups of cells in the former case but only one group in the latter case. The practical precision of the final results is about the same in the two cases. For these reasons, cross-comparisons are preferable, and furthermore are suitable for laboratories where two groups of cells are maintained, one of which (the TG) is used for tests by cross-comparison on cells sent by lower-echelon laboratories, while the other group (the RG) has a known mean emf from higher-echelon measurements and is used in cross-comparisons with the TG for the regular monitoring of the cells in both the RG and TG.

	1	2	3	4
1	....	F	F	F
2	R	....	F	F
3	R	R	....	F
4	R	R	R	....

Fig. 4. Pattern for intercomparison of 4 cells (F = 'forward', R = 'reverse').

(b) Intercomparison of One Group of n Cells with a Known Mean Emf M

Figure 4 shows a recommended pattern for the particular case of n = 4 cells. The number of pairwise comparisons is n(n-1) = 12, and of these, 6 are the 'forward' measurements and the other 6 the 'reverse' measurements. Let E_i = emf of the ith cell and let y_{ij} be the measurement in microvolts of emf difference in the ith row and jth column of Fig. 4. The model is:

$$y_{ij} = E_j - E_i + B + e_{ij} \quad (i, j = 1, 2....n; i \neq j) \tag{1}$$

B is a small 'bias' emf, for instance a thermal emf. As is the standard practice in dc measurements, taking measurements in both the forward and reverse directions eliminates systematic errors of this type. The e_{ij} are small random-error emfs. The more precise (i.e., internally consistent) the measurements are, the smaller the e_{ij} will be.

The observational quantities are the y_{ij}; the E's and the B are quantities to be determined from the observations, using the following equations:

$$E_k = M + (1/2n) \sum_{i \neq k}^{n} (y_{ik} - y_{ki}) \quad (k=1, 2..n) \tag{2}$$

$$B = (1/n(n-1)) \sum_{i \neq j}^{n} y_{ij} \tag{3}$$

After the E's and B have been calculated, the e_{ij} can be calculated using equation (1). A computational check is that the sum of these n(n-1) random error emfs must be zero.

The observational precision, s, is given by:

$$s^2 = \sum_{i \neq j}^{n} e_{ij}^2 / (n^2 - 2n), \tag{4}$$

where $n^2 - 2n$ is the number of degrees of freedom.

Example: A group of 4 cells has a known mean emf 1.018 154 47 V, and an intercomparison of these cells has provided the following 12 observations in microvolts:

	1	2	3	4
1	....	- 27.8	- 31.5	- 38.5
2	+ 27.9	....	- 3.8	- 11.3
3	+ 31.6	+ 3.8	....	- 7.4
4	+ 38.9	+ 11.2	+ 7.5	....

Applying equations (2) and (3) gives the following results:

$E_1 = 1.018\ 17900$ V

$E_2 = 1.018\ 15127$ V

$E_3 = 1.018\ 14750$ V

$E_4 = 1.018\ 14012$ V

and $B = 0.05$ μV.

The random error emfs are (in microvolts)

	1	2	3	4
1	....	− 0.125	− 0.050	+ 0.325
2	+ 0.125	.....	− 0.075	− 0.200
3	+ 0.050	− 0.025	....	− 0.075
4	− 0.025	0.000	+ 0.075	.....

and the observational precision is:

$$s = 0.158\ \mu V.$$

(c) Cross-comparison of one group (RG) of n cells with another group (TG) of m cells.

TEST GROUP \ REFERENCE GROUP	1	2	3	4	5	6
1	F	R	F	R	F	R
2	R	F	R	F	R	F
3	F	R	F	R	F	R
4	R	F	R	F	R	F

Fig. 5. Pattern for cross-comparison of six reference cells against four test cells.

Figure 5 shows a recommended pattern for the particular case $n = 6$, $m = 4$. The number of pairwise comparisons is $mn = 24$. Let E_i ($i = 1, 2...n$) be the emf of the ith cell in RG, and let D_i ($i = 1, 2...m$) be the emf of the ith cell in TG. Let y_{ij} be the measurement in microvolts of emf difference in the ith row and jth column of Fig. 5. The model is:

$$y_{ij} = (-1)^{(i + j)}(E_j - D_i) + B + e_{ij} \qquad (5)$$

where the B and the e_{ij} have the same meaning as in equation (1). Referring to Fig. 5, it is convenient to take all the F-measurements first, working from left to right in row 1, then top to bottom, then similarly taking all the R-measurements. Because only half the measurements are F and the other half R, this pattern may be called a 'half-matrix' design.

The equations giving the E's, D's and the B in terms of the observed y_{ij} are as follows:

There are four cases: m, n both even (as in Fig. 5), m odd and n even, m even and n odd, and both m and n odd. It is useful first to define quantities as follows, where the index i is taken from 1 to m and the index j from 1 to n:

$$S = \sum_{ij} y_{ij} \qquad (6)$$

$$T = \sum_{ij} y_{ij}(-1)^{i+j} \qquad (7)$$

$$U = \sum_{j} (-1)^{j} \sum_{i} y_{ij} \qquad (8)$$

$$V = \sum_{i} (-1)^{i} \sum_{j} y_{ij} \qquad (9)$$

and:

$$g_k = \sum_{i} y_{ik} (-1)^{i} \qquad (10)$$

$$G_k = (-1)^{k} g_k \qquad (11)$$

$$h_r = \sum_{j} y_{rj} (-1)^{j} \qquad (12)$$

$$H_r = (-1)^{r} h_r \qquad (13)$$

In the equations below for E_k and D_r, the suffix k extends from 1 to n, and the suffix r from 1 to m.

(i) *m, n both even.*

This is the simplest and also most common case, as most of the commercially available thermocontrolled enclosures contain an even number of cells.

$$E_k = M - (T/mn) + G_k/m \qquad (14)$$

$$D_r = M - (H_r/n) \qquad (15)$$

$$B = S/mn. \qquad (16)$$

For the other three cases, define the quantities:

$$L = 1/(n^2-1) \qquad (17)$$

$$L^1 = 1/(m^2-1) \qquad (18)$$

(ii) n even, m odd.

$$E_k = M - (T/mn) + (-1)^k((g_k/m) + (L^1/n)(S + (V/m))) \quad (19)$$

$$D_r = M - (H_r/n) \quad (20)$$

$$B = (mL^1/n)(S + (V/m)) \quad (21)$$

(iii) n odd, m even.

$$E_k = M - (T/mn) + (G_k/m) \quad (22)$$

$$D_r = M - (-1)^r((h_r/n) + (L/m)(S+(U/n))) \quad (23)$$

$$B = (nL/m)(S + (U/n)) \quad (24)$$

(iv) m, n both odd.

Define first the quantity:

$$W = nmS + nV + mU + T \quad (25)$$

$$E_k = M - (T/mn) + (G_k/m) + LL^1 (1+(-1)^k n)(W/mn) \quad (26)$$

$$D_r = M - (H_r/n) - (-1)^r LL^1W/n \quad (27)$$

$$B = WLL^1 \quad (28)$$

After the E's, D's and B have been calculated, the random errors e_{ij} can be calculated using equation (5). As a computational check, this sum must be zero. The observational precision s for any of the four cases above is given by:

$$s^2 = \sum_{ij} e_{ij}^2 / (nm-n-m) \quad (29)$$

s can be used to estimate the standard error of each of the calculated quantities (the E's, D's and B). Call s(E) the standard error of any E_i, s(D) the standard error of any D_i, and s(B) the standard error of B. Then:

(i) m, n both even.

$$s^2(E) = s^2(n-1)/mn \quad (30)$$

$$s^2(D) = s^2/n \quad (31)$$

$$s^2(B) = s^2/mn \quad (32)$$

(ii) n even, m odd.

$$s^2(E) = s^2(1/m)(1 - (2/n) + m^2L^1/n) \quad (33)$$

$$s^2(D) = s^2(1/n) \quad (34)$$

$$s^2(B) = s^2(mL^1/n) \quad (35)$$

(iii) n odd, m even.

$$s^2(E) = s^2(n-1)/mn \quad (36)$$

$$s^2(D) = s^2(1/n)(1 + L/m) \tag{37}$$

$$s^2(B) = s^2(nL/m) \tag{38}$$

(iv) m, n both odd.

$$s^2(E_k) = s^2(1/m)(1-(1/n) + LL^1(n^2+1+2n(-1)^k)/n) \tag{39}$$

$$s^2(D) = s^2(1/n)\ (1+mLL^1) \tag{40}$$

$$s^2(B) = s^2mnLL^1 \tag{41}$$

In practice, the redundancy of the observations will usually make the standard errors of the cell emfs less than 0.1 microvolt, which is often less than the variation of a standard cell from one week to the next. The standard error of B indicates whether there is a significant bias to the observations, e.g. due to a large thermal emf. This is worth knowing even though the taking of both F and R measurements will eliminate its effect from the final results.

Example

An RG of four cells is cross-compared with a TG of four cells. M for the RG = 1.018 1250 V. The 16 observations, made in the pattern of Fig. 5 with four columns instead of six, are as follows in microvolts:

		RG			
		1	2	3	4
	1	- 16.2	+ 19.4	- 18.8	+ 26.3
TG	2	- 8.3	+ 6.2	- 5.9	- 0.9
	3	- 12.7	+ 16.2	- 15.2	+ 23.2
	4	+ 19.6	- 21.8	+ 22.1	- 29.1

Applying equations (14), (15) and (16) gives the results:

E_1 = 1.018 1288 V

E_2 = 1.018 1260 V

E_3 = 1.018 1263 V

E_4 = 1.018 1189 V

and:

D_1 = 1.018 1452 V

D_2 = 1.018 1201 V

D_3 = 1.018 1418 V

D_4 = 1.018 1482 V

with B = 0.26 μV.

The random errors are (in microvolts):

		RG			
		1	2	3	4
	1	- 0.050	- 0.013	- 0.150	- 0.188
TG	2	+ 0.088	+ 0.050	- 0.013	+ 0.025
	3	+ 0.100	+ 0.138	+ 0.100	+ 0.063
	4	- 0.038	+ 0.075	- 0.038	- 0.150

so that s = 0.135 μV, s(E) = 0.058 μV, s(D) = 0.068 μV, s(B) = 0.034 μV.

Confidence limits for the estimated quantities (the E's, D's and B) may be obtained as follows. The desired confidence range is assumed to be 99% and the factor given below, which is derived from the t-distribution and depends on the number of degrees of freedom, is the multiplier of the standard error. The trend of the figures may be adequately seen from only five entries.

Degrees of freedom	Factor (±)
6	3.7
8	3.4
12	3.0
20	2.8
very many	2.6

In the above cross-comparison example, which had 8 degrees of freedom, the 99% confidence range for each of the D's is therefore ± 3.4 × 0.068 μV or ± 0.23 μV.

(d) Corrections for temperature.

For both intercomparisons and cross-comparisons, the mean emf M, known in advance, must always refer to a particular temperature, called the 'defined group temperature' (DGT), and preferably defined to a precision of 0.002° C or better. DGT is conveniently made a round number, e.g. 20.000 or 30.000. During an actual run of measurements, the group of cells with a known M will probably have a slightly different temperature from the DGT, and M must be given a corrected value, using the 'international' temperature-coefficient formula for saturated standard cells:

$$M_t = M_{20} + E_{corr} \tag{42}$$

where:

$$E_{corr} = -\ 40.6(t-20) - 0.95(t-20)^2 + 0.01(t-20)^3 \tag{43}$$

M_t is the corrected emf at temperature t, and M_{20} the emf at 20.000°C. M_t and M_{20} are in volts and E_{corr} (the correction emf) is in microvolts. A more useful version of (42) and (43), when a 30°C-controlled enclosure is used, is:

$$M_t = M_{30} + E_{corr}{}^1 \tag{44}$$

where:

$$E_{corr}{}^1 = -\ 56.6(t-30) - 0.65(t-30)^2 + 0.01(t-30)^3 . \tag{45}$$

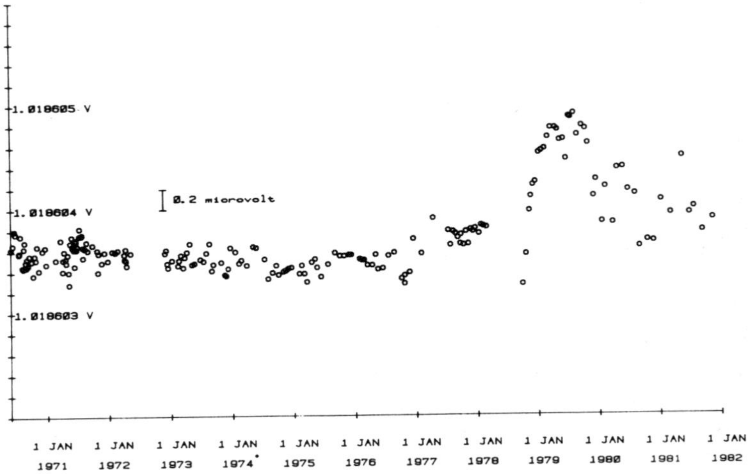

Fig. 6. History of DAP cells group 1.

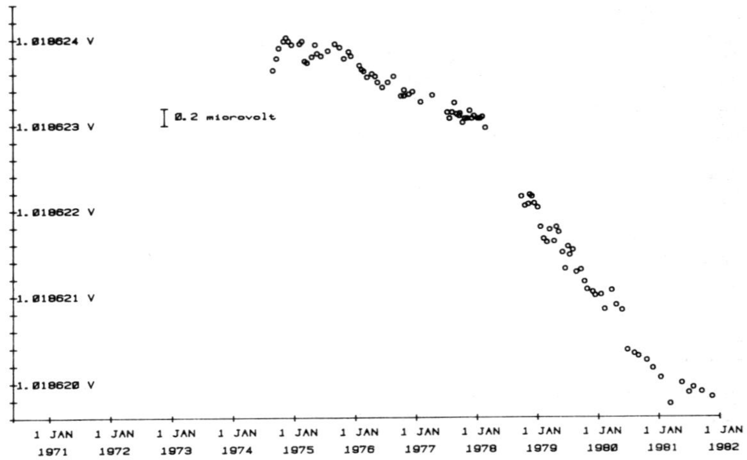

Fig. 7. History of DAP cells group 2.

6. Long-Term Variation in Standard-Cell Emfs.

Figure 6 shows the variation from May 1970 to December 1981 of the mean emf of a group of 7 DAP standard cells at 20.000°C. Each point in Fig. 6 was determined by reference to the Josephson-effect standard, and has an associated 99% confidence range of about 0.3 ppm. This group of cells, manufactured in the 1930's and 1940's, formed part of the DAP primary standard of emf before the Josephson-effect standard was developed. Fig. 7 shows the inherently larger drift in the mean emf of a group of four 20.000°C DAP cells of another much newer brand. Measurements on this group started in 1974 and were concurrent thereafter with those on the first group. In 1978 the laboratory was transferred to a new site, and this may have been the cause of the shift in value of all the cells of the older group.

Clients' cells submitted to the DAP for calibration are first allowed at least a week to settle after transport and, if not already in temperature-controlled enclosures, to adapt to the oil-bath temperature. The schedule of measurements is normally once a week for four to five weeks, but the duration is prolonged if the settling period is obviously too short, and any initial transient or erratic behaviour is excluded from the calculation of the reported mean emf.

A useful figure of merit that a high-echelon laboratory can apply to its clients' cells is the 'pace' of a cell, defined as the increase in its reported emf between two successive calibration times, divided by the interval between them. A cell submitted for calibration three times will therefore have two paces. Both pace and drift can be stated in microvolts per year, but the latter term, often evaluated by least-squares, connotes a larger number of such calibrations by the high-echelon laboratory. If, as in the case of the DAP, the interval is around two years, the drift rate will be known only for the relatively few clients' cells which have a calibration history extending over ten years or thereabouts. In the ten-year period that ended in February 1982 DAP clients' cells afforded 286 paces, grouped as a histogram in Fig. 8. The mean pace was -0.58 μV/year, about 44% fell in the range 0 to -1 μV/year and about 99% in the range ± 5 μV/year. Most of the cells were in 30° C-controlled enclosures, a few were in 28°C, 37°C and 25°C enclosures, and some were placed in the DAP oil-bath at 20°C for calibration.

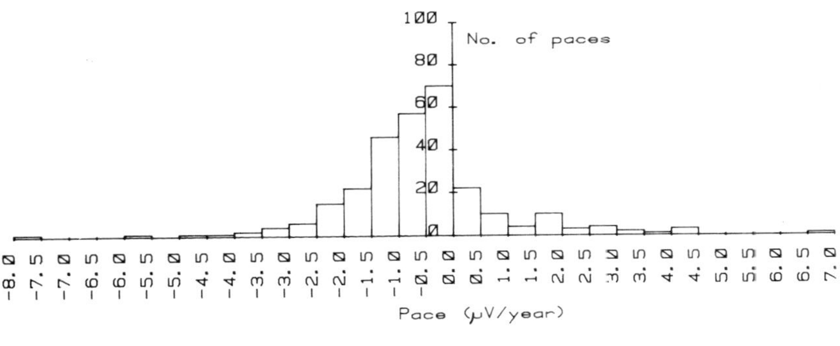

Fig. 8. Histogram of DAP clients' cells.

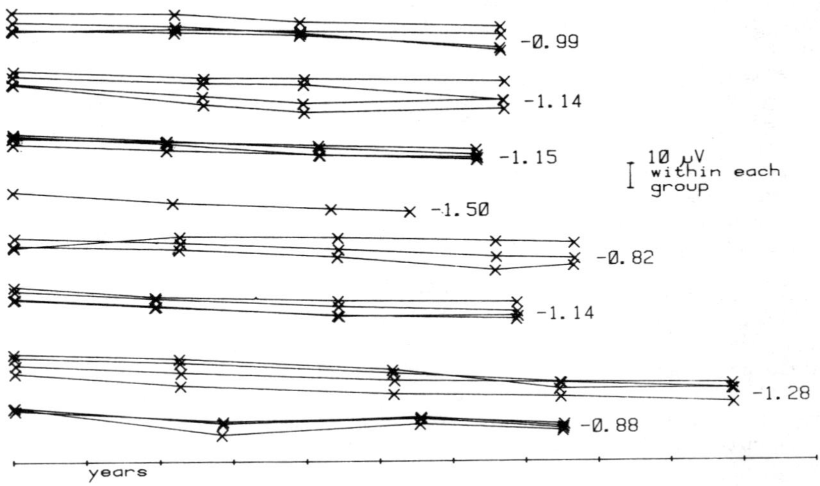

Fig. 9. DAP clients' cells, submitted four or more times (30°C).

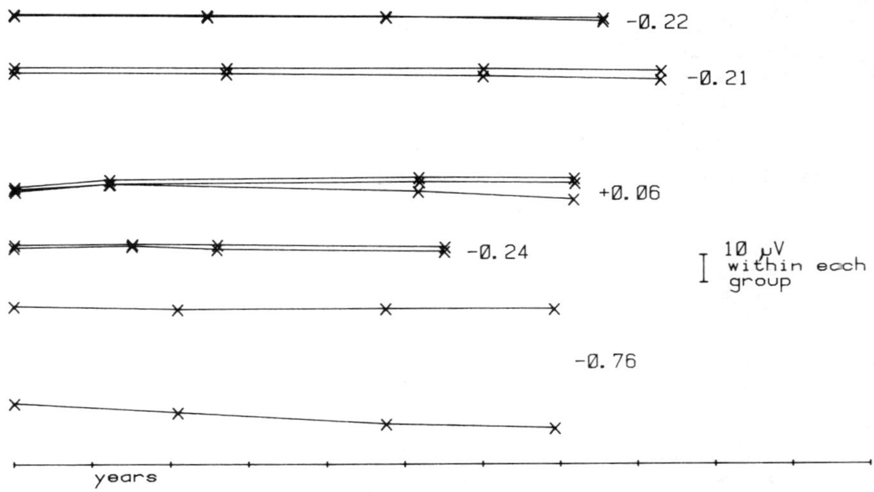

Fig. 10. DAP clients' cells, submitted four or more times (20°C).

The long-term variation of all clients' cells submitted to the DAP for calibration four or more times in the above-mentioned ten-year period is shown in Fig. 9 for 8 groups of 30°C cells (including a 'group' of only one cell), and in Fig. 10 for 5 groups of 20°C cells. The groups are shown separated vertically by arbitrary amounts for convenience in display, but within any particular group the excursion in emf of a cell and its separation in emf from others in its group are correctly described by the vertical scale as marked. Elapsed time in years is marked on the horizontal axis, with all the groups given the same initial calibration date for display convenience. The lines joining the calibration points are intended for easy reading and do not imply linear drifts between calibration times. The average drift rate for each group, in microvolts per year, is shown at the end of each group's trace. The drift rate averaged over the individual cells was -1.08 μV/year (standard deviation 0.39 μV/year) for the 30°C cells, and was -0.24 μV/year (0.46 μV/year) for the 20°C cells. This rather significant difference can probably be attributed to the fact that (i) all the 30°C cells were Muirhead D402's, whereas most of the 20°C cells were the more expensive Muirhead K30A's; (ii) the 20° cells were all considerably older and hence better aged than the 30°C cells. It should be noted that although a cell's temperature may vary slightly from one calibration time to the next, its emf as reported by the DAP, and all the emfs occurring in these graphs, have been corrected (using the international formula) to exactly the same value of temperature, e.g. '30.000°C' and 20.000°C. The reason for the quotes around the former value is that it is strictly an abbreviation for: 'the zero-deviation setting of the dial of a nominally 30°C enclosure'. The zero-deviation setting might actually correspond to (say) 30.05°C, but this would be of little importance as long as the temperature-measuring circuitry in the enclosure remained precise to a mK or so. Each enclosure surveyed here contained a thermistor and a reference resistor to measure its temperature, and undesired changes in these components (e.g. thermistor aging) would degrade the precision. This additional source of error in assessing long-term variation in emf does not exist for the 20°C cells, since their temperature is measured with high accuracy by platinum thermometers regularly checked against the DAP standards of temperature.

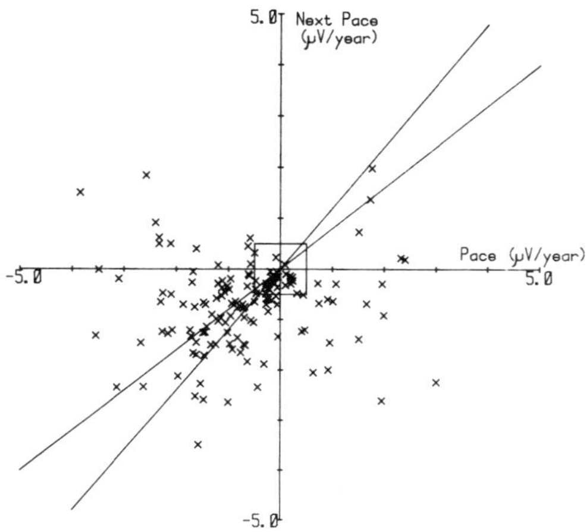

Fig. 11. Predictability of DAP clients' cells.

If a particular cell's paces were all approximately equal, its performance could justifiably be regarded as predictable. However, Fig. 11 shows that such a cell would be uncommon. Here, the abscissa of each point is the earlier pace of a cell submitted for calibration at least three times, and its ordinate is its immediately following pace. Thus if each cell had its own characteristic and unchanging pace, all the points in Fig. 11 would lie on the straight line in the first and third quadrants passing through the origin at 45° to either axis. As a relaxed requirement for predictability, we might stipulate that either or both of two conditions should hold: (a) a cell's immediately following pace must not be more than 20% different from its present pace, or (b) both paces should be less than 0.5 μV/year in absolute value. In case (a), a cell would be 'predictably variable', and in case (b), 'predictably good'. In Fig. 11, the points falling in the radial band denote successive paces obeying condition (a), and points in the square, condition (b). Out of a total of 143 points, only 49 fall in the radial band or square or both. The chance that a cell's currently observed pace will reliably indicate its next pace is therefore on the average only about one in three.

7. Electronic Standard Cells

The DAP has so far (Sept. 1982) examined only one type of electronic cell (Widdis). This is a battery-powered device, and like other types of electronic cell consists essentially of a temperature-compensated Zener diode selected for its very low noise. The Zener voltage is divided down to a nominal output of 1.018 61 V, with an output resistance comparable to the resistance of chemical cells, so that the unit can be compared with chemical cells using the same null-detector as is used for chemical-cell comparisons. A null-detector (galvanometer or electronic), used with this unit and a chemical cell, shows no more noise than it typically does with two chemical cells. The temperature-coefficient of the unit's output emf has been measured by the DAP to be less than ± 2 μV/$^{\circ}$C over the range 18°C to 26°C, and varies from positive through zero to negative within this range. Compared with chemical cells, therefore, the unit is just as quiet, has a much lower temperature-coefficient, and is presumably less prone to transportation errors.

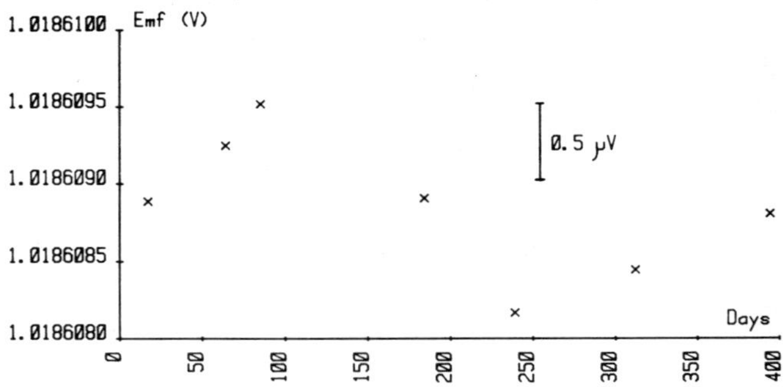

Fig. 12. Medium-term variation of an 'electronic' standard cell.

In part because of its presumed portability, one such unit has been used by the DAP as a transportable standard for the dissemination of the volt among the participating countries in the 1981-82 Asia-Pacific Metrology Programme. A platinum-film thermometer was secured inside the unit and connected to a front-panel socket, to enable the inside temperature to be conveniently monitored and so eliminate as far as possible the effect of even the small temperature-coefficient. The points in Fig.12 represent the emf of the unit as measured at the DAP before the commencement of the programme and during it, so that successive points are separated by intervals during which the unit was either travelling or undergoing measurement in another laboratory. The results available so far augur fairly well for the use of electronic standard cells as volt-disseminators to the ± 1 μV level.

8. Summary

We have listed the characteristics of saturated standard cells, and have described the instrumentation for maintaining and comparing them and two experimental designs for such comparisons. Their long-term behaviour, when properly maintained and used, is of particular interest to a standards laboratory and its clients, and a section is devoted to this. We have taken a necessarily brief but rewarding look at the performance of one brand of 'electronic' standard cell.

Acknowledgements

Mr W. Murray of the Division of Applied Physics did the painstaking measurements on many of the chemical cells surveyed above and on the electronic cell. Mr Murray and Mr P.R. Dencher of the DAP gave valuable assistance in installing the thermometer inside the electronic cell and constructing associated circuitry.

References

Arri, E., and Pogliano, U. (1979). *IEEE Trans. Instrum. Meas.* **IM-28**,311.
Braudaway, D.W., and Kleimann, R.E. (1974). *IEEE Trans. Instrum. Meas.* **IM-23**, 282.
Cutkosky, R.D., and Field, B.F. (1974). *IEEE Trans. Instrum. Meas.* **IM-23**, 295.
Denton, J.J. (1959). *Communication of National Physical Laboratory.*
Dunn, A.F. (1971). *IEEE Trans. Instrum. Meas.* **IM-20**, 2.
Dunn, A.F. (1974). *IEEE Trans. Instrum. Meas.* **IM-23**, 278.
Frenkel, R.B., Division of Applied Physics, CSIRO. 'Standard-cell Intercomparison using a Calculator-Controlled Galvanometer in an Integral-Control Feedback System' (to be published).
Froehlich, M. (1967). *Z. Instrumentenkd.* **75**, 393.
Froehlich, M., and Melchert, F. (1971). *Metrologia* **7**, 58.
Froehlich, M. (1974). *Metrologia* **10**, 35.
Hamer, W.J. (1967). *Instrum. Soc. Am.* Preprint No. M2-1-MESTIND-67.
Hamer, W.J. (1972). *J. Res. Nat. Bur. Stand.* **76A** (Phys. & Chem.), 185.
Hamer, W.J., Skapars, A., and Field, B.F. (1972).*J.Res.Nat.Bur.Stand.* **76A** (Phys. and Chem.), 321.
Harkness, S., and Dix, C.H. (1977). *Proc. EUROMEAS* **77**, 158.
Hartshorn, L., and Manning, F.A. (1954). *J. Sci. Instrum.* **31**, 115.
Harvey, I.K., Macfarlane, J.C., and Frenkel, R.B. (1976). *Metrologia* **12**, 55.

Harvey, I.K., Division of Applied Physics, CSIRO (unpublished).
Hill, J.J., Harkness, S., and Denton, J.J. (1976). *IEEE Trans. Instrum. Meas.* **IM-25**, 501.
Hirayama, H., and Shimazaki, K. (1972). *IEEE Trans. Instrum. Meas.* **IM-21**, 319.
Jennings, J.C.E., and O'Connor, R.J. (1964). *J.Sci.Instrum.(J.Phys.E)*, Series 2, **2**, 353.
John, J.A., Wolock, F.W., and David, H.A. (1972). *Nat.Bur.Stand.* Applied Mathematics Series 62.
Klein, K.D. (1978). *Mitt. Physikalisch-Technischen Bundesanstalt* **88**, 4.
McMartin, M.P., and Kusters, N.L. (1968). *IEEE Trans. Instru. Meas.* **IM-17**, 263. The derived commercial instrument used at the DAP is the Guildline DC potentiometer, Type 9930.
Preston, J.S. (1946). *J. Sci. Instrum.* **23**, 173. The derived commercial instrument used at the DAP is the Tinsley Photocell Galvanometer Type 5214.
Richardson, R.C., Division of Applied Physics, CSIRO. (unpublished).
Shida, K., and Hirayama, H. (1980). *Jpn. J. Appl. Phys.* **19**, 1567.
Sloggett, G.J. (1977). *Metrologia* **13**, 57.
Vinal, G.W., *'Primary Batteries'* (Wiley & Sons; Chapman & Hall), 1950, Ch. 6
Widdis, F.C. 'The Development, Use and Performance of the Cropico Electronic Standard Cell', Cropico Ltd., Croydon, UK. (1980 or later). The unit under test at the DAP is the 'Cropico ESC-1'.
Wilbur-Ham, J. (1973). *IEEE Trans. Instrum. Meas.* **IM-22**, 234.
Wolff, F.A. (1908). *Bull. Bur. Stand.* 5, 309. An account of Wolff's work on the temperature-coefficient, including the effect of subsequent changes in the unit of voltage and scale of temperature, is given in the article by Hamer (1972) previously referred to.

Calibration of Resistance Ratios

J.D. Thane
CSIRO Division of Applied Physics, Sydney, Australia 2070

Abstract

Ratio circuits used in resistance and voltage measurement systems provide multiple or sub-multiple extensions to a calibrated standard. Since ratios have no dimensions, they do not require calibration in terms of a national standard and may even be self-calibrating. Some fundamental methods of interchanging resistance elements to determine ratios are discussed and other techniques in commercial use are briefly described.

1. Introduction

Resistance-ratio circuits are employed in such instruments as bridges, volt ratio boxes, potentiometers and Kelvin-Varley dividers. They may be used to divide a voltage by some convenient factor for measurement by a low voltage instrument, or to provide a multiplying factor to extend the range of a Wheatstone or Kelvin bridge. The deviation from the nominal ratio value is a direct contribution to the overall error in the measuring system, so it is important to measure ratio values as accurately as possible. In many cases, such ratio circuits are conveniently accessible, so that a calibration using ratio techiques is possible, and precise resistance determinations may not be necessary. Many measurement textbooks provide a good treatment of resistance measurements, but very few discuss ratio techniques as a subject. The intention of this paper is to survey some of the more popular ratio-measuring concepts and show some practical examples of their use.

2. Ratio of a Two Resistor String

(a) Resistor Interchange

Two nominally equal resistors in series form a 2:1 divider, and the dividing ratio may be determined by comparison with a pair of nominally equal resistors in a bridge circuit as shown in Fig. 1. The deviation is determined firstly with the resistors A and B connected as shown and secondly with the resistors A and B interchanged. The deviation of the 2:1 divider is then the average of the two measured values.

(b) Series-Parallel Interchange

To extend this technique to ratios higher than 1:1, the resistors A and B must be networks that can be connected to produce different values of resistance, but with the same deviation from nominal in each configuration. In Fig. 2, which shows the test circuit for a 10:1 divider, the three A resistors and the three B resistors are all nominally equal.

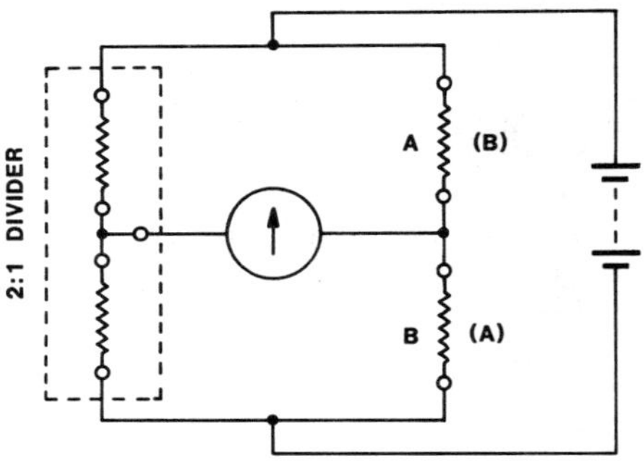

Fig. 1. Calibration by interchange of resistors A and B.

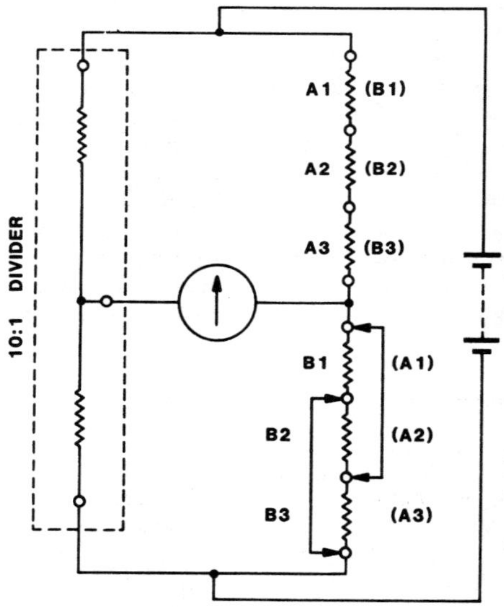

Fig. 2. Series-parallel interchange.

The resistance of A1, A2 and A3 in series will be nominally nine times the resistance of the parallel network B1, B2 and B3, and this forms the reference 10:1 divider in the first configuration. Interchanging the two groups with B1, B2 and B3 in series and with A1, A2 and A3 in parallel forms the second configuration for the reference 10:1 divider. The deviation of the test divider is again the average of the two readings. The principle of having a number of resistors in the A group and the B group may be extended. The ratios formed may be calculated from $r = n^2 + 1$, where r is the ratio and n is the number of resistors in each group.

(c) Calibration of Ratio Decades

While ratios of 1:1, 5:1 and 10:1 are useful for many ratio calibrations, using more resistors in the networks only provides inconvenient ratios such as 17:1 , 26:1 and so on. However, by using series-paralleling in strings of two to six resistors, ratios can be produced for each step of a single decade divider. A paper (Thompson 1978) describing this self-checking ratio method has been published and is recommended for a detailed discussion.

3. Ratios Obtained by Using 1:1 Resistance Comparisons

(a) Adjustable Ratios

Where high accuracy is required, the problem of determining ratio corrections can be avoided completely if individual resistances can be adjusted to a common value without the need for accurate, calibrated standards. Two examples of this technique, which is used in commercially available instruments, illustrate its usefulness.

Firstly, a Kelvin-Varley divider may be fitted with adjustable trimming resistances to each of the eleven resistors in the first, most critical decade. By means of an external, or preferably an internal, comparison circuit as shown in Fig. 3, all the individual resistances may be adjusted to the same magnitude giving what is sometimes referred to as a perfect divider. The principle is expanded in a seven decade Kelvin-Varley divider (John Fluke Manufacturing Co.) to give ratios in error by less than 1 step in the seventh dial for the first three decades.

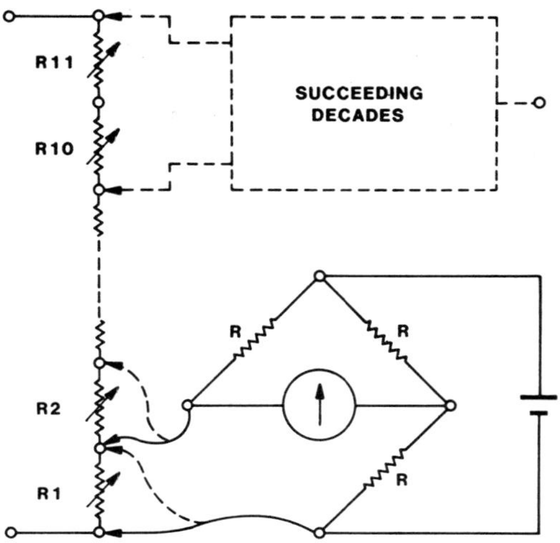

Fig. 3. A method of adjusting resistors R1 to R11 for equality.

Secondly, a 10:1 bridge with decade multipliers in the ratio arms can be arranged so that the next higher multiplier can be adjusted against the one in use by direct comparison with a reference 10:1 ratio. The circuits in Fig. 4 illustrate how the first and second multipliers (M1, M2) are adjusted to be 10 and 100 times the value of the dividing-ratio resistor, D. It is seen that no resistance-magnitude measurements are made, and the values of D, M1, M2 etc. are not relevant. Having adjusted the multiplier resistances to their ideal values, the 10:1 reference ratio is removed and the unknown and standard resistances are connected in their place in the usual manner. A megohm bridge (Keithley Instruments Inc.) uses this calibration principle to provide resistance measurements to 10^{14} ohms with an accuracy (standard deviation = 1σ) of 0.3%.

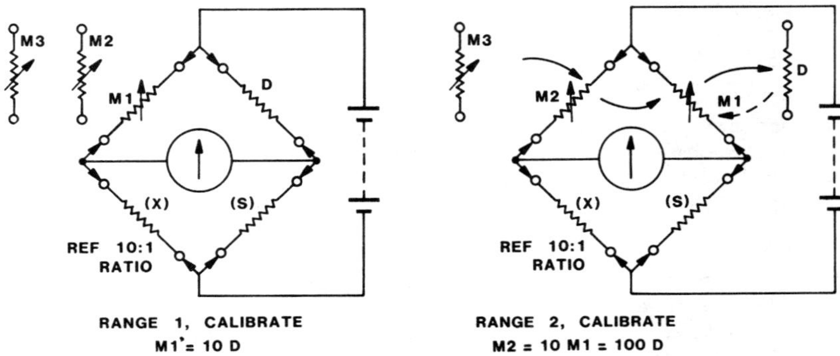

Fig. 4. Self-Calibrating Range Multipliers.

(b) Self-Calibrating Volt Ratio Box

A self-calibrating volt box can be constructed so that the ratios of input to output can be determined by a series of 1:1 resistance comparisons. Such a divider could consist of a string of resistors in binary-related magnitudes as shown in Fig. 5. One-to-one comparisons of R_2 versus R_1, R_3 versus $R_2 + R_1$ etc. as shown in Table 1, can be made to a few parts per million by resistance interchange or substitution techniques using equipment with an accuracy of only 1%.

Table 1. Steps in a self-calibrating voltage divider.

Resistor	Magnitude	Voltage	Comparison Test
R1	R	1	
R2	R	2	R2 Vs R1
R3	2R	4	R3 Vs R2 + R1
R4	4R	8	R4 Vs R3 + R2 + R1
R5	8R	16	R5 Vs R4 + R3 + R2 + R1

Corrections for each step can be progressively summed to provide a total correction for each terminal or range of the divider. Several self-calibrating voltage dividers are available which require only modest

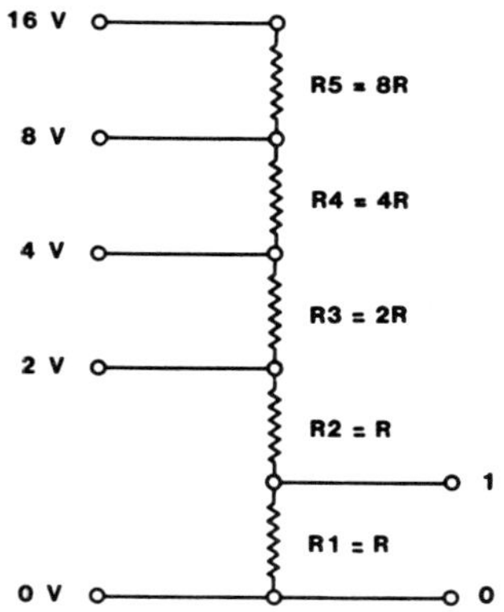

Fig. 5. Basic form of self-calibrating volt box.

comparison equipment (Leeds and Northrup Co.) or which are completely self-contained with a comparison circuit built in. (H. Tinsley and Co.) A detailed description and analysis of a self-calibrating divider is given Luppold (1969), section 6.

(c) Potentiometer Calibrations

High resolution potentiometers may be calibrated in a similar way by making 1:1 comparisons of the increment of each step in a decade with the increment provided by the whole of the following dial. The relative step deviations obtained in this manner can be progressively summed to provide the deviation or correction for each dial setting.

4. Current Comparator Bridge

The methods discussed so far rely on 1:1 and 10:1 resistance comparisons using interchange techniques for adjusting resistance ratios or for determining their deviations.

A method has been devised for comparing two resistances in a DC circuit in terms of transformer turns ratio. MacMartin and Kusters (1966) have detailed the development and performance of the instrument, which is known as the Direct Current Comparator Bridge and which is shown diagramatically in Fig. 6. The inherent accuracy of transformer ratios allows the ratio of unlike resistors to be measured without the need to determine bridge corrections as is the case with Wheatstone and Kelvin bridges. Another advantage resulting from the equipotential balance arrangement is that the current in the lower value resistor is proportionally higher than that in the other resistor. For example, if a 1000:1 ratio is used, a 0.001 ohm heavy current shunt at 100 amps can be compared with a 1 ohm standard at 0.1 amp.

A bridge constructed on the direct current comparator principle (Guildline Instruments) allows comparison of resistors between 0.1 milliohm and 10 kilohms and is useful for the calibration of potentiometers. However, the complexity of the instrument makes it relatively expensive and it is generally found only in large electrical standards laboratories.

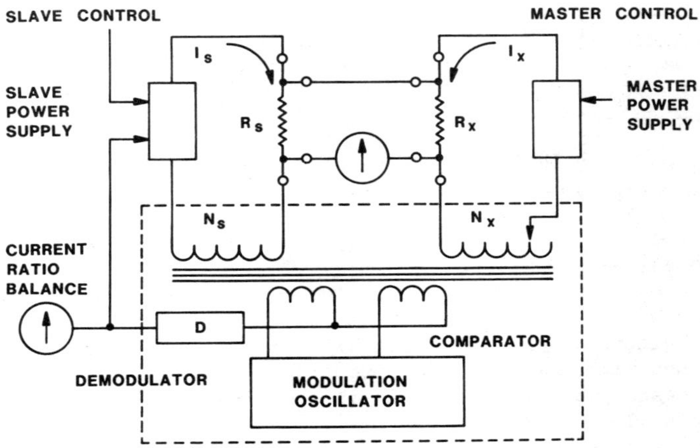

Fig. 6. D.C. Comparator Bridge.

5. Using Ratio Standards

Once the ratio standards are calibrated by using either comparison, adjustment or self-calibrating techniques, direct calibration of other ratio devices can be made. The circuit shown in Fig. 7 shows how the standard and test decade ratio devices should be connected.

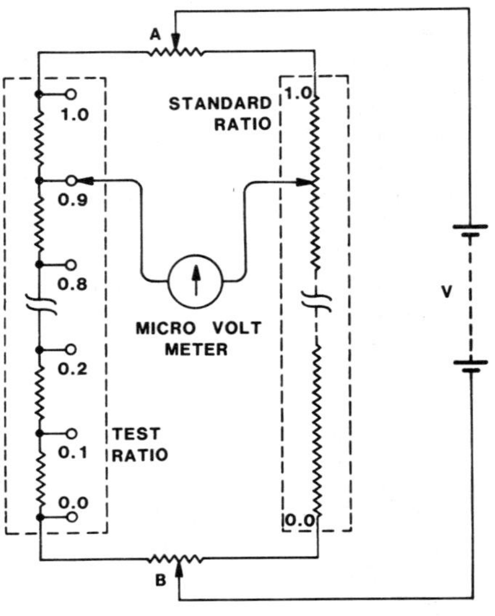

Fig. 7. Comparison of ratio devices.

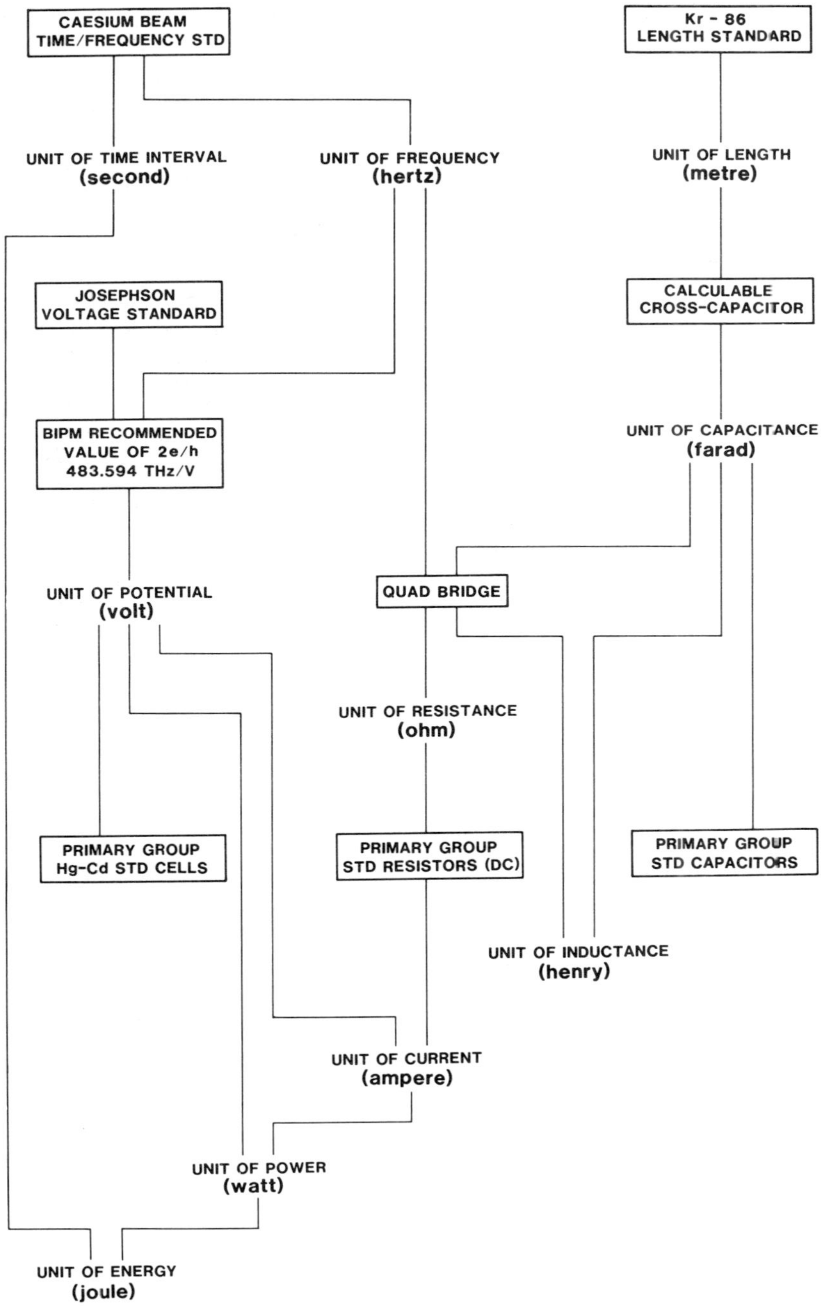

Fig. 1. Hierarchy of standards used in the realisation, maintenance and derivation of electrical units.

2. Standards Used to Realise and Maintain the Electrical Units

The hierarchy of standards used in the realisation and maintenance of the basic electrical units in the Division is shown in Fig. 1. SI base units of time interval (second) and length (metre) are realised directly in terms of their definitions from a caesium-beam time and frequency standard, and a krypton-86 length standard. The unit of frequency (hertz) is derived directly from the caesium-beam standard, and the unit of capacitance (farad) is realised in terms of the metre by the use of a calculable cross-capacitor. From frequency and capacitance, the unit of resistance (ohm) is realised by means of a special quad bridge. The unit of potential (volt) is maintained by the use of a voltage standard based on the superconducting ac Josephson effect, which relates voltage to frequency, and a BIPM* recommended value for 2e/h of 483.594 THz/V. The caesium-beam standard, the Kr-86 standard, the calculable capacitor and the Josephson standard are briefly described elsewhere (Macfarlane 1981, 1982). The units of capacitance and resistance are maintained by primary groups of standard capacitors and standard resistors respectively, in conjunction with the calculable capacitor. These primary groups, together with a primary group of cadmium standard cells maintained in terms of the Josephson voltage standard, serve as working standards during the routine calibration of similar standards from other laboratories. The unit of inductance (henry) is derived from the ohm and the farad, indirectly from the calculable capacitor, by means of a modified Maxwell bridge. The unit of current (ampere), which is an SI base unit, is derived in the Division from the volt and the ohm, as is the unit of power (watt). The unit of energy (joule) is derived from the units of power and time-interval.

3. Standards Used in Establishing the Values of AC Electrical Quantities in Terms of the Basic DC Units

The measurement of ac quantities is normally carried out by the use of ac measuring instruments calibrated in terms of a chain of secondary standards. Figure 2 shows the chain of standards and the way in which they relate the ac quantities to the basic dc units. Fundamental elements in the chain are the ac-dc transfer standards which are used to establish dc quantities equivalent to the ac quantities. Standard current transformers, standard voltage transformers and standard attenuators are used in the determination of ac quantities which are outside the range of basic ac-dc transfer standards and other ac measuring instruments.

From Fig. 2, ac voltage is transferred to dc through an ac-dc voltage transfer standard and the equivalent dc voltage so established is determined in terms of the basic unit by the use of dc ratio standards and standard cells. Similarly, ac current is transferred to dc through a current transfer standard and the equivalent dc current is determined in terms of the basic unit by the use of standard dc shunts, standard resistors and standard cells. Ac power is transferred to dc by using an ac-dc power transfer standard and the equivalent dc power is evaluated in terms of dc voltage and current. Ac energy is derived from ac power and the basic unit of time interval.

---

*International Bureau of Weights and Measures.

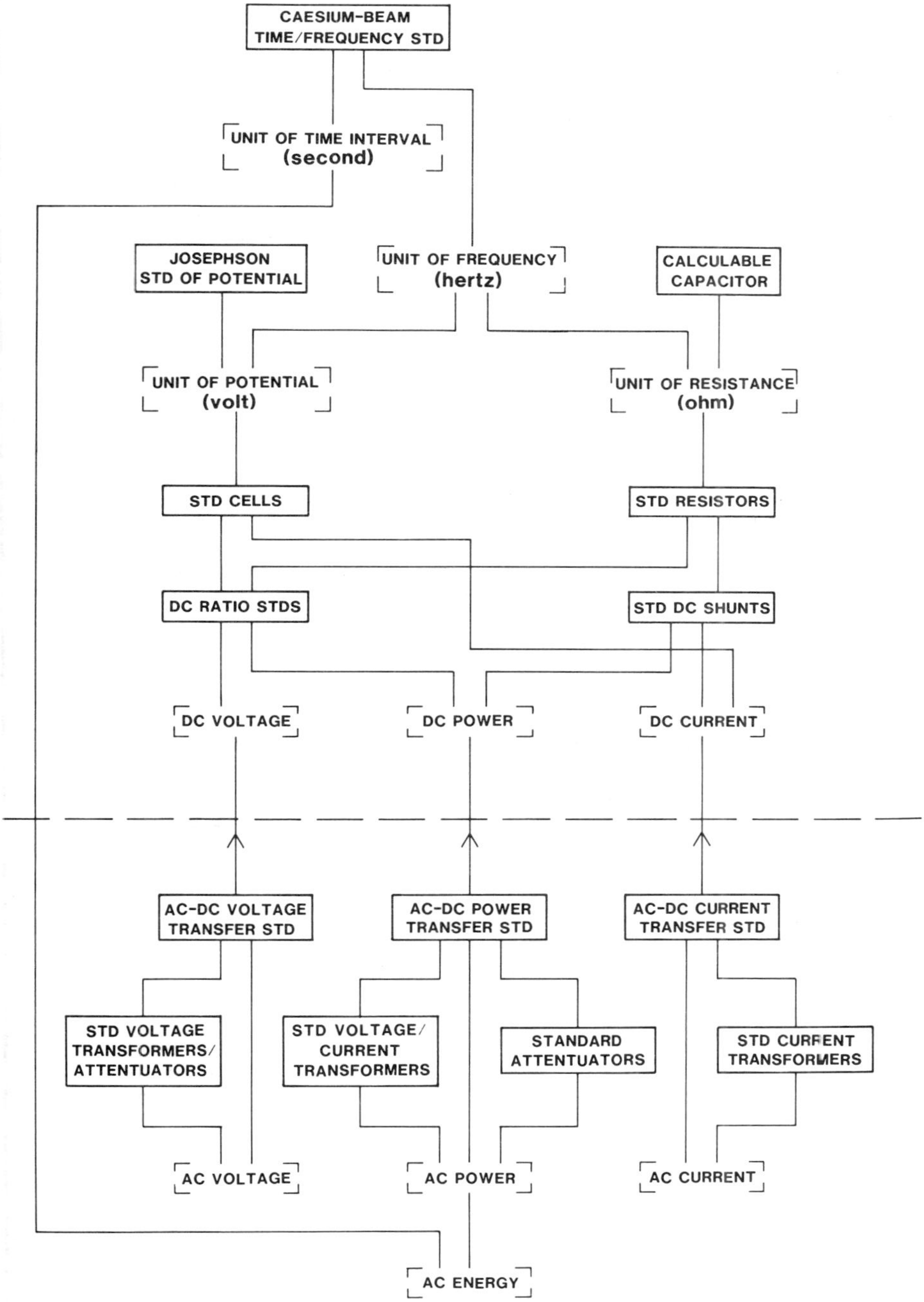

Fig. 2. Standards involved in the determination of ac quantities in terms of the basic electrical units.

Primary standards are maintained in the Division for the calibration of all of the standards referred to in Fig. 2. At frequencies up to 100 kHz ac-dc transfer standards for voltage and current are based on a group of thermal converters and a range of ac-dc shunts and series multipliers. Transfer standards for power, at power and audio frequencies, are based on thermal converters, standard ac-dc load resistors and more recently on a torque-compensated electrodynamometer. At radio and microwave frequencies thermal converters and a standard calorimeter are used as voltage and power transfer standards. Calibrations are provided for ac-dc transfer standards in the range 1-1200 V, 2.5 mA - 20 A; standard voltage and current transformers up to 550 kV and 22 kA respectively; and standard attenuators in the range 0-100 dB and at frequencies up to 12.4 GHz. Further details on some of the standards maintained in the Division for use at frequencies below 100 kHz, together with discussions on calibration and measurement technique, are given by Armitage (1982), Betts (1982), Franchimon (1982) and Rungis (1982).

4. Thermal AC-DC Transfer Standards

Thermal converters are widely used as ac-dc transfer standards both in commercial instruments and in national standards laboratories throughout the world. A typical converter is shown in Fig. 3. It consists basically of a straight-wire heater with a thermocouple attached to its mid-point. The thermocouple is attached by means of an electrically insulating bead, and the heater-thermocouple combination is mounted on four feed-through leads within an evacuated glass envelope. When used as an ac-dc transfer device, ac and dc excitations are applied sequentially to the heater and adjusted to achieve an equal heating effect. The temperature at the mid-point of the heater is indicated by a thermoelectric output emf from the thermocouple, and equality of heating on ac and dc is indicated by equal thermocouple outputs. Ideally, equal heating indicates equal dc and rms ac electrical inputs. Unfortunately, in practice, the heating effect of equal dc and rms ac excitations is not exactly the same and this gives rise to an ac-dc transfer error. At radio-frequencies the error may be as high as several percent due mainly to the presence of skin-effect, and a non-uniform distribution of inductance and stray capacitance. However, even at power and audio frequencies, where the above sources of error are not significant, there is still an error due to the presence of thermoelectric effects associated with the heater and its support leads.

The magnitude of the error due to thermoelectric effects is typically < 20 ppm for conventional thermal converters and may be < 5 ppm in converters with well chosen heater materials (Hermach and Williams 1966, Inglis 1978). Excitation currents and voltages in typical converters are in the range 2.5 - 10 mA and 0.25 - 1 V and errors at higher currents and voltages must be established for converters in combination with standard current shunts and series resistance multipliers. This is achieved by using special current shunts whose residual effects have been determined separately, and by assigning ac-dc voltage transfer errors to various multiplier-converter combinations on the basis of a build-up technique (Hermach and Williams 1966, Williams 1971, Inglis 1980).

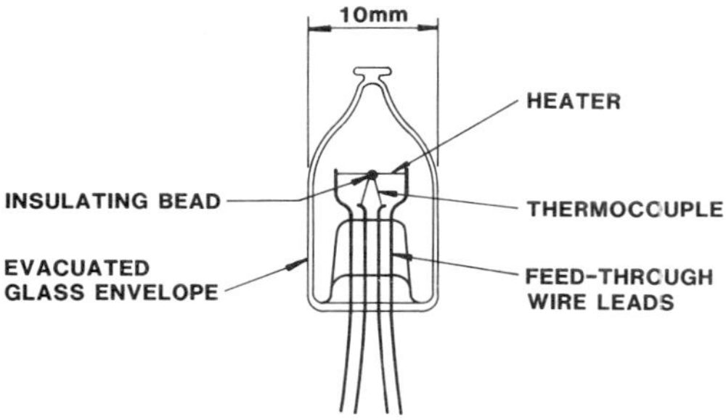

Fig. 3. Typical thermal converter.

Inherent in the voltage transfer build-up process is an accumulation of error with each succeeding step, and an uncertainty in the ac-dc transfer error of the basic converter of perhaps 5-20 ppm can lead to an inherent uncertainty contribution of ~ 25-100 ppm in multiplier-conver combinations at higher voltages. It is therefore important that the uncertainty in ac-dc transfer performance of the basic converter be kept to a minimum.

Several years ago, in response to pressure from external laboratories for higher accuracy calibrations, the Division initiated a program to investigate errors in thermal converters. The program is aimed primarily at improving the accuracy of the Division's basic thermal transfer standards at frequencies below 100 kHz. It is based on an absolute method for the determination of error variation with excitation level (Inglis 1978) and involves the measurement of differences in ac-dc transfer between thermal converters to a precision of < 0.1 ppm (Inglis 1982). The project has led to the development and accurate calibration of a number of special thermal converters. The special converters were designed and constructed with a view to minimising the effect of thermoelectric nonuniformities which arise in the heater during construction. These special thermal converters, in conjunction with a number of other special multi-junction converters form the Division's basic reference group. It is hoped that this work will lead to a basic uncertainty for ac-dc transfer of $\lesssim$ 1 ppm.

5. Power and Energy Standard

In the dissemination of a measuring capability for power and energy at power frequencies, the practice in the Division has been to provide an ac-dc transfer calibration and a zero power factor calibration for electrodynamometer wattmeter standards. This, together with the dissemination of the dc units, the unit of time interval, and the calibration of standard current and voltage transformers, has generally satisfied the requirements for power and energy measurement at power frequencies in Australia.

However, in recent years there has been an upsurge in the development and accuracy of electronic power and energy meters, and this has led to a growing demand for increased calibration accuracy and also for the acceptance of these instruments as secondary standards. Unlike electrodynamometers however, many of the electronic instruments are intended for ac use only and cannot be calibrated as ac-dc transfer standards. In order to adequately investigate the performance of these instruments, with a view to accepting and calibrating them as secondary standards, a project is under way in the Division to set up a new primary standard for power and energy. As an initial step in setting up the new standard, a survey of the main organisations in Australia requiring a precise power and energy measurement capability was carried out (Inglis and Johnson 1981) in order to establish a realistic accuracy objective. The survey showed a clear need for a standard capable of providing calibrations with an uncertainty of $< 0.02\%$.

The new standard consists of a commercial torque-compensated ac-dc dynamometer-type wattmeter, used in conjunction with a commercial wattmeter of the electronic time-division-multiplier type. Thermal transfer standards in combination with standard ac-dc load resistors provide an independent reference at unity power factor. Work is presently under way to evaluate the performance of the new standard and also to investigate the performance of a number of electronic wattmeters.

Conclusion

The maintenance and dissemination of the basic physical units in Australia is the responsibility of CSIRO. It maintains the volt and the ohm as the basic electrical units and disseminates these by the use of standard cells and standard resistors. Ac quantities of current, voltage, power and energy are determined in terms of the basic units by the use of ac-dc transfer standards and range extension devices. Primary standards are maintained at the CSIRO Division of Applied Physics and secondary standards, used by external laboratories in the determination of the ac quantities, are calibrated. Work is in progress in the Division to improve the accuracy of its basic ac-dc transfer standards and to establish a new and accurate power and energy standard.

References

Armitage, D.B. (1982). These proceedings.

Betts, P.J. (1982). These proceedings.

Franchimon, C.C. (1982). These proceedings.

Hermach, F.L. and Williams, E.S. (1966). *IEEE Trans. Instrum. Meas.*, IM-15, 260.

Inglis, B.D. (1978). *IEEE Trans. Instrum. Meas.*, IM-27, 440.

Inglis, B.D. (1980. *Metrologia.*, 16, 177.

Inglis, B.D. (1982). Asia/Pacific Met. Prog. Workshop on Metrology in Quality Assurance, Kuala Lumpur.

Inglis, B.D. (1982). *Metrologia*, 18, (in press).

Inglis, B.D. and Johnson, G.J. (1981). *J. Electr. Electron. Eng. Aust.*, 1, 273.

Macfarlane, J.C. (1981). CSIRO Inf. Serv. Sheet No. 33-1.

Macfarlane, J.C. (1982). These proceedings.

Rungis, J. (1982). These proceedings.

Williams, E.S. (1981). *J. Res. Nat. Bur. Stand.*, 75C, 145.

AC-DC Transfer of Voltage and Current

Charles C. Franchimon
CSIRO Division of Applied Physics, Sydney, Australia 2070

Abstract

A method for accurately calibrating ac voltage and current quantities, using ac-dc transfer devices, is discussed. Various types of ac-dc transfer instruments are reviewed. The modern method of using ac-dc thermal voltage and current converters is described in detail.

1. Introduction

Standards of dc voltage and current are disseminated by use of the Weston standard cells and standard resistors. In most standards laboratories groups of standard cells are maintained, which have been observed for a number of years to determine their drifts and their temperature coefficients. Using suitable resistive networks, dc voltages at any required level can be compared with the standard cell by potentiometric methods. The drift and temperature coefficient of standard resistors are also known accurately and thus dc currents can also be readily measured.

AC quantities are more difficult to determine as there is no direct means of comparing an ac quantity with the standard cell. Some means of transferring the standards of voltage, current and power between the ac and dc systems is therefore required. The ideal transfer device is one which responds in exactly the same way to rms ac and dc quantities. Thus the ac quantity can be equated to the dc quantity which in turn can be referred back to the standard cell (see Fig. 1).

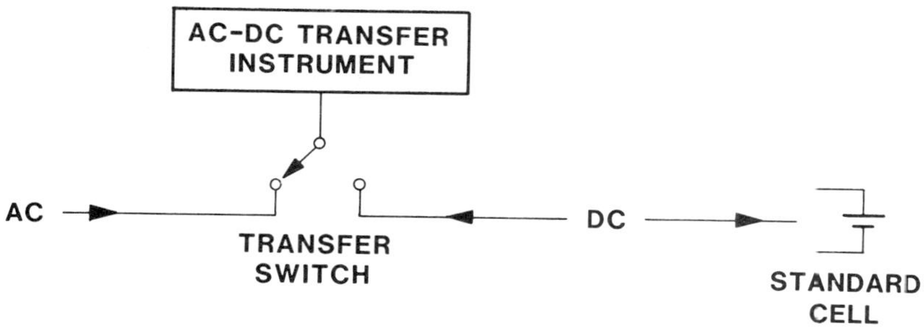

Fig. 1. Principle of ac-dc transfer

2. AC-DC Transfer Instruments

Any instrument responding to both ac and dc quantities car be used as an ac-dc transfer instrument. However, for accuracy and ease of operation it is convenient if the response to ac is the same or nearly the same as the response to dc i.e. rms responding. Various types of instruments fall into this category. The most commonly used are:

1. Electrodynamic instruments.
2. Electrostatic instruments.
3. Electrothermic instruments.

3. Electrodynamic Instruments

Electrodynamic instruments make use of the magnetic fields of two sets of coils, one fixed and the other movable. Each coil carries a current which is a function of the current or voltage to be measured and the reaction of the fixed and movable coils supplies the operating torque to the moving system. An electrodynamic instrument has a square law response and its indication is in terms of rms current or voltage.

Until about fifteen years ago this type of instrument was commonly used as an ac-dc transfer standard in industrial laboratories, while a specially compensated electrodynamometer was used as a national standard in the United States of America until about 1960. The main limitation of such an instrument is its frequency response. It is normally useful to only about 1 or 2 kHz. Sources of errors arise from electromagnetic coupling, capacitive currents, eddy currents and skin effects.

4. Electrostatic Instruments

In this type of instrument the deflecting torque results from the action of an electric field on charged electrodes. It is essentially a high-input-impedance, voltage-responding instrument. It has a square-law response and its indication is in terms of rms voltage. The frequency range of electrostatic voltmeters is normally limited by ancillary ranging dividers which allow the instrument to work near maximum sensitivity. Accurate electrostatic instruments tend to be rather fragile and cumbersome and are not commonly used in industrial type laboratories.

An electrostatic voltmeter was used as an ac-dc transfer standard at the National Physical Laboratory in Great Britain and at the CSIRO Division of Applied Physics in Australia. As a national standard its main disadvantage was the inherent ac-dc transfer error associated with dielectric films on the instrument's electrodes.

5. Electrothermic Instruments

Among the early electrothermic instruments suitable for ac and dc measurements were the hot-wire (or expansion) instruments.

The principle of these instruments was based on the linear expansion of a wire heated by an electric current. The resultant sag in the wire was mechanically translated into the movement of an instrument pointer. This type of instrument has been largely superseded by the thermal converter instrument.

A thermal converter (also called thermoelement or vacuojunction) consists of a thin-wire heater and a thermocouple which is attached to the midpoint of the heater by means of an insulating ceramic bead. The heater and thermocouple are normally encapsulated in an evacuated glass bulb. The heater wire is usually made of Nichrome or Evanohm and its dimensions are chosen to give an output from the thermocouple in the order of 7 mV at rated current. As the temperature of the bead is approximately proportional to the square of the current, the thermal converter has a square-law response.

High-frequency voltmeters and ammeters have been made using thermal converters by selecting suitable resistive multipliers or shunts and measuring the output of the thermocouple with a sensitive millivoltmeter.

Over the last thirty years or so this principle has been developed into the modern method of ac-dc transfer measurements with a sensitivity of a few parts per million. The output of the thermocouple, instead of being measured with a millivoltmeter, is opposed by an equal or near equal output voltage of a Lindeck potentiomenter element (see Fig. 2).

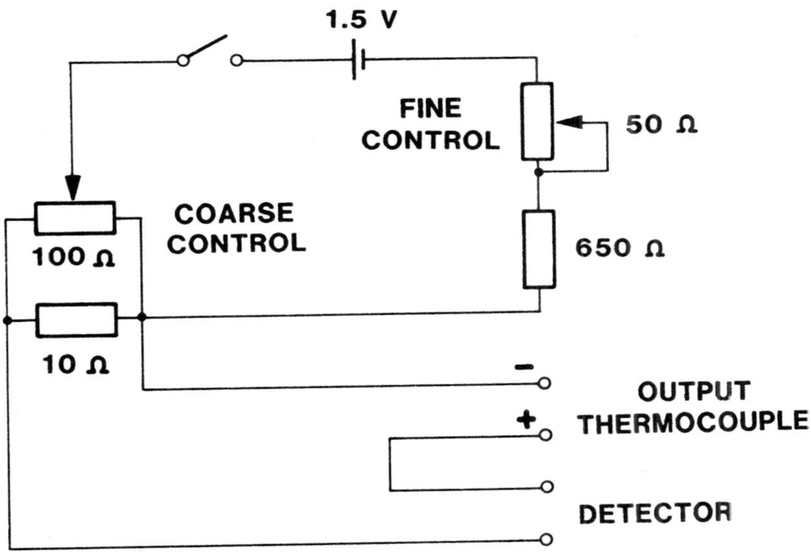

Fig. 2. Lindeck potentiometer element of the Hermach-Engelhard thermal converter.

The difference voltage is measured by a sensitive null detector. Thus, when an ac voltage (or current) is applied to the thermal converter, the potentiometer can be adjusted to give zero input to the null detector. If a dc voltage (or current) is then applied to the thermal converter from a variable dc source and the output of that source is adjusted to again obtain zero on the null detector an accurate comparison can be made between ac and dc voltages (or currents). (See Fig. 3).

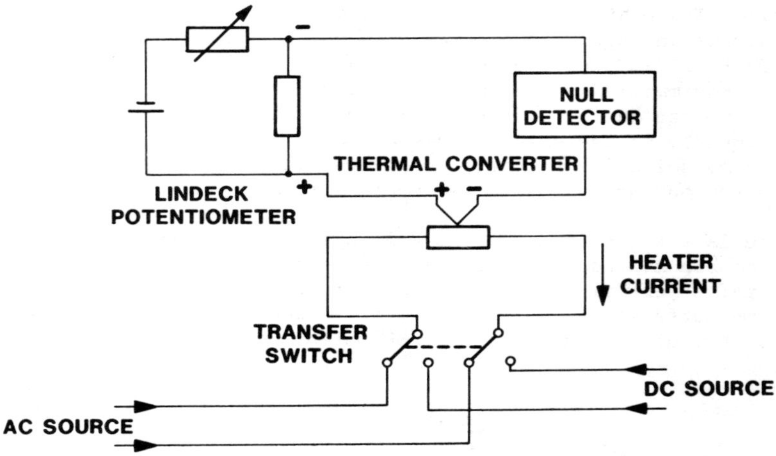

Fig. 3. Elementary thermal transfer instrument.

A thermal voltage converter consists of a thermal converter with a suitable multiplying resistor in series with its heater, while a thermal current converter consists of a thermal converter with a shunt resistor in parallel with its heater. These resistors are especially designed to have a low ac-dc transfer error.

The voltage ranges of commercial thermal transfer instruments usually extend from about 0.5 V to 1000 V and the current ranges from 5 mA to 20 A. Many such instruments are completely self-contained, with switched ranges, built-in potentiometer and null detector. Although the upper frequency limit is approximately 1 MHz, the instruments are most accurate in the audiofrequency range, where the uncertainties are typically 50 ppm.

For ac-dc voltage transfer tests at higher frequencies, a coaxial type thermal voltage converter is required. Two types of these are commercially available. One has the multiplying resistor and the thermal converter housed in the same tube, while the other type consists of a number of multiplying resistors, each in its own housing, which can be plugged into one of several thermal converters of different current ratings. The upper limit at which reasonable accuracy is maintained is 50 MHz.

In the Division of Applied Physics the ac-dc transfer standard consists of a commercial set of three coaxial thermal converters and five coaxial multiplying resistors. The current ratings of the thermal converters are 2.5, 5 and 10 mA and each can be connected to any of the five resistors to give voltage ranges of 0.5 V to 1200 V.

6. Limitations and Errors of Thermal Transfer Instruments

The output of thermal converters is usually a voltage in the order of 7 mV at rated current. At this level great care must be taken to avoid errors arising from spurious thermal emfs in the output circuit during measurements, especially if accuracies in terms of parts per million are to be achieved.

The main limitations on the high-frequency performance of thermal converters are due to the stray capacitances and inductances distributed throughout the heater and its support terminals, to skin effects and to the limited frequency range of multipliers and shunts.

When measuring ac quantities at very low frequencies, the heater and bead temperatures tend to follow the cyclic variations in the applied current. Therefore the thermal converter is only useful at frequencies greater than 25 Hz if reasonable accuracy is to be maintained.

Thomson and Peltier effects will cause changes in the distribution of temperature along the heater wire when dc is applied which may give rise to reversal errors. If these reversal errors are large enough, the sensitivity of the null detector may have to be reduced with a subsequent increase in the uncertainty of the measurements.

Second order Thomson and Peltier effects will affect the temperature of the bead, resulting in ac-dc transfer errors.

7. The Multijunction Thermal Converter

Because of the errors caused by Thomson and Peltier effects, a multijunction thermal converter was constructed with a large number of thermocouples distributed along a bifilar heater wire. Its advantages are that the Thomson and Peltier effects are averaged out and the reversal error is reduced to near zero due to the fact that the whole length of the heater is used. As the thermocouples are connected in series the output of a multijunction is higher than that of a single junction, while operating at a very much lower temperature.

The disadvantage of the multijunction thermal converter is its cost. It is many times more expensive to produce than a single junction.

8. Use of a Typical Commercial Thermal Voltage Converter to Calibrate an AC Voltmeter

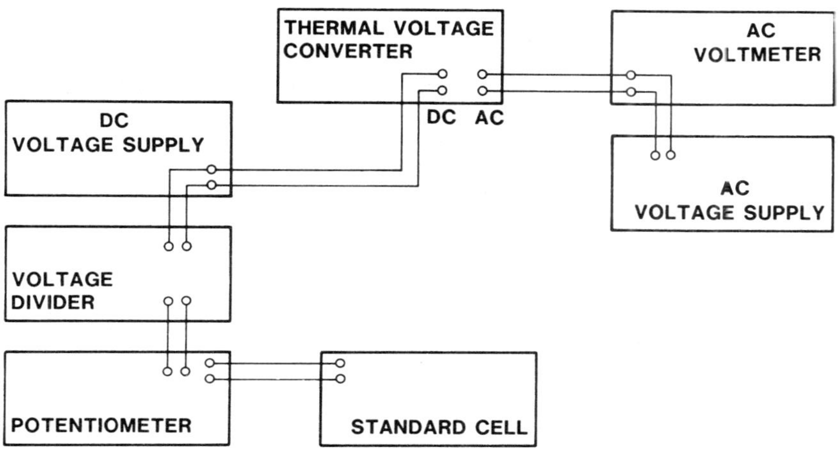

Fig. 4. AC voltmeter calibration

With the ac voltmeter connected in parallel with a thermal voltage converter (see Fig. 4), ac voltage is applied to give the required indication on the voltmeter. The thermal voltage converter is then used to determine accurately the ac voltage applied by comparing it with an equal dc voltage, which can be referred back to the standard cell by means of a resistive network and a potentiometer.

The voltage applied to the ac voltmeter is

$$E_{ac} = \left(1 + \frac{a_o}{100}\right) E_{dc}$$

where a_o = the ac-dc transfer correction (%) to the transfer instrument
E_{dc} = the average of forward and reverse voltages.

9. Use of a Typical Commercial Thermal Current Converter to Calibrate an AC Ammeter

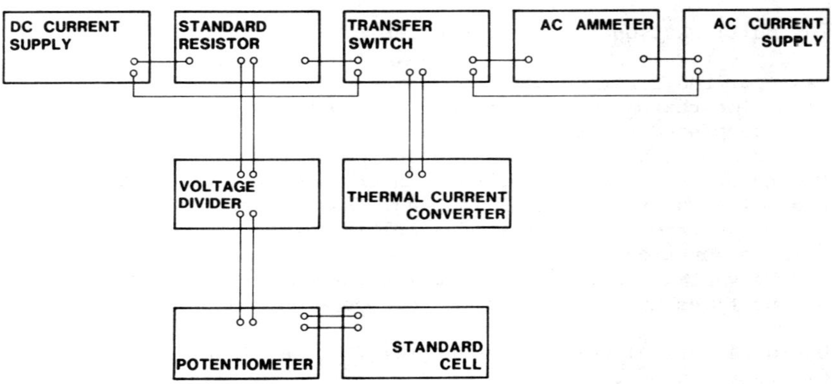

Fig. 5. AC ammeter calibration.

With the ac ammeter connected in series with a thermal current converter (see Fig. 5), ac current is applied to give the required indication on the ammeter. The thermal current converter is then used to determine accurately the ac current applied by comparing it with an equal dc current, which can be referred back to the standard cell by means of a resistive network and a potentiometer.

The current applied to the ac ammeter is

$$I_{ac} = \left(1 + \frac{a_o}{100}\right) I_{dc}$$

where a_o = the ac-dc transfer correction (%) to the transfer instrument
I_{dc} = the average of forward and reverse dc currents.

10. Comparing Two Typical Commercial Thermal Voltage Converters with Each Other

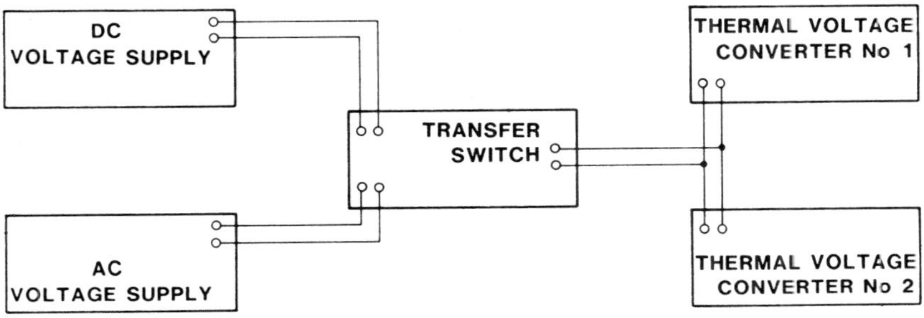

Fig. 6. Thermal voltage converter comparison.

With the transfer switch in the ac position and the two thermal voltage converters connected in parallel (see Fig. 6) the appropriate ac voltage is applied to both instruments. Using the internal reference circuits, both instruments are balanced to give zero deflection on the galvanometer.

With the transfer switch in the dc position, the dc supply is adjusted to give zero deflection on the galvanometer of one of the thermal voltage converters (say no. 1). The deflection on the galvanometer of thermal voltage converter no. 2 is then an indication of the difference in ac-dc transfer response of the two instruments and is given by

$$a_o = \frac{d_1 - d_2}{100\ k}$$

where a_o = the ac-dc transfer difference (%) between the two converters
d_1 = the deflection of the galvanometer on ac
d_2 = the average deflection on forward and reverse dc
k = the change in deflection for a 0.01% change in input.

11. Comparing Two Typical Commercial Thermal Current Converters with Each Other

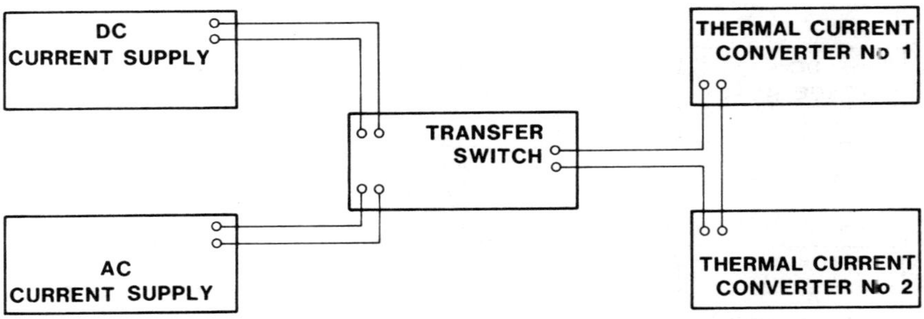

Fig. 7. Thermal current converter comparison.

With the transfer switch in the ac position and the two thermal current converters connected in series (see Fig. 7) the appropriate ac current is applied to both instruments. Using the internal reference circuits, both instruments are balanced to give zero deflection on the galvanometer.

With the transfer switch in the dc position, the dc supply is adjusted to give zero deflection on the galvanometer of one of the thermal current converters (say no. 1). The deflection on the galvanometer of thermal current converter no. 2 is then an indication of the difference in ac-dc transfer response of the two instruments and is given by

$$a_o = \frac{d_1 - d_2}{100\ k}$$

where a_o = the ac-dc transfer difference (%) between the two instruments
d_1 = the deflection of the galvanometer on ac
d_2 = the average deflection on forward and reverse dc
k = the change in deflection for a 0.01% change in input.

12. Conclusion

The use of thermal converters has improved many aspects of ac measurements. The frequency range, limited to 1 or 2 kHz with conventional indicating instruments, has been extended well into the MHz region. Accuracies in the order of 0.01% have been greatly improved, in some cases to a few parts per million.

Further research is being carried out to reduce errors arising from the Thomson and Peltier effects and to improve measurement techniques which will contribute to an improvement in the accuracy of ac measurements.

Bibliography

Harris, F.K. (1952). Electrical Measurements. John Wiley & Sons, Inc.

Hermach, F.L. (1952). *J. Res. Nat. Bur. Stand.* 48, 121.

Hermach, F.L. (1976). *IEEE Trans. Instrum. Meas.* 25, 489.

Hermach, F.L. and Williams E.S. (1966). *IEEE Trans. Instrum. Meas.* 15, 260.

Inglis, B.D. (1978). *IEEE Trans. Instrum. Meas.* 27, 440.

Inglis, B.D. (1981). *Metrologia,* 17, 111.

Klonz, M. and Wilkins, F.J. (1980). *IEEE Trans. Instrum. Meas.,* 29, 409.

Widdis, F.C. (1962). *Proc. Inst. Electr. Eng. Part C,* 109, 328.

Wilkins, F.J. (1972). *IEEE Trans. Instrum. Meas.,* 21, 334.

Wilkins, F.J. and Deacon, T.A. (1965). *Proc. Inst. Electr. Eng. Part 1,* 112, 794.

Williams, E.S. (1965). *Nat. Bur. Stand. Tech. Note* 257.

Williams, E.S. (1971). *J. Res. Nat. Bur. Stand.* 75C, 145.

The AC-DC Transfer Method of AC Power Standardization

D.B. Armitage
CSIRO Division of Applied Physics, Sydney, Australia 2070

Abstract

A method of disseminating standards of ac power is described which utilises a unity power-factor ac-dc transfer correction, a power factor correction and a dc correction. An ac-dc transfer wattmeter is calibrated by the standardising laboratory to determine the unity power-factor ac-dc transfer correction and also the zero power-factor (leading and lagging) error, from which a power-factor correction is derived. These corrections are combined with a dc correction measured by the user to establish an ac power correction. Thermal converters have been used in the dissemination of ac-dc transfer corrections for voltage and current standardisation. A transportable thermal ac-dc transfer wattmeter is proposed which utilises the proven suitability of thermal converters.

1. Introduction

The CSIRO Division of Applied Physics is charged with the maintenance and dissemination of dc and ac electrical units in Australia. The dc units are disseminated by the use of standard cells and standard resistors. The ac units of voltage, current and power are derived indirectly from the corresponding dc units by using ac-dc transfer instruments.

2. AC-DC Transfer Process

In the ac-dc transfer process, the difference between the ac and dc sensitivities of an appropriate instrument is determined by this Division , and reported in the form of a correction. The user may derive an ac calibration at any time by combining this correction algebraically with an appropriate dc correction which can be derived from in-house maintained standards of dc voltage and resistance.

The advantage of this process is that the ageing factors which affect an instrument's sensitivity equally on both ac and dc are accounted for by the user's own dc calibrations. Properties which change with time and necessitate re-calibration are the resistance of voltage-range multipliers, the resistance of current shunts, and the elastic properties of restoring-torque systems. When a history of these changes has been accumulated, the user can adjust the periods between re-calibrations to suit his own requirements. An advantage of the ac-dc transfer process is that even under these conditions an ac electrical quantity can be established with the best accuracy at a chosen time by performing a dc calibration at the rms magnitude required.

To account for the variations which occur only in the difference between the ac and dc sensitivities of an ac-dc transfer standard, this Division is required to standardise the ac-dc difference at about 5 yearly intervals. This long period is adequate because the variations in ac-dc difference only occur very slowly. Excluding catastrophes such as electrical overload or rough handling, they are caused by dielectric changes due to deterioration of insulation, shrinkage of varnishes, relaxation of stresses built into significant parts of the instrument and any other changes which alter the geometry of the instrument's circuit and so change associated inductances and capacitances. The ac-dc difference is largely independent of voltage and current on specific ranges. This is particularly so for thermal-converter-type instruments and for electrodynamometer instruments. As a result, at most two determinations (and often only one) are required on each range. A direct ac calibration would require enough calibration points on each range to determine the scale shape.

The ac-dc transfer process of disseminating ac standards is relatively straightforward in the case of voltage and current, there being in each case only one electrical component involved. In the case of power at unity power factor, voltage and current are both involved but this does not cause any difficulty. However, when power at other than unity power factor is required to be standardised by the ac-dc transfer process an additional correction is required to account for errors dependent on the power factor. The wattmeter ac correction is then

(DC Correction) + (AC-DC Transfer Correction) + (Power Factor Correction)

3. Power Factor Correction

A power factor correction can be derived from a measurement of the error of a wattmeter at zero power factor. Consider the zero-deflection wattmeter shown in Fig. 1. It is a dynamometer wattmeter provided with an electrical restoring torque system to counteract the wattmeter torque and so maintain its moving coil system in a constant position. The instantaneous torque on the moving system is

$$\tau = K_1 i_m i_f \left(\frac{dM}{d\Theta}\right) \tag{1}$$

where K_1 is a constant,
i_m is the instantaneous current in the moving coil
i_f is the instantaneous current in the fixed coil
M is the mutual inductance
and Θ is deflection angle between the moving and fixed coils.

If the moving-coil zero-deflection position is chosen at right angles to the fixed coils, M is zero and constant, Θ is constant and $\frac{dM}{d\Theta}$ is constant so that $\tau = K_2 i_m i_f$ without any error due to mutual inductance between the coils. If now the current in the moving coil i_m is proportional to the instantaneous voltage e across a load and the current in the fixed coil i_f is the instantaneous load current i, then

$$\tau = K_3 ei.$$

The average driving torque is $\frac{K_3}{T}\int_o^T ei\ dt$ and since the definition of average power is $P = \frac{1}{T}\int_o^T ei\ dt$ the restoring torque required to maintain zero deflection provides the measure of power without error due to mutual inductance at any power factor.

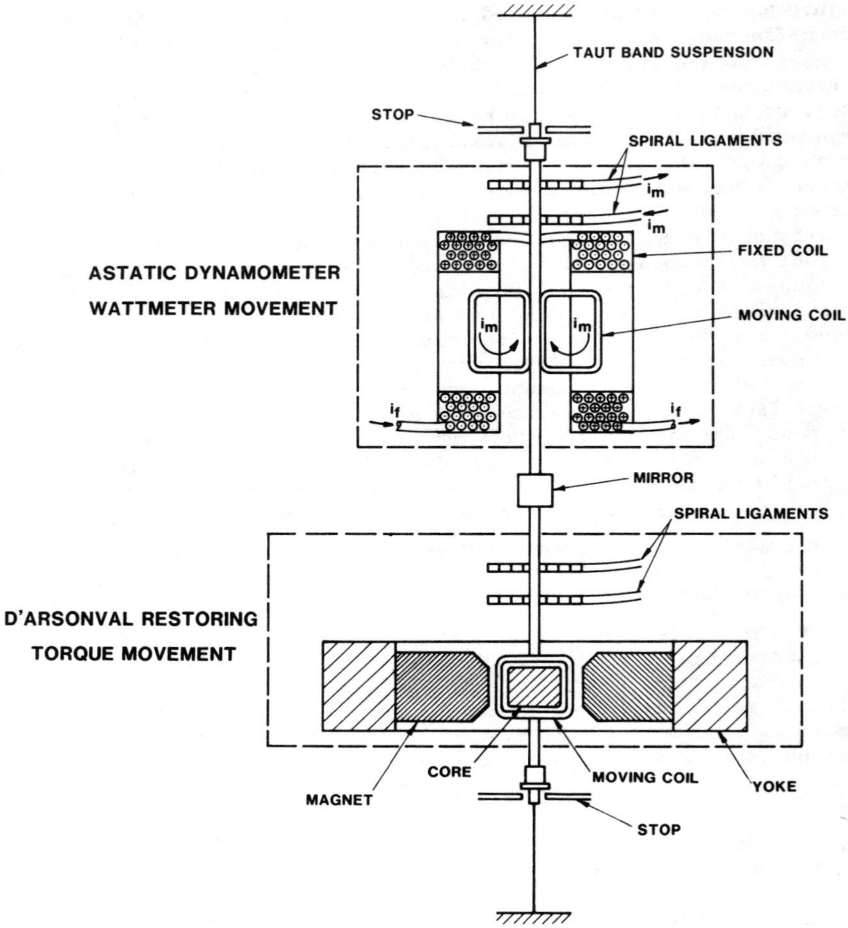

Fig. 1. Zero-deflection astatic wattmeter.

Other sources of power factor error, however, are present. They are due to inductance and capacitance in the voltage circuit, eddy currents in any metal including the wires of the wattmeter and capacitance currents in the fixed coils. There being no movement in the instrument these effects can all be lumped together and labelled as an equivalent phase-defect angle of the voltage circuit, i.e. they contribute to an equivalent phase angle by which the current in the voltage circuit deviates from the voltage applied to it. In Fig. 2 where a voltage represented by the vector V and a load current I_L lagging by a phase angle ϕ are applied to a wattmeter, this equivalent phase defect angle of the voltage circuit by which its current I_V lags the applied voltage is the angle α.

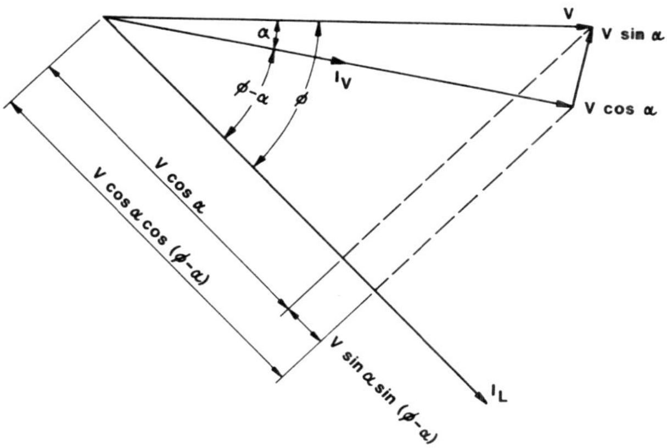

Fig. 2. Effect of a phase defect angle α in the voltage circuit of a wattmeter.

From the vector diagram, the effective power factor angle (i.e. the angle between the current I_V in the moving coil and the load current I_L in the fixed coils) is $(\phi - \alpha)$ and is less than the angle ϕ by which the load current lags the applied voltage V. The effective power factor in the wattmeter movement is greater than the applied power factor and the instrument will read high. For a leading load current, the effective power factor is less than the true power factor, and the wattmeter will indicate low. For the case where the load current lags the applied voltage the error in indication will be

$$k_\phi = VI_L \sin\alpha \sin(\phi - \alpha). \quad (2)$$

At unity power factor when $\phi = 0$

$$k_0 = -VI_L \sin^2\alpha$$

and at zero power factor lagging when $\phi = 90°$ lag

$$k_{90\ lag} = VI_L \sin\alpha \cos\alpha.$$

Since α is always less than a few minutes of angle

$$\sin\alpha \doteqdot \alpha \text{ and } \cos\alpha \doteqdot 1$$

and

$$k_0 \doteqdot -VI_L\,\alpha^2 \quad (3)$$

$$k_{90\ lag} \doteqdot VI_L\,\alpha. \quad (4)$$

Under the conditions considered above $k_{90\ lag}$ is the lagging zero power factor error and k_0 is the error at unity power factor.

Considering now the general phase angle error of equation (2), since α is the order of a few minutes of angle, $\sin\alpha \doteqdot \alpha$ and $\sin(\phi - \alpha) \doteqdot \sin\phi$, resulting in $k_\phi \doteqdot (VI_L\alpha)\sin\phi$.

Expressed as a fraction of the wattmeter indication $VI_L \cos\phi$, the power factor error at lagging phase angle ϕ is

$$E_{\phi\ lag} = \frac{(VI_L \alpha)\sin\phi}{VI_L \cos\phi}$$

$$= \frac{k_{90\ lag}}{D}\tan\phi.$$

where $k_{90\ lag}$ is the lagging zero power factor error and D is the wattmeter indication with voltage V and load current I_L at unity power factor. The corresponding wattmeter power factor correction at lagging phase angle ϕ is

$$C_{\phi\ lag} = -\frac{k_{90\ lag}}{D}\tan\phi.$$

Similarly, when the load current I_L leads the applied voltage by a phase angle ϕ,

$$C_{\phi\ lead} = -\frac{k_{90\ lead}}{D}\tan\phi$$

where $k_{90\ lead}$ is the leading zero power factor error and is of opposite sign to $k_{90\ lag}$.

4. Voltmeter and Ammeter Action

A contribution to ac-dc errors is made when voltage only or current only is applied to the wattmeter. These errors are present when making measurements at unity power factor and are taken into account in determining the ac-dc transfer correction at unity power factor. They are also present when determining the zero power factor error. They are measured separately and allowed for so that the zero power factor error is known in isolation.

5. Deflecting Wattmeter

In a deflecting wattmeter the mutual inductance between moving and fixed coils varies as shown in Fig. 3 with the deflection angle Θ. By appropriate proportioning and shaping of the fixed and moving coils, $\frac{dM}{d\Theta}$, which enters directly into the instantaneous torque equation (1), can be made substantially constant and variations are taken into account in the dc calibration of the instrument. Under normal conditions these variations in $\frac{dM}{d\Theta}$ do not give rise to an ac-dc difference.

The variation of mutual inductance itself however does give rise to impedance changes which alter the ac-dc error. The effect is exceedingly small at power frequencies, e.g., for a typical 5 ampere, 150 volt wattmeter, the contribution of M to the error at rated voltage and current and 50 Hz is only 0.01 watt even at zero power factor where M is greatest. For deflection wattmeters of large scale length, e.g. 30 cm, the limit of reading is 100 - 200 ppm of the full scale length which is equivalent to 0.07 watt. If the effect of mutual inductance variation is small enough, the process of deriving an ac calibration from a dc calibration using an ac-dc transfer correction and a power factor correction derived from a zero power factor error will be adequate. That the process is adequate requires confirmation for an instrument type by an experimental calibration in which the reference power is found by accurate measurement of ac voltage, current and power factor. However, for the limit of reading of the portable instruments in this category, the process is considered adequate.

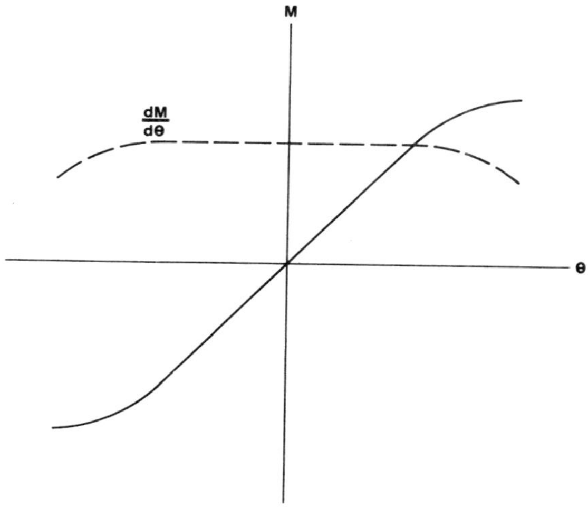

Fig. 3. Mutual inductance variation between two coils with angular displacement of their axes from the perpendicular position.

6. AC-DC Transfer and Zero Power Factor Testing

For many years now, this Division has tested deflection and zero-deflection wattmeters for ac-dc transfer correction and zero power factor error. This work has entailed the transportation of the customer's laboratory standard wattmeter to and from the Division with an attendant uncertainty in the effects of the travelling. This also applied to voltage and current laboratory standards before the appearance of the thermal voltage and current ac-dc transfer standards 25 years ago. The ruggedness and sensitivity of these latter instruments make them ideal for the purpose of disseminating ac-dc transfer difference, the actual ac calibrated laboratory standard never leaving the customer's laboratory.

To take advantage of these features, the Division has worked on a thermal converter wattmeter with the intention of using it to provide an ac-dc transfer reference and also a reference of zero power factor error which would be transportable with the same confidence and sensitivity now existing for voltage and current ac-dc transfer standards.

7. Conclusion

A laboratory can utilize the ac-dc transfer plus power factor correction method to provide a standard of ac power. The user has the choice of modern zero-deflection wattmeters with self-balancing systems of high sensitivity or deflecting wattmeters of a lesser sensitivity of indication. In order that these wattmeters may remain in the laboratory, a transportable thermal ac-dc power transfer instrument is being developed to transport the standard of unity power factor ac-dc transfer and zero power factor error from this Division.

The Use and Calibration of Instrument Transformers

Peter J. Betts
CSIRO Division of Applied Physics, Sydney, Australia 2070

Abstract

In the electrical supply industry, instrument transformers (ITs) are widely used where it is required to measure high voltages and currents. The nominal transformation ratio of an IT depends only on the number of turns on its windings; but, due to various losses, there are always errors in the transformation process. It is frequently necessary to correct for these errors, which requires calibration of the ITs. In this paper, the methods used in calibrating ITs, the required accuracy, and the required intervals between recalibrations are discussed.

1. Introduction

An instrument transformer (IT) is an alternating-current device used to extend the range of an electrical instrument such as a voltmeter, ammeter, wattmeter, watthour meter, or item of protection equipment. In addition to its use as a range extending device, a double wound IT may also serve to isolate an instrument from a high voltage circuit. It is usually impractical to manufacture precision indicating instruments with voltage ratings above about 500 V or current ratings above about 20 A, and as the electricity supply industry is required to measure or detect voltages and currents very much higher than these values, a convenient way to do so is to use an IT to transform such voltages and currents down to measurable values. Also, it may be required to use an IT to transform low voltages and currents up to values which can be accurately measured on available indicating instruments.

2. Transformer Ratio

A transformer used as a ratio standard is probably the most stable of all electrical ratio devices. To a first order of accuracy, the ratio is dependent only on numbers of turns wound on a ferromagnetic core. In the case of the voltage transformer (VT), a voltage applied to the primary winding causes a voltage to be induced in the secondary winding which is nearly equal to the applied voltage divided by the ratio of primary to secondary turns, provided the load connected to the secondary terminals is of high impedance.

In the case of the current transformer (CT), a current passed through the primary winding causes a current to be induced in the secondary winding which is nearly equal to the primary current multiplied by the ratio of primary to secondary turns, provided the load connected to the secondary terminals is of low impedance.

3. Transformer Error

Due to various losses, the actual transformation ratio of an IT is never exactly equal to the ratio as defined by numbers of turns. Therefore, where the highest accuracy of measurement is required, the error in the nominal transformation ratio must be determined. The following are three basic effects which give rise to internal losses in transformers.

The Effect of Loading

In properly designed ITs, by far the greatest contribution to error is that due to loading. The larger the volt-amperes delivered to the external load or burden connected to the secondary terminals, the greater will be the losses internal to the IT, and hence the greater the errors.

The Effect of Insulation-Leakage Currents

In applications where the voltage applied to an IT is high, or where high voltage is induced in one of the windings, insulation-leakage currents may contribute appreciably to the errors. Transformers which fall into this category are VTs with voltage ratings above about 30 kV and CTs with current ratings below about 0.5 A.

The Effect of Partial Coupling

When the primary turns of an IT are not uniformly distributed around the circumference of the core, the magnetic flux in the core may partly leak across an air path and therefore only partly couple with the secondary winding. This effect is usually associated with window type CTs where only one primary turn is used. Unless the primary conductor is subdivided into several uniformly distributed parallel paths, the error may change appreciably with changes in the primary conductor configuration.

4. Voltage Transformer Calibration

The error of a VT is normally defined as the error in the actual secondary voltage expressed as a fraction of the nominal secondary voltage. Voltage transformers are usually calibrated by comparing them directly with a higher-quality standard transformer of the same nominal transformation ratio. A simplified circuit for comparing two VTs is that of Fig. 1.

Rated burden is connected across the secondary winding of the VT under test, and a voltage is applied to the primary windings of both VTs. Since the actual transformation ratio of the standard VT is known, V1 may be used to measure the applied voltage. The voltmeter V2 measures the difference in voltage between the two secondary windings and is therefore an indication of the error of the VT under test, provided the error of the standard VT is negligible.

If it is required to know the error of the VT under test in both magnitude and phase, or if the error of the standard VT is not negligible, it will be necessary to replace the voltmeter V2 by some phase-sensitive measuring device such as a co-ordinate potentiometer or a VT testing set (eg. as described by Clothier and Medina 1954). Then the known error of the standard VT in complex terms may be added to the error indicated on the phase-sensitive device to obtain the error of the VT under test in complex terms.

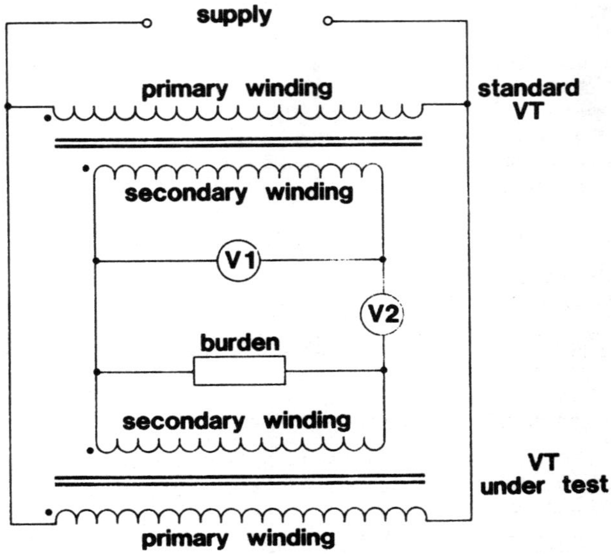

Fig. 1. Simplified circuit for comparing two VTs

At the CSIRO Division of Applied Physics, a more elaborate testing circuit is used in which the VT under test is compared with a capacitance divider in a bridge circuit (see Fig. 2). The capacitors are 3-terminal with air or compressed-gas dielectric, and are more stable at high voltages than a VT. The relative values of the capacitors may be determined by a 'build-up' method, which is readily carried out by switching the individual capacitance units into either arm of the bridge (Clothier and Medina 1956).

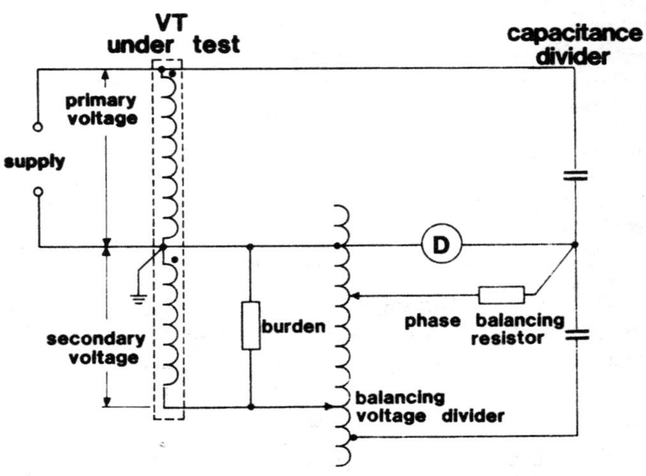

Fig. 2. CSIRO VT calibration circuit

5. Current Transformer Calibration

The error of a CT is normally defined as the error in the actual secondary current expressed as a fraction of the nominal secondary current. Current transformers are usually calibrated by comparing them with a higher quality standard of the same nominal transformation ratio in a circuit similar to that for calibrating VTs. A simplified circuit showing the principle involved is that of Fig. 3.

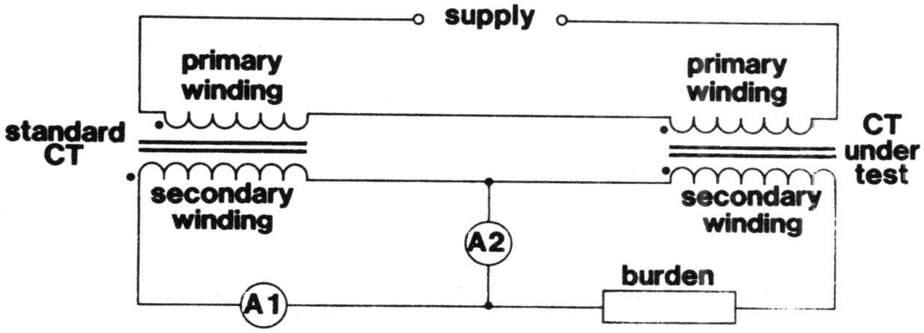

Fig. 3. Simplified circuit for comparing two CTs

Rated burden is connected into the secondary circuit of the CT under test and a current is passed through the primary windings of the two CTs in series. Since the actual transformation ratio of the standard CT is known, A1 may be used to measure the primary current. The ammeter A2 measures the difference in current between the two secondary windings and is therefore an indication of the error of the CT under test, provided the error of the standard CT is negligible. If it is required to measure the error of the CT under test in both magnitude and phase, or if the error of the standard CT is not negligible, the ammeter A2 must be replaced by a phase-sensitive measuring device such as a CT testing set.

At CSIRO, the traditional method of CT calibration is to compare the CT under test with a standard CT using an Arnold-type testing set as shown in Fig. 4 (Arnold 1934). In order to calibrate the primary-standard CTs of CSIRO, a build-up technique is adopted in which a CT of one ratio is compared with a composite connection of two CTs each of lower ratio (Betts *et al* 1982). More recently CSIRO has brought into use a current comparator circuit as shown in Fig. 5. This circuit incorporates a two-stage standard CT with errors which are very much smaller than can be obtained with CTs of conventional design (Kusters and Moore 1961).

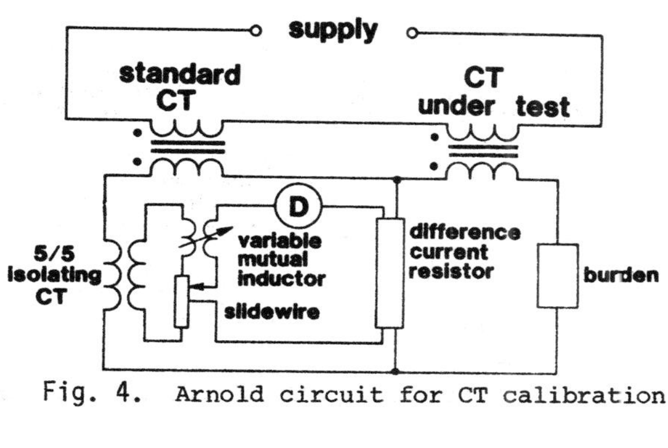

Fig. 4. Arnold circuit for CT calibration

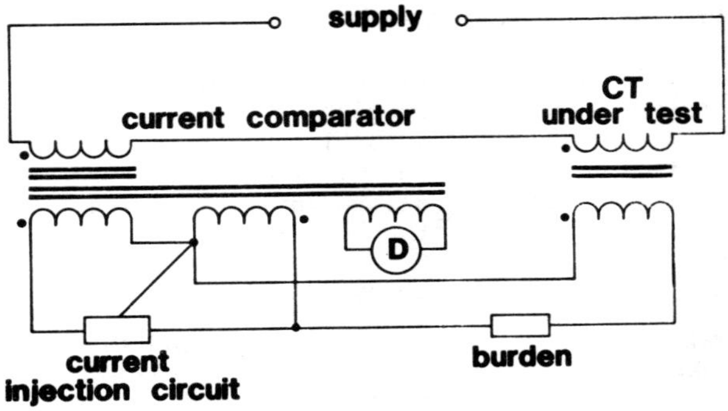

Fig. 5. Current comparator circuit for CT calibration

6. Accuracy

Instrument transformers which are used in applications where only a general indication is required, or used with protection devices, need to be accurate to only 1-5%. In tariff metering, overall accuracies of 1% may be required. An IT used in connection with such metering should not be in error by more than about 0.3%. Calibration of the ITs described above could be made in a simple circuit similar to Figs. 1 or 3, and the error of the standard transformer would be negligible if it were within about one tenth of the errors of the IT under test.

For the determination of generator efficiencies and other large plant efficiencies, overall accuracies of about 0.1% may be required in the electrical measurements. So as not to impose too great a requirement on the accuracy of the wattmeters used in these measurements, the range extending ITs should have errors which are known in both magnitude and phase with an uncertainty no greater than about 0.02%.

Calibration of ITs required for such precise measurements requires a standard which is known with even lower uncertainty. It is usually accepted that the improvement necessary in each step in the calibration chain giving traceability to the national standards, is a factor of about 3. Therefore a generating authority may require standard transformers with errors known with an uncertainty of 0.01% or better. Thus a national laboratory which calibrates these standards will require the errors of their own equipment to be known with an uncertainty of no more than about 0.003%. In order to be able to meet the demand for special calibrations, and to be able to carry out international comparisons, uncertainties better than 0.001% are aimed for at the CSIRO.

7. Recalibrations

Since the windings of an IT cannot be changed accidently, it might be expected that the actual transformation ratio would stay the same for all time. However, the effects that give rise to the errors occuring could themselves change. Hence there is a need for ITs to be calibrated periodically.

Where indications within the accuracy range 1-5% are required, an IT used in the measurement will probably never need recalibrating after its initial check by the manufacturer. However, ITs used in precision measurements will need regular checking. The intervals between checks vary according to the known stability of the IT.

A standard transformer used in the calibration of other ITs should be checked on one of its ratios at least once a year. If a multi-ratio standard has a 1 to 1 ratio, this ratio can be checked absolutely by comparing its primary directly with its secondary. Alternatively, if more than one multiratio standard is available, an intercomparison between standards on a ratio common to both could be made. Such a 'spot check' is carried out each time one of CSIRO's standards are used. A full calibration of a standard should only be necessary every 10 years unless the spot checks indicate that significant changes have occured. At CSIRO all standard CTs are fully calibrated every 5 years even though changes from one calibration to the next are rarely greater than 5 ppm. The relative values of the capacitance units used in CSIRO's VT calibration circuit are determined each time the units are used.

Instrument transformers in which leakage currents in the insulation are a major cause of error tend to be the least stable, since the characteristics of the insulation can change with humidity and temperature cycling or contamination of the insulating material. In severe cases, it might be necessary to recalibrate such ITs every 2 years.

Where a CT has a current rating above about 50 A, poor design or poor workmanship at the terminations, particularly where nickel or chromium plating has been used, often results in a CT which is stable for a few years and then starts to deteriorate. This deterioration is caused by oxides and other contaminants which when heavily coating the terminations, result in partial rectification at the contact faces. This in turn allows a small component of dc to flow in a winding, which changes the magnetic characteristics of the core. This effect is usually evident during calibration when the errors are observed to change and continue to change by large amounts for several minutes. It will then be necessary to demagnetize the core. Recalibration at intervals of less than 5 years will also be necessary with these CTs; or, in severe cases, short term stability may become so poor that the CT becomes useless for precision measurements. Where plain copper terminations are used, instabilities due to poor contacts are practically unknown.

Voltage transformers with ratings below about 10 kV and CTs rated within the range 1-25 A tend to be the most stable. Insulation leakage is of little importance to them, and the voltage developed in their windings is usually enough to break down any poor contacts in their terminations which might otherwise cause instabilities.

8. Conclusions

Instrument transformers are widely used devices in the electrical supply industry. They are relatively inexpensive, accurate and stable. However, like all measuring devices they need to be calibrated if they are to be used for precise measurements.

The accuracy obtainable from an IT is, up to a point, limited only by size. This is because the more iron and copper used in their construction, the lower will be their internal losses. Developments in the area of 2-stage and electronically aided ITs have made considerable advances in recent years (Fiander 1979). Such sophistication certainly enables greatly increased accuracies to be achieved, or enables a much smaller IT to be used, but the long term stability of these more complicated devices has not yet been fully established.

Reference

Arnold, A.H.M. (1934). Current Transformer Testing. *J. Inst. Electr. Eng.* **74**, 424

Betts, P.J, Clothier, W.K., and Smith, H.A. (1982). A Method for the Absolute Calibration of Current Transformers. *Proc. Inst. Electr. Eng.* (in press).

Clothier, W.K., and Medina, L. (1954). A Portable Voltage Transformer Testing Set. *Aust. J. Appl. Sci.* **5**, 145.

Clothier, W.K., and Medina, L. (1956). The Absolute Calibration of Voltage Transformers. *Proc. Inst. Electr. Eng. A*, **104**, 204.

Fiander, J.R. (1979). 2-Stage Current Transformers and their Application. Australian Institute of Physics, First National Conference on Applied Physics, Rockhampton.

Kusters, N.L., and Moore, W.J.H. (1961). The Current Comparator and its Application to the Absolute Calibration of Current Transformers. *Trans. Am. Inst. Electr. Eng. Part 3.* **80**, 94.

High-Voltage Measurements at the CSIRO Division of Applied Physics

J. Rungis
CSIRO Division of Applied Physics, Sydney, Australia 2070

Abstract

This paper discusses methods of measuring DC, AC and impulse voltages higher than 10 kV. The calibration systems used by the CSIRO Division of Applied Physics are presented and a mobile high-voltage calibration unit is described.

1. Introduction

Separate measurement methods have to be considered for the three types of voltages (DC, AC and impulse) encountered in a high-voltage laboratory. High-voltage in this paper means a voltage greater than 10 kV. High-voltage measurement methods can be classified into two types: a) direct, and b) indirect.

In the direct method, an object or substance is placed in the electric field and the influence of the field on it is a measure of the voltage. In indirect methods, a voltage divider is used to tap off a small fraction of the high-voltage which then can be measured by low-voltage measurement techniques.

2. Measurement Methods

(a) Direct Methods

(i) Spark gaps. The simplest of all high-voltage measuring methods is the breakdown of atmospheric air in a symmetric-gap arrangement. This usually consists of two spheres but crossed cylinders can also be used. The method consists of either setting the gap to a known spacing before sparkover or measuring the gap after sparkover. The voltage value corresponding to the gap and the sphere size is read from standard tables. The advantages of the method are that it is relatively simple to set up and easy to use and maintain. The disadvantages are that it is only a single-value and "one-shot" measurement method and cannot be used for continuous monitoring of a voltage. Other disadvantages or difficulties of operation are: uncertainties introduced by variable atmospheric conditions, particularly the presence of dust particles in the gap, the lack of spark-initiating electrons in a small gap, the uncertainty about the effects of humidity and the need for relatively large clearances around the spheres to prevent distortion of the field.

Spark gaps measure the peak value of all three types of voltage. The standards specify uncertainties of ±3% for AC and impulse voltages and ±5% for DC voltages. Better short-term consistency can be obtained if care is taken with the measurement methods; on the other hand anomalously low sparkover voltages can occur if the atmosphere is dirty or the flashover surfaces are damaged.

(ii) Electrostatic voltmeters. There are various designs of voltmeters which use the principle of electrostatic attraction to measure the applied voltage. Generally the mechanical force on one electrode is measured when the voltage is applied between the electrodes. However a metal object of known shape can be placed in the field, as in the ellipsoid and twin-sphere electrostatic voltmeters, and the force on it due to the field can be measured. The force of attraction is proportional to the square of the applied voltage for both AC and DC voltages. Because of their mechanical properties, electrostatic voltmeters measure the root-mean-square voltage of power and higher frequency voltages.

Their advantages are that they are continuous-reading instruments and have a very high input impedance, thus presenting a very small load to the source during the measurement. They can be made absolute by a choice of electrode geometry which permits the voltage to be calculated from the measured value of the force it produces. Other electrode arrangements can be used to produce a more nearly linear scale but these voltmeters then have to be calibrated.

A disadvantage of electrostatic voltmeters is that, because they have to measure relatively small forces, they are subject to interference from vibrations and draughts. Levelling is also very important as they are influenced by gravity. For these reasons, electrostatic voltmeters are usually employed at the lower end of the high-voltage scale (below 50 kV). Larger units have been developed as measurement standards and are not used for routine measurements.

(iii) Particle acceleration and molecular effects. High-voltages can be used to accelerate electrons and other particles and conversely the study of the effects and properties of accelerated particles can be used as a measure of the accelerating voltage. Effects that can be used include: the deflection of electrons or ions by an electric or magnetic field; measurement of the X-rays emitted or diffraction patterns produced by accelerated electrons; resonance peaks in nuclear reaction cross-sections dependent on the energy of the incident particle. All these methods require large amounts of hardware, as the particles have to be accelerated in vacuum, but they are useful for the internal calibration of particle accelerators. However, if the calibration is to be transferred to some external meter, direct access to the high-voltage source is required. This is not always possible without special access provisions, as accelerators often use compressed gas insulation to reduce size.

The effect of an electric field on the optical properties of molecules can be used to measure high-voltages. The most developed of these methods is the Kerr effect in which the refractive index of a liquid varies with the strength of a transversely applied electric field. A similar method is the Pockels effect in which the refractive index of a crystal lattice varies in both transverse and longtitudinal directions with applied electric field.

In both effects the molecular relaxation time is fast compared to the voltages to be measured, so that the effects can be used to measure voltages from DC to impulse.

A similar phenomenon is the Stark effect which is the splitting of a spectral line into components when an electric field is applied to the substance giving rise to the line.

All these molecular methods have the disadvantage that they depend on the local field strength and not directly on the applied voltage which is the quantity to be measured. This means that electrode spacing has to be known accurately and the effects of fringe fields due to the edges and viewing ports has to be kept small. The advantages of the methods are that they relate the electric field strength to molecular and atomic parameters, which should be constant and reproducible if sample purity is not a problem, and which eventually should be calculable for simple substances.

(iv) Generating voltmeters. These voltmeters really fall into a class somewhere between direct and indirect reading voltmeters. They are indirect in the sense that they are not connected to the high-voltage source but direct because they measure the voltage (or actually the electric-field strength) directly. This is achieved by having a rotating electrode which provides a cyclicly varying capacitance. The current from this capacitance which is proportional to the voltage across the plates of the probe, is rectified and displayed on a meter. AC voltages can also be measured with this type of meter if the motor drive speed is synchronized with the supply frequency and is half its value. Generating voltmeters place no load on the voltage source but care has to be taken that the probe does not disturb the field distribution.

(v) Other types. Some very empiric voltmeters have been constructed which depend on the ionization of air at a sharp point. One such device reported is called the ionic-wind voltmeter and depends on the velocity of ions from a point source being proportional to the applied voltage after initial ionization takes place. These devices, although simple to construct are difficult to calibrate and lack stability. The production of ozone can cause corrosion and is a health hazard if adequate ventilation is not provided.

Air ionization or corona onset, however, can be useful as a voltage indicator. This principle has been used to measure the gradient along an insulator string. A small point-point gap is attached across successive insulators on a string. The gap spacing is such that corona will occur at the points at some voltage in the region of interest. The total voltage input to the string is monitored and the values at which corona starts, for each location of the gap, are noted.

(b) Indirect Methods

All the measurement methods considered in this section will be of the divider type: resistive, capacitive and inductive. No consideration will be given to the low voltage measurement methods. There voltages range from microvolts to a few hundred volts and can be measured with adequate accuracy.

A problem common to all dividers is that they have to be calibrated. This can usually be achieved quite easily at low-voltage but their behaviour then has to be ascertained to the full operating voltage. This can be done by comparison with a divider of known characteristics or by comparison with another type of instrument, preferably of higher maximum working voltage.

(i) Resistance dividers. The only dividers that can be used for DC measurement are resistance dividers. These same dividers, with some restrictions, can be used for power-frequency AC measurements, but for impulse measurements totally different dividers are required because of the high frequencies involved.

The most successful high-voltage resistance divider design is due to Park. This consists of a helix of 1 MΩ wire-wound resistors, usually with a maximum rating of 1 kV/1 MΩ or 1 mA. The resistors are encapsulated and the metal outer case is joined to one end of the resistor, thus providing shielding. This shielding is necessary to prevent corona on the resistors. Adequate insulation has to be provided for the resistor supports as they provide a shunt or leakage path in parallel with the resistor chain. However the biggest disadvantages of wire-wound resistance dividers of this type are the thermal and electrostatic forces on the wire, which cause resistance changes, and strains in the wire and insulation. In practice, it has been found that AC operation has to be limited to 50% of maximum DC operation because of voltage-induced strains in the wire. Uncertainties of 0.05% or less on DC can be achieved with this type of divider. A less accurate (~ 2%) but more robust divider can be constructed from a chain of carbon resistors.

For the measurement of impulse voltages, much lower resistance ohmic dividers are used and special care has to be taken with the physical layout of the high and low-voltage arms to keep inductance to an acceptably low value, and capacitance grading may be required at higher voltages.

(ii) Capacitance dividers. For both AC and impulse voltages, capacitance dividers can be used to measure voltage although, as in the case of resistance dividers, the designs and values for the two types of voltage are quite different.

For power frequencies it is found that currents of the order of 10 mA down to 1 mA are convenient in dividers for voltage measurement. In capacitance terms this means a value of 100 pF for a voltage of 300 kV. Larger capacitances result in too much load on the supply and smaller capacitances mean a loss of sensitivity.

Various types of capacitors can be used for the top and bottom arms of the divider, but it is convenient to have capacitors of the same type, and thus similar temperature coefficients, to reduce the temperature coefficient of the complete divider. The simplest method is to form a high voltage capacitor between a high-voltage terminal and a plate near earth potential. This air capacitor can then be placed in series with a higher value capacitor, thus providing a capacitance divider. This method has severe limitations because the air capacitor so formed usually has a very small capacitance value and it can be greatly influenced by surrounding objects.

The most commonly used high-voltage capacitors are paper-oil insulated foil capacitors, usually made of lower voltage units in series, in oil, in a paper resin-impregnated tube. These dividers have uncertainties of no better than 1 to 2%, because of dimensional instabilities and quite large temperature and voltage coefficients.

Much better accuracies can be achieved by the use of compressed-gas capacitors. If these are not subjected to large (~ 10°C) temperature fluctuations, capacitance and divider stabilities of better than 0.01% can be achieved.

Purely capacitive dividers are only suitable for impulse voltage measurements at low levels where lead inductance can be reduced sufficiently by the use of short co-axial leads. For larger high-voltage circuits and for fast impulses, lead inductance becomes very significant and oscillations will occur in the capacitive-inductive circuit so formed. This problem can be overcome by constructing a damped capacitive voltage divider where the LC oscillations are damped by a resistive element. Best results are achieved by having distributed resistive-capacitive elements instead of trying to achieve adequate damping with one resistor at the input of the divider.

(iii) Inductive dividers. These are better known as voltage transformers. They have very good long-term stability and therefore are very useful as ratio standards for the measurement of AC voltages. Their main disadvantage is that they become rather large and expensive at higher voltages. To overcome this problem for voltages above about 200 kV, capacitive voltage transformers have been constructed. They consist of a high-voltage (usually paper-oil) capacitor as the top arm of a divider and a magnetic voltage transformer as the lower arm. Because of drifts in the high-voltage capacitors and because of the large size of the units, in-situ calibration techniques have been developed.

3. High-Voltage Measurements in the Division of Applied Physics

(i) DC voltages. The laboratory has two DC voltage sources, both of which are rectified and smoothed AC circuits. The maximum voltages are 100 kV and 1 MV. The latter consists of two 500 kV units that can be used separately. The voltage is measured with wire-wound resistance dividers and digital voltmeters of adequate accuracy. The laboratory has two commercial dividers rated at 100 kV each and our own design modular units consisting of two 100 kV and two 150 kV sections to give a maximum measurement capacity of 500 kV, with an uncertainty of ±0.05%.

Each divider section is tested to full operating voltage against a combination of the other sections. In the absence of another calibrated divider or some other calibrating instrument, the procedure provides an estimate of resistance change due to thermal loading and voltage stress, because, in the fully loaded divider, each resistor is dissipating the maximum load of 1 W, but, in the combined divider, each resistor is only dissipating 10 to 20% of the maximum load. Dividers are compared in an equal ratio bridge or sequential readings of voltage output are taken and averaged with a computer-controlled data acquisition system. A sensitivity of 0.005% can be achieved. The main contribution to measurement noise comes from the supply-voltage fluctuations.

(ii) AC voltages. The main test transformer has a maximum output of 600 kV and a tertiary winding with a full voltage output of 250 V which can be used to monitor the transformer voltage. However, as the transformer ratio will change with load, this is only used for monitoring and control purposes. The output voltage of the transformer is measured with a paper-oil capacitance divider which has a high-voltage capacitance of 200 pF and ratios of 1000, 2000 and 4000:1. This voltage measurement has an uncertainty of about 2%.

Much better voltage measurement accuracy can be achieved using a voltage transformer. However the transformer with the highest working voltage in this laboratory has a maximum range of only 100 kV: the voltage and phase errors of this transformer are known with an uncertainty of 0.02%.

Compressed-gas capacitors are used to extend this accuracy to higher voltages. The laboratory has two gas capacitors, one 330 kV and the other 550 kV working voltage. Their capacitance values are stable to better than 0.005% over their full voltage range. A current comparator bridge is used to compare the gas capacitors and also to compare capacitance dividers with voltage transformers as shown in Figs. 1 and 2.

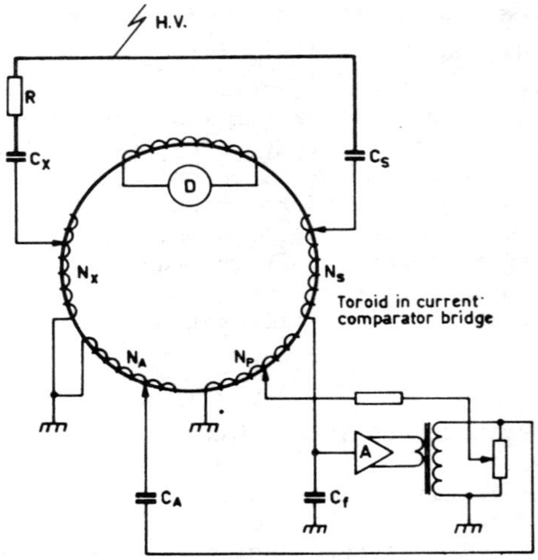

Fig. 1. Comparison of capacitors with a current comparator bridge.

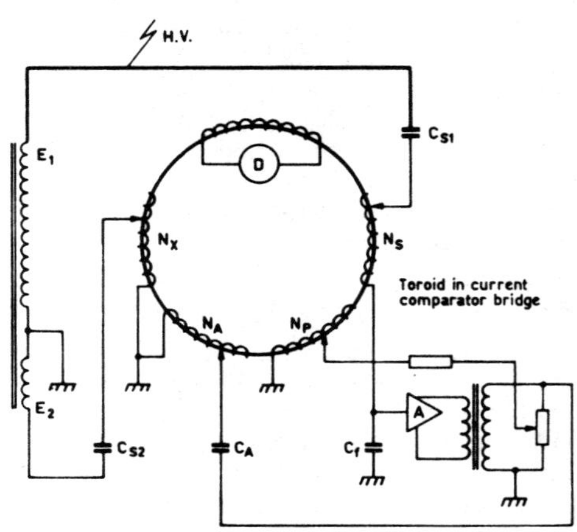

Fig. 2. Voltage transformer calibration with a current comparator bridge.

(iii) Impulse voltages. Located in the laboratory is a 12-stage 2.4 MV impulse generator belonging to the Electricity Commission of NSW. The Commission routinely uses sphere gaps for calibration, and resistance and capacitance dividers for monitoring impulse voltages. The Division has a 1 MV damped-capacitive voltage divider.

4. Mobile High-Voltage Calibration Unit

The calibration of high-voltage measurement apparatus presents special difficulties. Because of its large size, it is difficult to transport it to a central laboratory for calibration, and often it cannot be taken out of service for a prolonged period because substitute apparatus is not available or is too expensive.

To remedy this situation a mobile calibration system has been constructed. *In-situ* calibration has the advantage of taking into account environmental factors and proximity effects, and removing the uncertainties due to mechanical shocks to customers' equipment during transport. Disadvantages are the usually higher electrical noise and larger temperature fluctuations in visited laboratories and lack of the support facilities normally available at the base laboratory.

The main item carried is a gas-capacitor, which usually is the 330 kV model but the 550 kV one can be accommodated, together with a current-comparator bridge. This enables the calibration of capacitors and voltage transformers to be performed. A 100 kV voltage transformer is also carried as the basic high-voltage ratio standard. A low voltage audio-frequency bridge is used to monitor the capacitor after transit and before and after the application of high-voltage.

For impulse voltage measurements the unit can carry a 1 MV damped-capacitive divider and associated measuring equipment and for DC voltages a 500 kV resistive divider. Provision is also made to carry additional equipment such as a partial discharge detector and radio-influence voltage-measuring equipment, as required.

All the equipment is mounted in a 7-tonne air-conditioned truck. After removal of the larger pieces of equipment the truck is used as a screened measurement laboratory, as the steel-sheet walls are all electrically connected, as are the wired-glass windows. The truck is designed to be used in other measurement laboratories, as it does not carry its own power supplies.

5. Bibliography

Allibone, T.E. and Saunderson, J.C. (1980). "The measurement of DC Voltages up to 1 MV using Sphere-gaps and Rod-gaps". *Gaseous Dielectrics II*, Christophorou, L.G. ed. Pergamon Press.

Altar, W. and Garbury, M. (1949). "Absolute speed gauge for high voltage particles". *Phys. Rev.*, 76, 496.

Anderson, W.E., Davis, R.S., Petersons, O. and Moore, W.J.M. (1978). "An international comparison of high voltage capacitor calibrations". *IEEE Trans. Power Appar. and Syst.*, PAS-97, 1217.

Bettenhausen, L.H. and Gallagher, W.J. (1964). "An alpha-particle voltmeter". *Nucleonics*, June, 70.

Brooks, H.B., Defandorf, F.M. and Silsbee, F.B. (1938). "An absolute electrometer for the measurement of high alternating voltages". *J. Res. Nat. Bur. Stand.*, 20, 253.

Crepaz, S. (1973). "A high voltage measuring device based on the Pockels effect". Association des ingenieurs electriciens Sorti de l'Institut electrotechnique Montifiore. *Bulletin Scientifique*, 86, 141.

Defandorf, F.M. (1948). "The measurement of high voltage". *J. Wash. Acad. Sci.*, 38, 33.

Hara, K., Kobayashi, T., Matsui, T., Nakase, T. and Yonezaki, G. (1971). "Stark Voltmeter as an Electromotive Force Reference". *Jp. J. Phys.*, 10, 1066.

Harris, F.K., Sze, W.C., Kusters, N.L., Petersons, O. and Moore, W.J.M. (1964). "An international comparison of voltage-transformer calibrations to 350 kV". *IEEE Trans. Commun. Electron.*, 70, 13.

Hebner, R.E., Malewski, R.A. and Cassidy, E.C. (1977). "Optical methods of electrical measurement at high voltage levels". *Proc. IEEE*, 65, 1524.

Hillhouse, D.L. and Peterson, A.E. (1973). "A 300 kV compressed gas standard capacitor with negligible voltage dependance". *IEEE Trans. Instrum. Meas.*, IM22, 408.

Hillhouse, D.L., and Petersons, O. and Sze, W.C. (1978). "A prototype system for on-site calibration of coupling capacitor voltage transformers (CCVTs)". *IEEE Trans. Power Appar. Syst.*, PAS-98, 1026.

Howard, P.R. (1955). "The measurement of high voltages". *Electrical Times*, 14 April, 601, 19 May, 819, 14 July, 45.

Kusters, N.L., and Petersons, O. (1963). "A transformer-ratio-arm bridge for high voltage capacitance measurements". *IEEE Trans. Commun. Electron.*, 606, 69.

Kusters, N.L., and Petersons, O. (1963). "The voltage coefficient of precision capacitors". *IEEE Trans. Commun. Electron.*, 69, 612.

Medina, L. (1959). "A spark gap with crossed cylinder electrodes for voltage measurements in the range 20 to 140 kVp". *Aust. J. Appl. Sci.*, 10, 404.

National Physical Laboratory. (1954). Precision electrical measurements; proceedings of a symposium. H.M.S.O. London.

Park, J.H. (1962). "Special shielded resistor for high voltage DC measurements". *J. Res. Nat. Bur. of Stand.*, 66C, 19.

Paul, J.C. (1970. "Design and construction of rotating voltmeter for the measurement of high dc voltage". *J. Phys. E.*, 3, 321.

Rungis, J. (1973). "Optimizing an ellipsoid voltmeter". *J. Phys. E.*, 6, 365.

Rungis, J., and Brown D.E. (1973). "Construction of an ellipsoid voltmeter". *J. Phys. E.*, 6, 724.

Rungis, J., and Brown, D.E. (1975). "Voltage induced capacitance fluctuations in a compressed gas, high voltage capacitor". *J. Phys. E.*, 8, 16.

Rungis, J., and Brown, D.E. (1979). "Mobile high voltage calibration unit". *Aust. Phys.*, November, 160.

Rungis, J., and Brown, D.E. (1981). "Experimental study of factors affecting capacitance of high voltage compressed-gas capacitors". *Proc. Inst. Electr. Eng. Part A*, 128, 273.

Rungis, J., and Brown, D.E. (1982). "Stacking of high voltage resistance dividers". *Proc. Inst. Electr. Eng. Part A*, 129, 119.

Schwab, A.J. (1972). "High voltage measurement techniques". The M.I.T. Press, Cambridge, Massachusetts.

Shoupp, W.E., Jennings, B., and Jones, W. (1949). "An absolute calibration of the Li^7 (p,n) threshold voltage". *Phys. Rev.* 76, 502.

Smith, W.E. and Rungis, J. (1975). "Twin adhering conducting spheres in an electric field - an alternative geometry for an electrostatic voltmeter". *J. Phys. E.*, 8, 379.
Thomas, R.J. (1970). "High-impulse current and voltage measurement". *IEEE Trans. Instrum. Meas.*, IM-19, 102.
Thornton, W.M. (1931). "High voltage precision measurements". *Inst. Electr. Eng.*, 69, 1273.